FUNDAMENTALS
OF
SOIL DYNAMICS

D0945202

FUNDAMENTALS
OF
SOIL DYNAMICS

BRAJA M. DAS

The University of Texas at El Paso

ELSEVIER

New York • Amsterdam • Oxford

Elsevier Science Publishing Co., Inc.
52 Vanderbilt Avenue, New York, New York 10017

Sole distributors outside the United States and Canada:
Elsevier Science Publishers B. V.
P. O. Box 211, 1000 AE Amsterdam, The Netherlands

Library of Congress Cataloging in Publication Data

Das, Braja M., 1941–
    Fundamentals of soil dynamics.

    Includes bibliographical references and index.
    1. Soil dynamics. I. Title.
TA710.D258 1983   624.1'5136   82-9321
                                            AACR2
ISBN 0-444-00705-9

Manufactured in the United States of America

To
Janice and Valerie

CONTENTS

PREFACE

During the past two decades, considerable progress in the area of soil dynamics has been made. Soil dynamics courses have been added or expanded for graduate-level study in most universities. The knowledge gained from the intensive research conducted all over the world has gradually filtered into actual planning, design, and construction process of foundations of structures.

This text on soil dynamics has been developed partly out of my notes prepared for teaching an introductory course for graduate students. In developing this text, simplicity of presentation for clear understanding of beginners has been my main consideration. For that reason, a major portion of the text has been assigned to the treatment of fundamental concepts of the subjects under consideration. Also, more recently developed materials have been drawn out of published literature and incorporated in the text.

Both systems of units (i.e., English and SI) have been used throughout the text. A number of worked-out problems have also been included, which I believe are essential for the student. A list of references has also been included at the end of each chapter.

During the preparation of the text, I received constant inspiration from my wife and Professor Paul C. Hassler of The University of Texas at El Paso. Professor Hassler and Sands H. Figuers, a graduate student in the Department of Geological Sciences at The University of Texas at El Paso, helped me in obtaining several reference materials needed for the preparation of the manuscript.

Thanks are due to my wife, Janice, for typing the entire manuscript and for her help in preparation of the figures and tables.

Braja M. Das

El Paso, Texas
March 1982

1

FUNDAMENTALS OF VIBRATION

Considerable advancements in the area of soil dynamics have been made recently. These advancements include a better understanding of the behavior of soils subjected to dynamic loading conditions. Based on the analytical and experimental evaluations, new design criteria for foundations subjected to dynamic loading have been developed. This text is intended to present the fundamentals of soil dynamics as related to the design of foundations, lateral earth pressure on retaining structures, soil liquefaction and evaluation of liquefaction potential of natural soil deposits, and bearing capacity of shallow foundations.

Satisfactory design of foundations for vibrating equipments is based on displacement considerations. Displacement due to vibratory loading can be classified under two major divisions:

1. cyclic displacement due to the elastic response of the soil–foundation system to the vibratory loading, and
2. permanent displacement due to compaction of soil below the foundation.

In order to consider the first condition described above, it is essential to know the nature of the unbalanced forces in a foundation such as shown in Figure 1.1 (Richart, 1962). Note that a foundation can vibrate in any or all six possible modes. For ease of analysis, each mode is considered separately and design is carried out by considering the displacement due to each mode separately. The mathematical considerations of the displacement of foundations can be made by treating soil as a viscoelastic material. This can be explained with the aid of Figure 1.2a, which shows a foundation subjected to a vibratory loading in the vertical direction. The parameters for the vibration of the foundation can be evaluated by treating the soil as equivalent to a spring and a dashpot which supports the foundation as

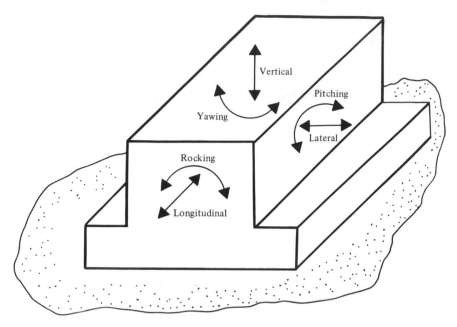

FIGURE 1.1 Six modes of vibration for a foundation. [Richart, F. E., Jr. (1962), "Foundation Vibration," *Transactions*, *ASCE*, 127, Part I, Fig. 1, p. 865.]

FIGURE 1.2 A lumped parameter vibrating system.

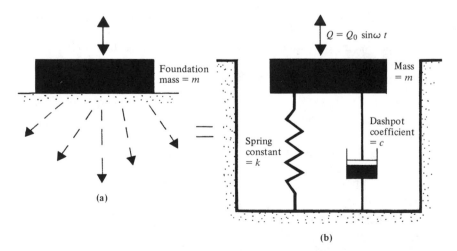

shown in Figure 1.2b. This is usually referred to as a *lumped parameter vibrating system.*

In order to solve the vibration problems of lumped parameter systems, one needs to know the fundamentals of vibration engineering. Therefore, a brief review of the mathematical solutions of simple vibration problems is presented. More detailed discussion regarding other approaches to solving foundation vibration problems and evaluation of basic parameters such as the spring constant and damping coefficient are presented in Chapters 5 and 6.

1.1 FUNDAMENTAL DEFINITIONS

Free Vibration. Vibration in system under the action of forces inherent in the system itself and in the absence of external impressed forces.

Forced Vibration. Vibration of a system caused by an external force.

Degree of Freedom. The number of independent coordinates required to describe the solution of a vibrating system. For example, the position of the mass m in Figure 1.3a can be described by a single coordinate z, so it

FIGURE 1.3 Degrees of freedom for vibrating system.

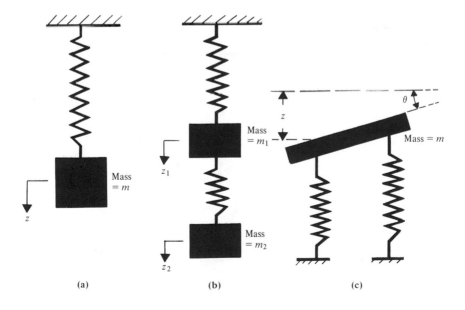

is a *single degree of freedom system*. In Figure 1.3b, two coordinates (z_1 and z_2) are necessary to describe the motion of the system; hence this system has *two* degrees of freedom. Similarly, in Figure 1.3c, two coordinates (z and θ) are necessary, and the number of degrees of freedom is two.

1.2 SINGLE DEGREE OF FREEDOM SYSTEM

1.2.1 Free Vibration of a Spring–Mass System

Figure 1.4 shows a foundation resting on a spring. Let the spring represent the elastic properties of the soil. The load W represents the weight of the foundation plus that which comes from the machinery supported by the foundation. If the area of the foundation is equal to A, the intensity of load transmitted to the subgrade can be given by

$$q = W/A \tag{1.1}$$

Due to the load W, a static deflection z_s will develop. By definition,

$$k = W/z_s \tag{1.2}$$

where k is the spring constant for the elastic support.

The coefficient of subgrade reaction k_s can be given by

$$k_s = q/z_s \tag{1.3}$$

If the foundation is disturbed from its static equilibrium position, the system will vibrate. The equation of motion of the foundation when it has been disturbed through a distance z can be written from Newton's second law of motion as

$$(W/g)\ddot{z} + kz = 0$$

FIGURE 1.4 Free vibration of a mass–spring system.

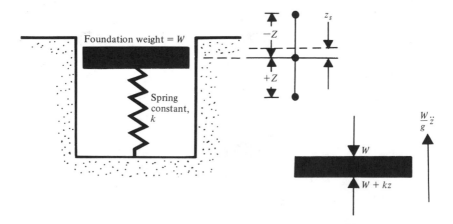

or

$$\ddot{z} + (k/m)z = 0 \tag{1.4}$$

where g is the acceleration due to gravity, $\ddot{z} = d^2z/dt^2$, t is time, and m is mass $= W/g$.

In order to solve Eq. (1.4), let

$$z = A_1\cos\omega_n t + A_2\sin\omega_n t \tag{1.5}$$

where A_1 and A_2 are both constants and ω_n is the undamped natural circular frequency.

Substitution of Eq. (1.5) into Eq. (1.4) yields

$$-\omega_n^2(A_1\cos\omega_n t + A_2\sin\omega_n t) + (k/m)(A_1\cos\omega_n t + A_2\sin\omega_n t) = 0$$

or

$$\omega_n = \pm\sqrt{k/m} \tag{1.6}$$

The unit of ω_n is in radians per second (rad/sec). Hence,

$$z = A_1\cos\left(\sqrt{k/m}\,t\right) + A_2\sin\left(\sqrt{k/m}\,t\right) \tag{1.7}$$

In order to determine the values of A_1 and A_2, one must, substitute the proper boundary conditions. At time $t = 0$, let,

$$\text{Displacement } z = z_0$$

and

$$\text{Velocity} = dz/dt = \dot{z} = v_0$$

Substituting the first boundary condition in Eq. (1.7),

$$z_0 = A_1 \tag{1.8}$$

Again, from Eq. (1.7),

$$\dot{z} = -A_1\sqrt{k/m}\,\sin\left(\sqrt{k/m}\,t\right) + A_2\sqrt{k/m}\,\cos\left(\sqrt{k/m}\,t\right) \tag{1.9}$$

Substituting the second boundary condition in Eq. (1.9),

$$\dot{z} = v_0 = A_2\sqrt{k/m}$$

or

$$A_2 = v_0/\sqrt{k/m} \tag{1.10}$$

Combination of Eqs. (1.7), (1.8), and (1.10) gives

$$z = z_0\cos\left(\sqrt{\frac{k}{m}}\,t\right) + \frac{v_0}{\sqrt{k/m}}\sin\left(\sqrt{\frac{k}{m}}\,t\right) \tag{1.11}$$

Now let

$$z_0 = Z\cos\alpha \tag{1.12}$$

and

$$\frac{v_0}{\sqrt{k/m}} = Z \sin \alpha \qquad (1.13)$$

Substitution of Eqs. (1.12) and (1.13) into Eq. (1.11) yields

$$z = Z \cos(\omega_n t - \alpha) \qquad (1.14)$$

where

$$\alpha = \tan^{-1}\left(v_0 / z_0 \sqrt{k/m}\right) \qquad (1.15)$$

$$Z = \sqrt{z_0^2 + \left(v_0 / \sqrt{k/m}\right)^2} = \sqrt{z_0^2 + (m/k) v_0^2} \qquad (1.16)$$

The relation for the displacement of the foundation given by Eq. (1.14) can be represented graphically as shown in Figure 1.5. At time

$$t = 0, \qquad z = Z \cos(-\alpha) \qquad = Z \cos \alpha$$

$$t = \frac{\alpha}{\omega_n}, \qquad z = Z \cos\left(\omega_n \frac{\alpha}{\omega_n} - \alpha\right) \qquad = Z$$

$$t = \frac{\frac{1}{2}\pi + \alpha}{\omega_n}, \; z = Z \cos\left(\omega_n \frac{\frac{1}{2}\pi + \alpha}{\omega_n} - \alpha\right) = 0$$

$$t = \frac{\pi + \alpha}{\omega_n}, \qquad z = Z \cos\left(\omega_n \frac{\pi + \alpha}{\omega_n} - \alpha\right) \quad = -Z$$

$$t = \frac{\frac{3}{2}\pi + \alpha}{\omega_n}, \; z = Z \cos\left(\omega_n \frac{\frac{3}{2}\pi + \alpha}{\omega_n} - \alpha\right) = 0$$

$$t = \frac{2\pi + \alpha}{\omega_n}, \; z = Z \cos\left(\omega_n \frac{2\pi + \alpha}{\omega_n} - \alpha\right) = Z$$

$$\vdots$$

From Figure 1.5, it can be seen that the nature of displacement of the foundation is sinusoidal. The magnitude of maximum displacement is equal to Z. This is usually referred to as the *single amplitude*. The peak-to-peak displacement amplitude is equal to $2Z$, which is sometimes referred to as the *double amplitude*. The time required for the motion to repeat itself is called as the *period* of the vibration. Note that, in Figure 1.5, the motion is repeating itself at points A, B, and C. The period T of this motion can therefore be given by

$$T = 2\pi / \omega_n \qquad (1.17)$$

The *frequency of oscillation f* is defined as the number of cycles in unit time, or

$$f = 1/T = \omega_n / 2\pi \qquad (1.18)$$

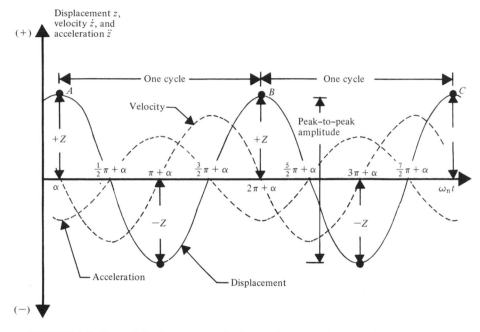

FIGURE 1.5 Plot of displacement, velocity, and acceleration for the free vibration of a mass–spring system. Note: velocity leads displacement by $\frac{1}{2}\pi$ rad; acceleration leads velocity by $\frac{1}{2}\pi$ rad.

It has been shown in Eq. (1.6) that, for this system, $\omega_n = \pm\sqrt{k/m}$. Thus,

$$f = f_n = (1/2\pi)\sqrt{k/m} \tag{1.19}$$

The term f_n is generally referred to as the *undamped natural frequency*. Since $k = W/z_s$ and $m = W/g$, Eq. (1.19) can also be expressed as

$$f_n = (1/2\pi)\sqrt{g/z_s} \tag{1.20}$$

Table 1.1 gives values of f_n for various values of z_s.

The variation of the velocity and acceleration of the motion with time can also be represented graphically. From Eq. (1.14), the expressions for the velocity and the acceleration can be obtained as

$$\dot{z} = -(Z\omega_n)\sin(\omega_n t - \alpha) = Z\omega_n\cos(\omega_n t - \alpha + \tfrac{1}{2}\pi) \tag{1.21}$$

and

$$\ddot{z} = -Z\omega_n^2\cos(\omega_n t - \alpha) = Z\omega_n^2\cos(\omega_n t - \alpha + \pi) \tag{1.22}$$

The nature of variation of the velocity and acceleration of the foundations is also shown in Figure 1.5.

Example 1.1 A mass is supported by a spring. The static deflection of the spring due to the mass is 0.015 in. Find the natural frequency of vibration.

TABLE 1.1 Undamped Natural Frequency[a]

z_s (mm)	Undamped natural frequency f_n (cps)
0.02	111.47
0.05	70.5
0.10	49.85
0.20	35.25
0.50	22.29
1.0	15.76
2	11.15
5	7.05
10	4.98

[a] Note: $g = 9.81$ m/sec^2.

Solution: From Eq. (1.20),

$$f_n = (1/2\pi)\sqrt{g/z_s},$$

$$g = 32.2 \text{ ft/sec}^2, \quad z_s = 0.015 \text{ in.}$$

So,

$$f_n = (1/2\pi)\sqrt{(32.2)(12)/0.015} = 25.54 \text{ cps (cycles/sec)}$$

Example 1.2 For a machine foundation, given weight of the foundation = 45 kN (kilonewtons) and spring constant $= 10^4$ kN/m, determine

a. the natural frequency of vibration
b. the period of oscillation

Solution

a. $f_n = \dfrac{1}{2\pi}\sqrt{\dfrac{k}{m}} = \dfrac{1}{2\pi}\sqrt{\dfrac{10^4}{(45/9.81)}} = 7.43$ cps

b. From Eq. (1.18),

$$T = 1/f_n = 1/7.43 = 0.135 \text{ sec}$$

1.2.2 Forced Vibration of a Spring–Mass System

Figure 1.6 shows a foundation which has been idealized to a simple spring–mass system. Weight W is equal to the weight of the foundation itself and that supported by it; the spring constant is k. This foundation is being subjected to an alternating force $Q_0 \sin(\omega t + \beta)$. This type of problem is generally encountered with foundations supporting reciprocating engines, and so on. The equation of motion for this problem can be given by

$$m\ddot{z} + kz = Q_0 \sin(\omega t + \beta) \tag{1.23}$$

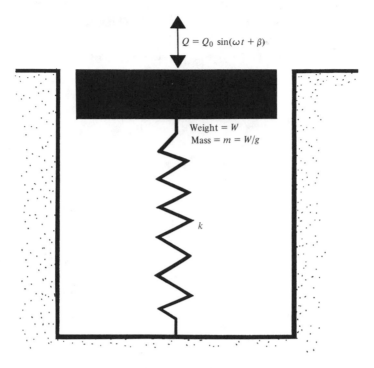

FIGURE 1.6 Forced vibration of a mass–spring system.

Let $z = A_1 \sin(\omega t + \beta)$ be a particular solution to Eq. (1.23) ($A_1 = $ const). Substitution of this into Eq. (1.23) gives

$$-\omega^2 m A_1 \sin(\omega t + \beta) + k A_1 \sin(\omega t + \beta) = Q_0 \sin(\omega t + \beta)$$

$$A_1 = (Q_0/m)/(k/m - \omega^2) \qquad (1.24)$$

Hence, the particular solution to Eq. (1.23) is of the form

$$z = A_1 \sin(\omega t + \beta) = \frac{Q_0/m}{(k/m - \omega^2)} \sin(\omega t + \beta) \qquad (1.25)$$

The complementary solution of Eq. (1.23) must satisfy

$$m\ddot{z} + kz = 0$$

As shown in the preceding section, the solution to the above equation may be given as

$$z = A_2 \cos \omega_n \cdot t + A_3 \sin \omega_n \cdot t \qquad (1.26)$$

where

$$\omega_n = \sqrt{k/m}, \qquad A_2, A_3 = \text{const}$$

Hence, the general solution to Eq. (1.23) is obtained by adding Eqs. (1.25) and (1.26), or

$$z = A_1 \sin(\omega t + \beta) + A_2 \cos \omega_n \cdot t + A_3 \sin \omega_n \cdot t \qquad (1.27)$$

Now, let the boundary conditions be as follows: At time $t = 0$,

$$z = z_0 = 0 \qquad (1.28)$$
$$dz/dt = \text{velocity} = v_0 = 0 \qquad (1.29)$$

From Eqs. (1.27) and (1.28),

$$A_1 \sin \beta + A_2 = 0$$

or

$$A_2 = -A_1 \sin \beta \qquad (1.30)$$

Again, from Eq. (1.27),

$$dz/dt = \dot{z} = A_1 \omega \cos(\omega t + \beta) - A_2 \omega_n \sin \omega_n \cdot t + A_3 \omega_n \cos \omega_n \cdot t$$

Substituting the boundary condition given by Eq. (1.29) in the above equation

$$A_1 \omega \cos \beta + A_3 \omega_n = 0$$

or

$$A_3 = -(A_1 \omega / \omega_n) \cos \beta \qquad (1.31)$$

Combining Eqs. (1.27), (1.30), and (1.31),

$$z = A_1 [\sin(\omega t + \beta) - \cos(\omega t) \cdot \sin \beta - (\omega/\omega_n) \sin(\omega_n t) \cdot \cos \beta] \qquad (1.32)$$

For a real system, the last two terms inside the brackets in Eq. (1.32) will vanish due to damping, leaving the only term for steady-state solution.

If the force function is in phase with the vibratory system (i.e., $\beta = 0$), then

$$z = A_1 \left(\sin \omega t - \frac{\omega}{\omega_n} \sin \omega_n t \right)$$

$$= \frac{Q_0/m}{k/m - \omega^2} \left(\sin \omega t - \frac{\omega}{\omega_n} \sin \omega_n t \right)$$

$$= \frac{Q_0/k}{1 - \omega^2/\omega_n^2} \left(\sin \omega t - \frac{\omega}{\omega_n} \sin \omega_n t \right) \qquad (1.33)$$

However, $Q_0/k = z_s = $ static deflection. If one lets $1/(1 - \omega^2/\omega_n^2)$ be equal to M [equal to the magnification factor or $A_1/(Q_0/k)$], Eq. (1.33) reads as

$$z = z_s M (\sin \omega t - (\omega/\omega_n) \sin \omega_n t) \qquad (1.34)$$

The nature of variation of the magnification factor M with ω/ω_n is shown in Figure 1.7a. Note that the magnification factor goes to infinity when

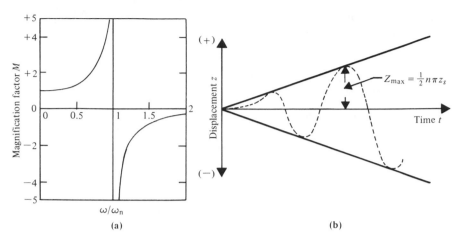

FIGURE 1.7 Forced vibration of a mass–spring system: **(a)** variation of magnification factor with ω/ω_n; **(b)** variation of displacement with time at resonance $(\omega = \omega_n)$.

$\omega/\omega_n = 1$. This is called the *resonance condition*. For resonance condition, the right-hand side of Eq. (1.34) yields $0/0$. Thus, applying *L'Hôpital's rule*,

$$\lim_{\omega \to \omega_n}(z) = z_s \frac{(d/d\omega)\left[\sin \omega t - (\omega/\omega_n)\sin \omega_n t\right]}{(d/d\omega)\left(1 - \omega^2/\omega_n^2\right)}$$

or

$$z = \tfrac{1}{2}z_s\left(\sin \omega_n t - \omega_n t \cos \omega_n t\right) \tag{1.35}$$

The velocity as resonance condition can be obtained from Eq. (1.35) as

$$\dot{z} = \tfrac{1}{2}z_s\left(\omega_n \cos \omega_n t - \omega_n \sin \omega_n t + \omega_n^2 t \sin \omega_n t\right)$$
$$= \tfrac{1}{2}\left(z_s \omega_n^2 t\right)\sin \omega_n t \tag{1.36}$$

Since the velocity is equal to zero at the point where the displacement is at maximum, for *maximum displacement*

$$\dot{z} = 0 = \tfrac{1}{2}\left(z_s \omega_n^2 t\right)\sin \omega_n t$$

or

$$\sin \omega_n t = 0, \quad \text{i.e.,} \quad \omega_n t = n\pi \tag{1.37}$$

where n is an integer.

For the condition given by Eq. (1.37), the displacement equation (1.35) yields

$$|z_{max}|_{res} = \tfrac{1}{2}n\pi z_s \tag{1.38}$$

where z_{max} = maximum displacement.

It may be noted that, when n tends to ∞, $|z_{max}|$ is also infinite, which points out the danger to the foundation. The nature of variation of z/z_s vs time for the *resonance condition* is shown in Figure 1.7b.

Maximum Force on Foundation Subgrade

The maximum and minimum force on the foundation subgrade will occur at the time when the amplitude is maximum, i.e., when velocity is equal to zero. This can be derived from displacement equation (1.33):

$$z = \frac{Q_0}{k}\frac{1}{\left(1 - \omega^2/\omega_n^2\right)}\left(\sin \omega t - \frac{\omega}{\omega_n}\sin \omega_n t\right)$$

Thus, the velocity at any time is

$$\dot{z} = \frac{Q}{k}\frac{1}{\left(1 - \omega^2/\omega_n^2\right)}\left(\omega \cos \omega t - \omega \cos \omega_n t\right)$$

For maximum deflection, $\dot{z} = 0$, or

$$\omega \cos \omega t - \omega \cos \omega_n t = 0$$

Since ω is not equal to zero,

$$\cos \omega t - \cos \omega_n t = 2 \sin\left(\tfrac{1}{2}\omega_n - \omega\right)t \sin\left(\tfrac{1}{2}\omega_n + \omega\right)t = 0$$

Thus,

$$\left(\tfrac{1}{2}\omega_n - \omega\right)t = n\pi; \qquad t = 2n\pi/(\omega_n - \omega) \tag{1.39}$$

or

$$\left(\tfrac{1}{2}\omega_n + \omega\right)t = m\pi; \qquad t = 2m\pi/(\omega_n + \omega) \tag{1.40}$$

where m and $n = 1, 2, 3, \ldots$.

Equation (1.39) is not relevant (beating phenomenon). Substituting Eq. (1.40) into Eq. (1.33) and simplifying it further,

$$z = z_{max} = \frac{Q_0}{k} \cdot \frac{1}{\left(1 - \omega/\omega_n\right)} \cdot \sin\left(\frac{2\pi m\omega}{\omega_n + \omega}\right) \tag{1.41}$$

In order to determine the maximum dynamic force, the maximum value of z_{max} given in Eq. (1.41) is required:

$$z_{max(max)} = (Q_0/k)(1 - \omega/\omega_n)^{-1} \tag{1.42}$$

So,

$$F_{dynam(max)} = k\left[z_{max(max)}\right] = k(Q_0/k)(1 - \omega/\omega_n)^{-1} = Q_0(1 - \omega/\omega_n)^{-1} \tag{1.43}$$

Hence, the total force on the subgrade will vary between the limits

$$W - Q_0(1 - \omega/\omega_n)^{-1} \qquad \text{and} \qquad W + Q_0(1 - \omega/\omega_n)^{-1}$$

Example 1.3 A machine foundation can be idealized as a mass–spring system. This foundation can be subjected to a force which can be given as Q (lb) $= 8000 \sin \omega t$. Given

$f = 800$ cycles/min

Weight of the machine + foundation $= 40{,}000$ lb

Spring constant $= 400{,}000$ lb/in.,

determine the maximum and minimum force transmitted to the subgrade.

Solution:

$$\text{Natural angular frequency} = \omega_n = \sqrt{\frac{k}{m}} = \sqrt{400{,}000 \Big/ \left(\frac{40{,}000}{32.2 \times 12}\right)}$$

$$= 62.16 \text{ rad/sec}$$

$$F_{\text{dynam}} = Q_0 (1 - \omega/\omega_n)^{-1}$$

But

$$\omega = 2\pi f = 2\pi(800/60) = 83.78 \text{ rad/sec}$$

Thus

$$F_{\text{dynam}} = 8000 \Big/ \left(1 - \frac{83.78}{62.16}\right) = 23{,}000 \text{ lb}$$

Maximum force on the subgrade $= 40{,}000 + 23{,}000 = 63{,}000$ lb
Minimum force on the subgrade $= 40{,}000 - 23{,}000 = 17{,}000$ lb

1.2.3 Free Vibration with Viscous Damping

In the case of undamped free vibration as explained in Section 1.2.1, vibration would continue once the system had been set in motion. However, in practical cases, all vibrations undergo a gradual decrease of amplitude with time. This characteristic of vibration is referred to as *damping*. Figure 1.2b (see p. 2) shows a foundation supported by a spring and a dashpot. The dashpot represents the *damping characteristic* of the soil. The dashpot coefficient is equal to c. For free vibration of the foundation (i.e., the force $Q = Q_0 \sin \omega t$ on the foundation is zero), the differential equation of motion can be given by

$$m\ddot{z} + c\dot{z} + kz = 0 \tag{1.44}$$

Let $z = Ae^{rt}$ be a solution to Eq. (1.44), where A is a constant. Substitution of this into Eq. (1.44) yields

$$mAr^2 e^{rt} + cAre^{rt} + kAe^{rt} = 0$$

or

$$r^2 + (c/m)r + k/m = 0 \tag{1.45}$$

The solutions to Eq. (1.45) can be given as

$$r = -\frac{c}{2m} + \sqrt{\frac{c^2}{4m^2} - \frac{k}{m}} \qquad (1.46)$$

There are three general conditions that may be developed from Eq. (1.46):

1. If $c/2m > \sqrt{k/m}$, both roots of Eq. (1.45) are real and negative. This is referred to as an *overdamped* case.
2. If $c/2m = \sqrt{k/m}$, $r = -c/2m$. This is called the *critical damping* case. Thus, for this case,

$$c = c_c = 2\sqrt{km} \qquad (1.47a)$$

3. If $c/2m < \sqrt{k/m}$, the roots of Eq. (1.45) are complex:

$$r = -\frac{c}{2m} \pm i\sqrt{\frac{k}{m} - \frac{c^2}{4m^2}}$$

This is referred to as a case of *underdamping*.

It is possible now to define a *damping ratio D*, which can be expressed as

$$D = c/c_c = c/(2\sqrt{km}) \qquad (1.47b)$$

Using the damping ratio, Eq. (1.46) can be rewritten as

$$r = -\frac{c}{2m} \pm \sqrt{\frac{c^2}{4m^2} - \frac{k}{m}} = \omega_n\left(-D \pm \sqrt{D^2 - 1}\right) \qquad (1.48)$$

where $\omega_n = \sqrt{k/m}$.

For the *overdamped condition* ($D > 1$),

$$r = \omega_n\left(-D \pm \sqrt{D^2 - 1}\right)$$

For this condition, the equation for displacement (i.e., $z = Ae^{rt}$) may be written as

$$z = A_1\exp\left[\omega_n t\left(-D + \sqrt{D^2 - 1}\right)\right] + A_2\exp\left[\omega_n t\left(-D - \sqrt{D^2 - 1}\right)\right] \qquad (1.49)$$

where A_1 and A_2 are two constants. Now, let

$$A_1 = \tfrac{1}{2}(A_3 + A_4) \qquad (1.50)$$

and

$$A_2 = \tfrac{1}{2}(A_3 - A_4) \qquad (1.51)$$

Substitution of Eqs. (1.50) and (1.51) into Eq. (1.49) and rearrangement gives

$$z = e^{-D\omega_n t}\left\{\tfrac{1}{2}A_3\left[\exp\left(\omega_n\sqrt{D^2-1}\,t\right)+\exp\left(-\omega_n\sqrt{D^2-1}\,t\right)\right]\right.$$
$$\left.+\tfrac{1}{2}A_4\left[\exp\left(\omega_n\sqrt{D^2-1}\,t\right)-\exp\left(-\omega_n\sqrt{D^2-1}\,t\right)\right]\right\}$$

or

$$z = e^{-D\omega_n t}\left[A_3\cosh\left(\omega_n\sqrt{D^2-1}\,t\right)+A_4\sinh\left(\omega_n\sqrt{D^2-1}\,t\right)\right]\qquad\text{(for }D>1)$$

$$(1.52)$$

Equation (1.52) shows that the system which is overdamped will *not oscillate at all*. The variation of z with time will take the form shown in Figure 1.8a.

FIGURE 1.8 Free vibration of a mass–spring–dashpot system: **(a)** overdamped case; **(b)** critically damped case; **(c)** underdamped case.

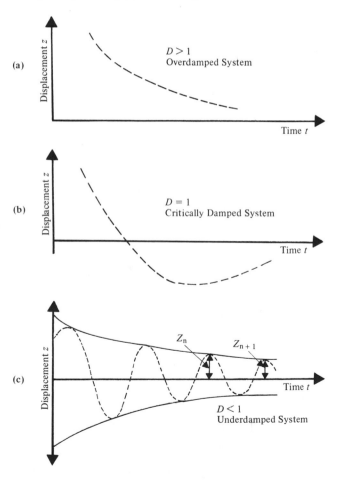

The constants A_3 and A_4 in Eq. (1.52) can be evaluated by knowing the initial conditions. Let, at time $t = 0$, displacement $= z = z_0$ and velocity $= dz/dt = v_0$. From Eq. (1.52) and the first boundary condition,

$$z = z_0 = A_3 \tag{1.53}$$

Again, from Eq. (1.52) and the second boundary condition,

$$dz/dt = v_0 = \left(\omega_n\sqrt{D^2 - 1}\, A_4\right) - D\omega_n A_3$$

or

$$A_4 = \frac{v_0 + D\omega_n A_3}{\omega_n\sqrt{D^2 - 1}} = \frac{v_0 + D\omega_n z_0}{\omega_n\sqrt{D^2 - 1}} \tag{1.54}$$

Substituting Eqs. (1.53) and (1.54) into Eq. (1.52),

$$z = e^{-D\omega_n t}\left[z_0\cosh\left(\omega_n\sqrt{D^2 - 1}\, t\right) + \frac{v_0 D\omega_n z_0}{\omega_n\sqrt{D^2 - 1}}\sinh\left(\omega_n\sqrt{D^2 - 1}\, t\right)\right] \tag{1.55}$$

For a *critically damped condition* ($D = 1$), from Eq. (1.48),

$$r = -\omega_n \tag{1.56}$$

Given this condition, the equation for displacement ($z = Ae^{rt}$) may be written as,

$$z = (A_5 + A_6 t)e^{-\omega_n t} \tag{1.57}$$

where A_5 and A_6 are two constants. This is similar to the case of the overdamped system except for the fact that the sign of z changes only once. This is shown in Figure 1.8b.

The values of A_5 and A_6 in Eq. (1.57) can be determined by using the initial conditions of vibration. Let, at time $t = 0$,

$$z = z_0, \qquad dz/dt = v_0$$

From the first of the above two conditions and Eq. (1.57),

$$z = z_0 = A_5 \tag{1.58}$$

Similarly, from the second condition and Eq. (1.57),

$$dz/dt = v_0 = -\omega_n A_5 + A_6 = -\omega_n z_0 + A_6$$

or

$$A_6 = v_0 + \omega_n z_0 \tag{1.59}$$

A combination of Eqs. (1.57)–(1.59) yields

$$z = [z_0 + (v_0 + \omega_n z_0)t]e^{-\omega_n t} \tag{1.60}$$

Lastly, for the *underdamped condition* ($D < 1$),

$$r = \omega_n\left(-D \pm i\sqrt{1 - D^2}\right)$$

Thus, the general form of the equation for the displacement ($z = Ae^{rt}$) can be expressed as

$$z = e^{-D\omega_n t}\left[A_7\exp\left(i\omega_n\sqrt{1-D^2}\,t\right) + A_8\exp\left(-i\omega_n\sqrt{1-D^2}\,t\right)\right] \quad (1.61)$$

where A_7 and A_8 are two constants.

Equation (1.61) can be simplified to the form

$$z = e^{-D\omega_n t}\left[A_9\cos\left(\omega_n\sqrt{1-D^2}\,t\right) + A_{10}\sin\left(\omega_n\sqrt{1-D^2}\,t\right)\right] \quad (1.62)$$

where A_9 and A_{10} are two constants.

The values of the constants A_9 and A_{10} in Eq. (1.62) can be determined by using the following initial conditions of vibration. Let, at time $t = 0$,

$$z = z_0 \quad \text{and} \quad dz/dt = v_0$$

The final equation with the above boundary conditions will be of the form

$$z = e^{-D\omega_n t}\left[z_0\cos\left(\omega_n\sqrt{1-D^2}\,t\right) + \frac{v_0 + D\omega_n z_0}{\omega_n\sqrt{1-D^2}}\cdot\sin\left(\omega_n\sqrt{1-D^2}\,t\right)\right]$$

$$(1.63)$$

Equation (1.63) can further be simplified as

$$z = Z\cos(\omega_d t - \alpha) \quad (1.64)$$

where

$$Z = e^{-D\omega_n t}\sqrt{z_0^2 + \left(\frac{v_0 + D\omega_n z_0}{\omega_n\sqrt{1-D^2}}\right)^2} \quad (1.65)$$

$$\alpha = \tan^{-1}(v_0 + D\omega_n z_0)/\omega_n z_0\sqrt{1-D^2} \quad (1.66)$$

$$\omega_d = \text{damped natural circular frequency} = \omega_n\sqrt{1-D^2} \quad (1.67)$$

The effect of damping is to decrease gradually the amplitude of vibration with time. In order to evaluate the magnitude of decrease of the amplitude of vibration with time, let Z_n and Z_{n+1} be the two *successive* positive or negative maximum values of displacement at times t_n and t_{n+1} from the start of the vibration as shown in Figure 1.8c. From Eq. (1.65),

$$\frac{Z_{n+1}}{Z_n} = \frac{\exp(-D\omega_n t_{n+1})}{\exp(-D\omega_n t_n)} = \exp\left[-D\omega_n(t_{n+1} - t_n)\right] \quad (1.68)$$

However, $t_{n+1} - t_n$ is the period of vibration T,

$$T = 2\pi/\omega_d = 2\pi/\omega_n\sqrt{1-D^2} \quad (1.69)$$

Thus, combining Eqs. (1.68) and (1.69),

$$\delta = \ln(Z_n/Z_{n+1}) = 2\pi D/\sqrt{1-D^2} \quad (1.70)$$

The time δ is called the *logarithmic decrement*.

If the damping ratio D is small, Eq. (1.70) can be approximated as

$$\delta = \ln(Z_n/Z_{n+1}) = 2\pi D \qquad (1.71)$$

Example 1.4 For a machine foundation, given weight $= 60$ kN, spring constant $= 11,000$ kN/m, and $c = 200$ kN-sec/m, determine:

a. whether the system is overdamped, underdamped, or critically damped
b. the logarithmic decrement
c. the ratio of two successive amplitudes

Solution: a. From Eq. (1.47),

$$c_c = 2\sqrt{km} = 2\sqrt{11,000(60/9.81)} = 518.76 \text{ kN-sec/m}$$

$$c/c_c = D = 200/518.76 = 0.386 < 1$$

Hence, the system is *underdamped*.
b. From Eq. (1.70),

$$\delta = 2\pi D/\sqrt{1-D^2} = 2\pi(0.386)/\sqrt{1-(0.386)^2} = 2.63$$

c. Again, from Eq. (1.70),

$$Z_n/Z_{n+1} = e^\delta = e^{2.63} = 13.87$$

Example 1.5 For Example 1.4, determine the damped natural frequency.

Solution: From Eq. (1.67),

$$f_d = \sqrt{1-D^2}\,f_n$$

where f_d = damped natural frequency.

$$f_n = \frac{1}{2\pi}\sqrt{\frac{k}{m}} = \frac{1}{2\pi}\sqrt{\frac{11,000\times 9.81}{60}} = 6.75 \text{ cps}$$

Thus,

$$f_d = \left(\sqrt{1-(0.386)^2}\right)(6.75) = 6.23 \text{ cps}$$

1.2.4 Steady-State Forced Vibration with Viscous Damping

Figure 1.2b (see p. 2) shows the case of a foundation resting on a soil which can be approximated to an equivalent spring and dashpot. This foundation is being subjected to a sinusoidally varying force $Q = Q_0 \sin \omega \cdot t$. The differential equation of motion for this system can be given by

$$m\ddot{z} + kz + c\dot{z} = Q_0 \sin \omega t \qquad (1.72)$$

The transient part of the vibration is damped out quickly; so, considering the particular solution for Eq. (1.72) for the steady-state motion, let

$$z = A_1 \sin \omega t + A_2 \cos \omega t \qquad (1.73)$$

where A_1 and A_2 are two constants.

Substituting Eq. (1.73) into Eq. (1.72),

$$m\left(- A_1 \omega^2 \sin \omega t - A_2 \omega^2 \cos \omega t\right) + k\left(A_1 \sin \omega t + A_2 \cos \omega t\right)$$

$$+ c\left(A_1 \omega \cos \omega t - A_2 \omega \sin \omega t\right) = Q_0 \sin \omega t \qquad (1.74)$$

Collecting *sine* and *cosine* functions in Eq. (1.74) separately,

$$\left(- mA_1 \omega^2 + kA_1 - cA_2 \omega\right)\sin \omega t = Q_0 \sin \omega t \qquad (1.75a)$$

$$\left(- mA_2 \omega^2 + A_2 k + cA_1 \omega\right)\cos \omega t = 0 \qquad (1.75b)$$

From Eq. (1.75a),

$$A_1\left(\frac{k}{m} - \omega^2\right) - A_2\left(\frac{c}{m}\omega\right) = \frac{Q_0}{m} \qquad (1.76)$$

and from Eq. (1.75b),

$$A_1\left(\frac{c}{m}\omega\right) + A_2\left(\frac{k}{m} - \omega^2\right) = 0 \qquad (1.77)$$

Solution of Eqs. (1.76) and (1.77) will give the following relations for the constants A_1 and A_2:

$$A_1 = \frac{(k - m\omega^2)Q_0}{(k - m\omega^2)^2 + c^2\omega^2} \qquad (1.78)$$

and

$$A_2 = \frac{- c\omega Q_0}{(k - m\omega^2)^2 + c^2\omega^2} \qquad (1.79)$$

By substituting Eqs. (1.78) and (1.79) into Eq. (1.73) and simplifying, one can obtain

$$z = Z \cos(\omega t + \alpha) \qquad (1.80)$$

where

$$\alpha = \tan^{-1}\left(- \frac{A_1}{A_2}\right) = \frac{k - m\omega^2}{c\omega} = \frac{1 - \left(\omega^2/\omega_n^2\right)}{2D(\omega/\omega_n)} \qquad (1.81)$$

$$Z = \sqrt{A_1^2 + A_2^2} = \frac{(Q_0/k)}{\sqrt{\left[1 - \left(\omega^2/\omega_n^2\right)\right]^2 + 4D^2\left(\omega^2/\omega_n^2\right)}} \qquad (1.82)$$

where $\omega_n = \sqrt{k/m}$ is the undamped natural frequency and D is the damping ratio.

Equation (1.82) can be plotted in a nondimensional form as $Z/(Q_0/k)$ against ω/ω_n. This is shown in Figure 1.9. In this figure, note that the maximum values of $Z/(Q_0/k)$ do not occur at $\omega = \omega_n$, as occurs in the case of forced vibration of a spring–mass system (Section 1.2.2). Mathematically, this can be shown as follows: From Eq. (1.82),

$$\frac{Z}{(Q_0/k)} = \frac{1}{\sqrt{\left[1-\left(\omega^2/\omega_n^2\right)\right]^2 + 4D^2\left(\omega^2/\omega_n^2\right)}} \qquad (1.83)$$

For maximum value of $Z/(Q_0/k)$,

$$\frac{\partial\left[Z/(Q_0/k)\right]}{\partial(\omega/\omega_n)} = 0 \qquad (1.84)$$

From Eqs. (1.83) and (1.84),

$$\frac{\omega}{\omega_n}\left(1 - \frac{\omega^2}{\omega_n^2}\right) - 2D^2\left(\frac{\omega}{\omega_n}\right) = 0$$

FIGURE 1.9 Plot of $Z/(Q_0/k)$ against ω/ω_n.

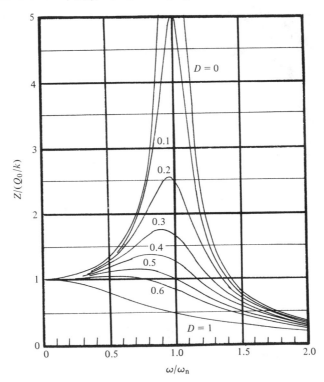

or

$$\omega = \omega_n \sqrt{1 - 2D^2} \qquad (1.85)$$

Hence,

$$f_m = f_n \sqrt{1 - 2D^2} \qquad (1.86)$$

where f_m is the frequency at *maximum amplitude* (the *resonant frequency for vibration with damping*) and f_n is the natural frequency $= (1/2\pi)\sqrt{k/m}$. Hence, the *amplitude of vibration at resonance* can be obtained by substituting Eq. (1.85) into Eq. (1.82):

$$Z_{res} = \frac{Q_0}{k} \frac{1}{\sqrt{[1 - (1 - 2D^2)]^2 + 4D^2(1 - 2D^2)}}$$

$$= \frac{Q_0}{k} \frac{1}{2D\sqrt{1 - D^2}} \qquad (1.87)$$

Maximum Dynamic Force Transmitted to the Subgrade

For vibrating foundations, it is sometimes necessary to determine the dynamic force transmitted to the foundation. This can be given by summing the spring force and the damping force caused by relative motion between mass and dashpot; that is,

$$F_{dynam} = kz + c\dot{z} \qquad (1.88a)$$

From Eq. (1.80),

$$z = Z\cos(\omega t + \alpha)$$

Therefore,

$$\dot{z} = -\omega Z \sin(\omega t + \alpha)$$

and

$$F_{dynam} = kZ\cos(\omega t + \alpha) - c\omega Z \sin(\omega t + \alpha) \qquad (1.88b)$$

If one lets

$$kZ = A\cos\phi \qquad \text{and} \qquad c\omega Z = A\sin\phi,$$

then Eq. (1.88) can be written as

$$F_{dynam} = A\cos(\omega t + \phi + \alpha) \qquad (1.89)$$

where

$$A = \sqrt{(A\cos\phi)^2 + (A\sin\phi)^2} = Z\sqrt{k^2 + (c\omega)^2} \qquad (1.90)$$

Hence, the *magnitude* of *maximum dynamic force* will be equal to $Z\sqrt{k^2 + (c\omega)^2}$.

Example 1.6 A machine and its foundation weighs 140 kN. The spring constant and the damping ratio of the soil supporting the soil may be taken as 12×10^4 kN/m and 0.2, respectively. Forced vibration of the foundation is caused by a force which can be expressed as

$$Q(\text{kN}) = Q_0 \sin \omega t$$
$$Q_0 = 46 \text{ kN}, \qquad \omega = 157 \text{ rad/sec}$$

Determine:

   a. the undamped natural frequency of the foundation
   b. amplitude of motion
   c. maximum dynamic force transmitted to the subgrade

Solution:

a.

$$f_n = \frac{1}{2\pi}\sqrt{\frac{k}{m}} = \frac{1}{2\pi}\sqrt{\frac{12 \times 10^4}{(140/9.81)}} = 14.59 \text{ cps}$$

b. From Eq. (1.82),

$$Z = \frac{Q_0/k}{\sqrt{\left(1 - \omega^2/\omega_n^2\right)^2 + 4D^2\left(\omega^2/\omega_n^2\right)}}$$

$$\omega_n = 2\pi f_n = 2\pi(14.59) = 91.67 \text{ rad/sec}$$

So

$$Z = \frac{46/(12 \times 10^4)}{\sqrt{\left[1 - (157/91.67)^2\right]^2 + 4(0.2)^2 \times (157/91.67)^2}}$$

$$= \frac{3.833 \times 10^{-4}}{\sqrt{3.737 + 0.469}} = 0.000187 \text{ m} = 0.187 \text{ mm}$$

c. From Eq. (1.90), the dynamic force transmitted to the subgrade

$$A = Z\sqrt{k^2 + (c\omega)^2}.$$

From Eq. (1.47),

$$c = 2D\sqrt{km} = 2(0.2)\sqrt{(12 \times 10^4)(140/9.81)} = 523.46 \text{ kN-sec/m}$$

Thus,

$$F_{\text{dynam}} = 0.000187\sqrt{(12 \times 10^4)^2 + (523.46 \times 157)^2} = 27.20 \text{ kN}$$

1.2.5 Rotating-Mass Type Excitation

In many cases of foundation equipment, vertical vibration of foundations is produced by counter-rotating masses as shown in Figure 1.10a. Since horizontal forces on the foundation at any instant cancel, the net vibrating force on the foundation can be determined to be equal to $2m_e e\omega^2 \sin\omega t$ (where m_e = mass of each counter-rotating element, e = eccentricity, and ω = angular frequency of the masses). In such cases, the equation of motion with viscous damping [Eq. (1.72)] can be modified to the form

$$m\ddot{z} + kz + c\ddot{z} = Q_0\sin\omega t \qquad (1.91)$$

$$Q_0 = 2m_e e\omega^2 = U\omega^2 \qquad (1.92)$$

$$U = 2m_e e \qquad (1.93)$$

and m is the mass of the foundation including $2m_e$.

FIGURE 1.10 (a) Rotating mase type excitation; **(b)** plot of $Z/(U/m)$ against ω/ω_n.

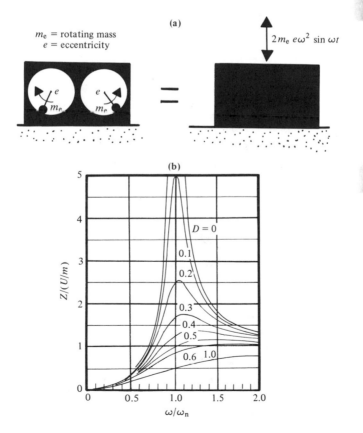

Equations (1.91)–(1.93) can be similarly solved by the procedure presented in Section 1.2.4.

The solution for displacement may be given as

$$z = Z\cos(\omega t + \alpha)$$ (1.94)

where

$$Z = \frac{(U/m)(\omega/\omega_n)^2}{\sqrt{(1 - \omega^2/\omega_n^2)^2 + 4D^2(\omega^2/\omega_n^2)}}$$ (1.95)

$$\alpha = \tan^{-1}\left\{[1 - (\omega^2/\omega_n^2)]/2D\left(\frac{\omega}{\omega_n}\right)\right\}$$ (1.96)

In Section 1.2.4, a nondimensional plot for the amplitude of vibration was given in Figure 1.9 [i.e., $Z(Q_0/k)$ vs ω/ω_n]. This was for a vibration produced by a sinusoidal forcing function ($Q_0 = $ const). For rotating mass type of excitation, a similar type of nondimensional plot for the amplitude of vibration can also be prepared. This is shown in Figure 1.10b, which is a plot of $Z/(U/m)$ vs ω/ω_n. Also proceeding in the same manner [as in Eq. (1.86) for the case where $Q_0 = $ const], the angular resonant frequency for rotating mass type excitation can be obtained as

$$\omega = \omega_n/\sqrt{1 - 2D^2}$$ (1.97)

or

$$f_m = \text{damped resonant frequency} = f_n/\sqrt{1 - 2D^2}$$ (1.98)

The amplitude at damped resonant frequency can be given [similar to Eq. (1.87)] as

$$Z_{res} = (U/m)(2D\sqrt{1 - D^2})^{-1}$$ (1.99)

1.3 TWO DEGREES OF FREEDOM SYSTEMS

1.3.1 Free Vibration of a Mass–Spring System (Coupled Translation)

A mass–spring system with two degrees of freedom is shown in Figure 1.11. If the masses m_1 and m_2 are displaced from their static equilibrium positions, the system will start to vibrate. The equations of motion of the two masses can be given as

$$m_1\ddot{z}_1 + k_1 z_1 + k_2(z_1 - z_2) = 0$$ (1.100)

$$m_2\ddot{z}_2 + k_3 z_2 + k_2(z_2 - z_1) = 0$$ (1.101)

where m_1 and m_2 are the masses of the two bodies, k_1, k_2, and k_3 are the spring constants, and z_1 and z_2 are the displacements of masses m_1 and m_2,

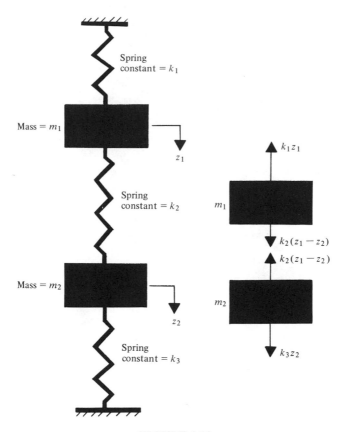

FIGURE 1.11

respectively. Now let

$$z_1 = A \sin(\omega t + \alpha) \qquad (1.102)$$

and

$$z_2 = B \sin(\omega t + \alpha) \qquad (1.103)$$

where A, B, and α are arbitrary constants.

Substitution of Eqs. (1.102) and (1.103) into Eqs. (1.100) and (1.101), respectively, yields

$$\left(k_1 + k_2 - m_1 \omega^2\right) A - k_2 B = 0 \qquad (1.104)$$

and

$$- k_2 A + \left(k_2 + k_3 - m_2 \omega^2\right) B = 0 \qquad (1.105)$$

For nontrivial solutions of ω in Eqs. (1.104) and (1.105),

$$\begin{vmatrix} \left(k_1 + k_2 - m_1 \omega^2\right) & -k_2 \\ -k_2 & \left(k_2 + k_3 - m_2 \omega^2\right) \end{vmatrix} = 0$$

or

$$\omega^4 - \left(\frac{k_1 + k_2}{m_1} + \frac{k_2 + k_3}{m_2} \right)\omega^2 + \frac{k_1 k_2 + k_2 k_3 + k_3 k_1}{m_1 m_2} = 0 \quad (1.106)$$

The roots of the above equation are

$$\omega^2 = \frac{1}{2} \left\{ \left(\frac{k_1 + k_2}{m_1} + \frac{k_2 + k_3}{m_2} \right) \right.$$

$$\left. \mp \left[\left(\frac{k_1 + k_2}{m_1} - \frac{k_2 + k_3}{m_2} \right)^2 + 4\frac{k_2^2}{m_1 m_2} \right]^{1/2} \right\} \quad (1.107)$$

or

$$\begin{matrix} \omega_1 \\ \omega_2 \end{matrix} = \frac{1}{\sqrt{2}} \left\{ \left(\frac{k_1 + k_2}{m_1} + \frac{k_2 + k_3}{m_2} \right) \right.$$

$$\left. \mp \left[\left(\frac{k_1 + k_2}{m_1} - \frac{k_2 + k_3}{m_2} \right)^2 + 4\frac{k_2^2}{m_1 m_2} \right]^{1/2} \right\}^{1/2} \quad (1.108)$$

In the above equation, ω_1 and ω_2 are the natural frequencies of the system. ω_1 is the first mode and ω_2 is the second. The general equation of motion of the two masses can now be written as

$$z_1 = A_1 \sin(\omega_1 t + \alpha_1) + A_2 \sin(\omega_2 t + \alpha_2) \quad (1.109)$$

and

$$z_2 = B_1 \sin(\omega_1 t + \alpha_1) + B_2 \sin(\omega_2 t + \alpha_2) \quad (1.110)$$

The ratios of A_1/B_1 and A_2/B_2 can also be determined from Eqs. (1.104) and (1.105) as

$$\frac{A_1}{B_1} = \frac{k_2}{k_1 + k_2 - m_1 \omega_1^2} = \frac{k_2 + k_3 - m_2 \omega_1^2}{k_2} \quad (1.111)$$

and

$$\frac{A_2}{B_2} = \frac{k_2}{k_1 + k_2 - m_1 \omega_2^2} = \frac{k_2 + k_3 - m_2 \omega_2^2}{k_2} \quad (1.112)$$

Example 1.7 Refer to Figure 1.11. At time $t = 0$, let the mass m_1 be displaced through a vertical distance of 5 mm and released. Determine the displacement equations of m_1 and m_2 with time. Assume $m_1 = m_2 = m$ and $k_1 = k_2 = k$.

Solution: From Eq. (1.108), the natural frequencies ω_1 and ω_2 can be obtained as

$$\begin{matrix} \omega_1 \\ \omega_2 \end{matrix} = \frac{1}{\sqrt{2}} \left\{ \left(\frac{2k}{m} + \frac{2k}{m} \right) \mp \left[\left(\frac{2k}{m} - \frac{2k}{m} \right)^2 + \frac{4k^2}{m^2} \right]^{1/2} \right\}^{1/2} = \frac{1}{\sqrt{2}} \left(\frac{4k}{m} \mp \frac{2k}{m} \right)^{1/2}$$

or

$$\omega_1 = \sqrt{k/m} \qquad \text{and} \qquad \omega_2 = \sqrt{3k/m}$$

Again, from Eq. (1.111),

$$A_1/B_1 = k/[2k - m_1(k/m)] = 1 \qquad \text{or} \qquad A_1 = B_1$$

Similarly, using Eq. (1.112),

$$A_2/B_2 = -1 \qquad \text{or} \qquad A_2 = -B_2$$

Thus, the general equations of motion for the two masses [Eqs. (1.109) and (1.110)] are,

$$z_1 = A_1 \sin\left(\sqrt{k/m}\, t + \alpha_1\right) + A_2 \sin\left(\sqrt{3k/m}\, t + \alpha_2\right) \qquad \text{(E1.7a)}$$

and

$$z_2 = A_1 \sin\left(\sqrt{k/m}\, t + \alpha_1\right) - A_2 \sin\left(\sqrt{3k/m}\, t + \alpha_2\right) \qquad \text{(E1.7b)}$$

Based on the initial condition at time $t = 0$, (1) $z_1 = 5$ mm, $z_2 = 0$; and (2) $\dot{z}_1 = 0$, $\dot{z}_2 = 0$. From Eq. (E1.7a),

$$5 = A_1 \sin\alpha_1 + A_2 \sin\alpha_2 \qquad \text{(E1.7c)}$$

From Eq. (E1.7b),

$$0 = A_1 \sin\alpha_1 - A_2 \sin\alpha_2 \qquad \text{(E1.7d)}$$

Combining Eqs. (E1.7c, d),

$$A_1 = 2.5/\sin\alpha_1 \qquad \text{(E1.7e)}$$

and

$$A_2 = 2.5/\sin\alpha_2 \qquad \text{(E1.7f)}$$

From Eq. (E1.7a) and the second initial condition (i.e., $\dot{z}_1 = 0$ at time $t = 0$),

$$\dot{z}_1 = 0 = A_1\omega_1\cos\alpha_1 + A_2\omega_2\cos\alpha_2 \qquad \text{(E1.7g)}$$

Similarly, from Eq. (E1.7b) and the second initial condition ($\dot{z}_2 = 0$ at time $t = 0$),

$$\dot{z}_2 = 0 = A_1\omega_1\cos\alpha_1 - A_2\omega_2\cos\alpha_2 \qquad \text{(E1.7h)}$$

Combining Eqs. (E1.7g, h),

$$2A_1\omega_1\cos\alpha_1 = 0 \qquad \text{and} \qquad 2A_2\omega_2\cos\alpha_2 = 0$$

or

$$\alpha_1 = \alpha_2 = \tfrac{1}{2}\pi \qquad \text{(E1.7i)}$$

Substituting Eq. (E1.7i) into Eqs. (E1.7e, f),

$$A_1 = A_2 = 2.5$$

With $A_1 = A_2 = 2.5$, Eqs. (E1.7a,b) can be rewritten as

$$z_1 \text{ (mm)} = 2.5\cos\left(\sqrt{k/m}\, t\right) + 2.5\cos\left(\sqrt{3k/m}\, t\right)$$

$$z_2 \text{ (mm)} = 2.5\cos\left(\sqrt{k/m}\, t\right) - 2.5\cos\left(\sqrt{3k/m}\, t\right)$$

Example 1.8 Consider a drop hammer foundation (Figure 1.12a). This hammer foundation can be approximated as a mass–spring system as shown in Figure 1.12b. Determine the amplitude of displacement for the anvil and the foundation. Given

   a. weight of the anvil and the frame $= W_1 = 130,000$ lb
   b. weight of the foundation $= W_2 = 200,000$ lb
   c. the spring constant for the pad between the foundation and the anvil $= k_2 = 150 \times 10^6$ lb/ft

FIGURE 1.12

Weight of anvil and frame $= W_1$
Weight of foundation $= W_2$

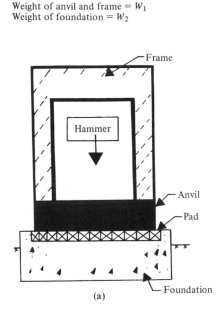

(a)

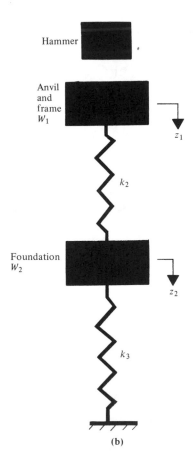

(b)

d. the spring constant for the soil supporting the foundation $= k_3 = 22 \times 10^6$ lb/ft
e. weight of the hammer $= W_h = 8000$ lb
f. velocity of the hammer before impact $= 10$ ft/sec

Solution: At the impact of the hammer, the initial conditions are as follows:

$$z_1 = 0 \qquad \dot{z}_1 = v_0$$
$$z_2 = 0 \qquad \dot{z}_2 = 0$$

According to the theory of conservation of momentum,

$$m_h v_{h(before)} = m_h v_{h(after)} + m_1 v_0 \qquad (E1.8a)$$

where

m_h = mass of the hammer

m_1 = mass of the anvil and frame

$v_{h(before)}$ = velocity of the hammer before impact

$v_{h(after)}$ = velocity of the hammer after impact

v_0 = velocity of the anvil and frame

Also,

$$n = \frac{v_0 - v_{h(after)}}{v_{h(before)}} \qquad (E1.8b)$$

where n is the coefficient of restitution. The value of n may vary between 0.2 and 0.5. For this problem, let $n = 0.4$. Combining Eqs. (E1.8a,b),

$$v_0 = \frac{1+n}{1+m_1/m_h} \cdot v_{h(before)} \qquad (E1.8c)$$

or

$$v_0 = \frac{1+0.4}{1 + \left(\dfrac{130{,}000 \times 32.2}{8{,}000 \times 32.2} \right)} \cdot 10 = 0.812 \text{ ft/sec}$$

Comparing the system given in Figure 1.12b with that given in Figure 1.11, it can be seen that they are equivalent if k_1 in Figure 1.11 is equal to zero. The initial boundary conditions for the motion will be as follows: At time $t = 0$,

$$z_1 = 0, \qquad z_2 = 0$$

Using the initial boundary conditions and Eqs. (1.109)–(1.112) with $k_1 = 0$, the following equations for displacement can be obtained.

$$z_1 = P_1 \left[\left(\frac{k_2}{m_1} - \omega_2^2 \right) \frac{\sin \omega_1 t}{\omega_1} - \left(\frac{k_2}{m_1} - \omega_1^2 \right) \frac{\sin \omega_2 t}{\omega_2} \right] \qquad (E1.8d)$$

and

$$z_2 = P_2\left(\frac{\sin\omega_1 t}{\omega_1} - \frac{\sin\omega_2 t}{\omega_2}\right) \qquad (E1.8e)$$

where

$$P_1 = \frac{v_0}{\left(\omega_1^2 - \omega_2^2\right)} \qquad (E1.8f)$$

$$P_2 = \frac{\left(k_2/m_1 - \omega_2^2\right)\left(k_2/m_1 - \omega_1^2\right)}{\left(k_2/m_1\right)\left(\omega_1^2 - \omega_2^2\right)} v_0 \qquad (E1.8g)$$

The values ω_1 and ω_2 can be calculated by using Eq. (1.108). Note that, that equation, $\omega_1 < \omega_2$. According to Barkan (1962, p. 207), it will be sufficient for these types of problems to approximate Eqs. (E1.8d, e) as follows:

$$z_1 = P_1\left(\frac{k_2}{m_1} - \omega_2^2\right)(\sin\omega_1 t/\omega_1) \qquad (E1.8h)$$

and

$$z_2 = P_2(\sin\omega_1 t/\omega_1) \qquad (E1.8i)$$

Calculation of ω_1 and ω_2 [Eq. (1.108)]

$$\begin{aligned}
\frac{\omega_1}{\omega_2} = \frac{1}{\sqrt{2}} &\left\{\left(\frac{0 + 150\times10^6}{(130\times10^3)/32.2} + \frac{150\times10^6 + 22\times10^6}{(200\times10^3)/32.2}\right)\right. \\
&\mp \left[\left(\frac{0 + 150\times10^6}{(130\times10^3)/32.2} - \frac{150\times10^6 + 22\times10^6}{(200\times10^3)/32.2}\right)^2\right. \\
&\left.\left. + \frac{4(150\times10^6)^2}{(130\times200\times10^6)/(32.2\times32.2)}\right]^{1/2}\right\}^{1/2} \\
&= (1/\sqrt{2})\left[64.84\times10^3 \mp (89.49\times10^6 + 3589\times10^6)^{1/2}\right]^{1/2} \\
&= (1/\sqrt{2})(64.84\times10^3 \mp 60.65\times10^3)^{1/2}
\end{aligned}$$

or

$$\omega_1 = 45.78 \text{ rad/sec}, \qquad \omega_2 = 250.53 \text{ rad/sec}$$

Calculation of Maximum Displacement of Anvil

From Eq. (E1.8h),

$$\text{maximum displacement} = \left[P_1\left(k_2/m_1 - \omega_2^2\right)\right]/\omega_1$$

$$P_1 = \frac{v_0}{\omega_1^2 - \omega_2^2} = \frac{0.812}{(45.78)^2 - (250.53)^2} = -\frac{0.812}{60669.5}$$

Therefore,

$$\text{maximum displacement} = \frac{0.812}{60669.5} \left[\frac{\dfrac{150 \times 10^6}{(130 \times 10^3)/32.2} - (250.53)^2}{45.78} \right]$$

$$= 0.00749 \text{ ft} = 0.0898 \text{ in.}$$

Maximum Displacement of Foundation

From Eq. (E1.8i),

maximum displacement of foundation

$$= \frac{\left(k_2/m_1 - \omega_2^2\right)\left(k_2/m_1 - \omega_1^2\right)v_0}{\left(k_2/m_1\right)\left(\omega_1^2 - \omega_2^2\right)\omega_1}$$

$$= \frac{\left[\dfrac{150 \times 10^6}{(130 \times 10^3)/32.2} - (45.78)^2\right]\left[\dfrac{150 \times 10^6}{(130 \times 10^3)/32.2} - (250.53)^2\right]}{\dfrac{150 \times 10^6}{(130 \times 10^3)/32.2}\left[(45.78)^2 - (250.53)^2\right]45.78}$$

$$= 0.00707 \text{ ft} = 0.0848 \text{ in.}$$

Note. Only the procedure to determine the amplitudes has been shown. However, the amplitudes may not be within the desired limit and the foundation may have to be redesigned. Tolerable amplitudes of displacement for various machine foundations are discussed in Chapter 6.

1.3.2 Coupled Translation and Rotation of a Mass–Spring System (Free Vibration)

Figure 1.13 shows a mass–spring system which will undergo translation and rotation. The equations of motion of the mass m can be given as

$$m\ddot{z} + k_1(z - l_1\theta) + k_2(z + l_2\theta) = 0 \tag{1.113}$$

$$mr^2\ddot{\theta} - l_1 k_1(z - l_1\theta) + l_2 k_2(z + l_2\theta) = 0 \tag{1.114}$$

where

θ = angle of rotation of the mass m

$\ddot{\theta} = d^2\theta/dt^2$

r = radius of gyration of the body about the center of gravity (*Note:* $mr^2 = J$ = mass moment of inertia about the center of gravity.)

k_1, k_2 = spring constants

z = distance of translation of the center of gravity of the body

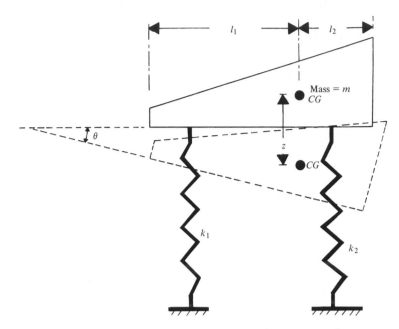

FIGURE 1.13 Coupled translation and rotation of a mass–spring system: free vibration.

Now, let

$$k_1 + k_2 = k_z \tag{1.115}$$

and

$$l_1^2 k_1 + l_2^2 k_2 = k_\theta \tag{1.116}$$

So, the equations of motion can be written as

$$m\ddot{z} + k_z z + (l_2 k_2 - l_1 k_1)\theta = 0 \tag{1.117}$$

$$mr^2\ddot{\theta} + k_\theta \theta + (l_2 k_2 - l_1 k_1)z = 0 \tag{1.118}$$

If $l_1 k_1 = l_2 k_2$, Eq. (1.117) is independent of θ and Eq. (1.118) is independent of z. This means that the two motions (i.e., translation and rotation) can exist independently of each one another (uncoupled motion); that is,

$$m\ddot{z} + k_z z = 0 \tag{1.119}$$

and

$$mr^2\ddot{\theta} + k_\theta \theta = 0 \tag{1.120}$$

The natural circular frequency ω_z of translation can be obtained by

$$\omega_z = \sqrt{k_z/m} \tag{1.121}$$

Similarly, the natural circular frequency of rotation ω_θ can be given by

$$\omega_\theta = \sqrt{k_\theta / mr^2} \tag{1.122}$$

However, if $l_1 k_1$ is not equal to $l_2 k_2$, the equations of motion (coupled motion) can be solved as follows: Let

$$k_z / m = E_1 \tag{1.123}$$

$$(l_2 k_2 - l_1 k_1)/m = E_2 \tag{1.124}$$

$$k_\theta / m = E_3 \tag{1.125}$$

Combining Eqs. (1.117), (1.118), (1.123)–(1.125),

$$\ddot{z} + E_1 z + E_2 \theta = 0 \tag{1.126}$$

$$\ddot{\theta} + (E_3/r^2)\theta + (E_2/r^2)z = 0 \tag{1.127}$$

For solution of the above equations, let

$$z = Z \cos \omega t \tag{1.128}$$

and

$$\theta = \Theta \cos \omega t \tag{1.129}$$

Substitution of Eqs. (1.128) and (1.129) into Eqs. (1.126) and (1.127) results in

$$(E_1 - \omega^2)Z + E_2 \Theta = 0 \tag{1.130}$$

and

$$(E_3/r^2 - \omega^2)\Theta + (E_2/r^2)Z - 0 \tag{1.131}$$

For nontrivial solutions of Eqs. (1.130) and (1.131),

$$\begin{vmatrix} E_1 - \omega^2 & E_2 \\ E_2/r^2 & E_3/r^2 - \omega^2 \end{vmatrix} = 0 \tag{1.132}$$

or

$$\omega^4 - (E_3/r^2 + E_1)\omega^2 + (E_1 E_3 - E_2^2)/r^2 = 0 \tag{1.133}$$

The natural frequencies ω_1, ω_2 of the system can be determined from Eq. (1.133) as

$$\begin{matrix} \omega_1 \\ \omega_2 \end{matrix} = \frac{1}{\sqrt{2}} \left\{ \left(\frac{E_3}{r^2} + E_1 \right) \mp \left[\left(\frac{E_3}{r^2} - E_1 \right)^2 + 4 \frac{E_2^2}{r^2} \right]^{1/2} \right\}^{1/2} \tag{1.134}$$

Hence, the general equations of motion can be given as

$$z = Z_1 \cos \omega_1 \cdot t + Z_2 \cos \omega_2 \cdot t \tag{1.135}$$

and

$$\theta = \Theta_1 \cos \omega_1 \cdot t + \Theta_2 \cos \omega_2 \cdot t \tag{1.136}$$

The amplitude ratios can also be obtained from Eqs. (1.130) and (1.131) as

$$\frac{Z_1}{\Theta_1} = -\frac{E_2}{E_1 - \omega_1^2} = \frac{-\left(E_3/r^2 - \omega_1^2\right)}{E_2/r^2} \tag{1.137}$$

and

$$\frac{Z_2}{\Theta_2} = -\frac{E_2}{E_1 - \omega_2^2} = \frac{-\left(E_3/r^2 - \omega_2^2\right)}{E_2/r^2} \tag{1.138}$$

PROBLEMS

1.1. Define the following terms:
   a. spring constant
   b. coefficient of subgrade reaction
   c. undamped natural circular frequency
   d. undamped natural frequency
   e. period
   f. resonance
   g. critical damping coefficient
   h. damping ratio
   i. damped natural frequency

1.2. A machine foundation can be idealized to a mass–spring system as shown in Figure 1.4 (p. 4). Given

   weight of machine + foundation = 400 kN

   spring constant = 100,000 kN/m

   determine the *natural frequency of undamped free vibration of this foundation and the natural period.*

1.3. Refer to Problem 1.2. What would be the static deflection z_s of this foundation?

1.4. Refer to Example 1.3 (p. 13). Let, for this foundation motion, time $t = 0$, $z = z_0 = 0$, $\dot{z} = v_0 = 0$.
   a. Determine the natural period T of the foundation.
   b. Plot the dynamic force on the subgrade of the foundation due to the forced part of the response for time $t = 0$ to $t = 2T$.
   c. Plot the dynamic force on the subgrade of the foundation due to the free part of the response for $t = 0$ to $t = 2T$.
   d. Plot the total dynamic force on the subgrade [i.e., algebraic sum of (b) and (c)].
   [*Hint:* Refer to Eq. (1.33).

$$\text{Force due to forced part} = k\left(\frac{Q_0/k}{1 - \omega^2/\omega_n^2}\right)\sin \omega t$$

$$\text{Force due to free part} = k\left(\frac{Q_0/k}{1 - \omega^2/\omega_n^2}\right)\left(-\frac{\omega}{\omega_n}\sin \omega_n t\right)$$

1.5. Refer to Figure 1.4. Instead of one spring, let two springs with spring constants k_1 and k_2 be attached in *series* to the foundation of mass m. Determine the natural frequency of undamped free vibration.

1.6. Repeat Problem 1.5 assuming that the two springs are attached in parallel.

1.7. Derive Eqs. (1.97) and (1.99).

1.8. A body weighs 30 lb. A *spring* and a *dashpot* are attached to the body in the manner shown in Figure 1.2b (p. 2). The spring constant is 15 lb/in. The dashpot has a resistance of 0.15 lb at a velocity of 2.5 in./sec. Determine the following for free vibration:
  a. damped natural frequency of the system
  b. damping ratio
  c. the ratio of successive amplitudes of the body (Z_n/Z_{n+1})
  d. amplitude of the body 5 cycles after it is disturbed assuming that at time $t = 0$, $z = 1$ in.

1.9. A machine foundation can be identified as a mass–spring system. This is subjected to a forced vibration. The vibrating force can be expressed as

$$Q = Q_0 \sin \omega t \qquad Q_0 = 15{,}000 \text{ lb} \qquad \omega = 3100 \text{ rad/min}$$

Given

$$\text{weight of machine} + \text{foundation} = 65{,}000 \text{ lb}$$
$$\text{spring constant} = 5000 \text{ kip/in.}$$

Determine the maximum and minimum force transmitted to the subgrade.

1.10. Repeat Problem 1.9 if

$$Q_0 = 200 \text{ kN} \qquad \omega = 6000 \text{ rad/min}$$
$$\text{weight of machine} + \text{foundation} = 400 \text{ kN}$$
$$\text{spring constant} = 120{,}000 \text{ kN/m}$$

1.11. A foundation weighs 800 kN. The foundation and the soil can be approximated as a mass–spring–dashpot system as shown in Figure 1.2b. Given

$$\text{spring constant} = 200{,}000 \text{ kN/m}$$
$$\text{dashpot coefficient} = 2340 \text{ kN-sec/m}$$

Determine
  a. critical damping coefficient c_c
  b. damping ratio
  c. logarithmic decrement
  d. damped natural frequency

1.12. The foundation given in Problem 1.11 is subjected to a vertical force $Q = Q_0 \sin \omega t$ in which

$$Q_0 = 25 \text{ kN} \qquad \omega = 100 \text{ rad/sec}$$

Determine
a. the amplitude of the vertical vibration of the foundation
b. maximum dynamic force transmitted to the subgrade

1.13. Redo Example 1.8 given

$$W_1 = 90{,}000 \text{ lb} \qquad k_2 = 140 \times 10^6 \text{ lb/ft}$$

$$W_2 = 180{,}000 \text{ lb} \qquad k_3 = 32 \times 10^6 \text{ lb/ft}$$

$$\text{weight of hammer} = 5000 \text{ lb}$$

$$\text{velocity of hammer before impact} = 8 \text{ ft/sec}$$

1.14. Refer to Example 1.7.
a. Plot a graph showing the variation of z_1 with $0 \leqslant \sqrt{k/m}\ t \leqslant 6\pi$.
[*Hints:* Plot $2.5 \cos\sqrt{k/m}\ t$ vs $\sqrt{k/m}\ t$; also plot $2.5 \cos\sqrt{3k/m}\ t$ vs $\sqrt{k/m}\ t$; add the coordinates to obtain z_1 vs $\sqrt{k/m}\ t$.
b. Repeat (a) for the expression of z_2.

REFERENCES

Barkan, D. D. (1962). *Dynamics of Bases and Foundations*, McGraw-Hill, New York.

Richart, F. E., Jr. (1962). "Foundation Vibration," *Trans. ASCE* 127 (Part I), 863–898.

2

STRESS WAVES
IN BOUNDED ELASTIC MEDIUM

If a load is suddenly applied to a body, the part of the body closest to the source of disturbance will be affected first. The deformation of the body due to the load will gradually spread throughout the body via *stress waves*. The nature of propagation of stress waves in a bounded elastic medium is the subject of discussion in this chapter.

2.1 LONGITUDINAL ELASTIC WAVES IN A BAR

Figure 2.1 shows a rod, the cross-sectional area of which is equal to A. Let the Young's modulus and the unit weight of the material which constitutes the rod be equal to E and γ, respectively. Now, let the stress along section $a-a$ of the rod increase by σ. The stress increase along the section $b-b$ can then be given by $\sigma + (\partial\sigma/\partial x)\Delta x$. Based on Newton's second law,

$$\sum \text{force} = (\text{mass})(\text{acceleration})$$

Thus, summing the forces in the x direction,

$$-\sigma A + \left(\sigma + \frac{\partial\sigma}{\partial x}\Delta x\right) A = \frac{(A\Delta x\,\gamma)}{g}\frac{\partial^2 u}{\partial t^2} \qquad (2.1)$$

where $A\Delta x\,\gamma$ = weight of the rod of length Δx, g is the acceleration due to gravity, u is displacement in the x direction, and t is time.

Equation (2.1) is based on the assumptions that (1) the stress is uniform over the entire cross-sectional area and (2) the cross section remains plane during the motion. Simplification of Eq. (2.1) gives

$$\partial\sigma/\partial x = \rho\left(\partial^2 u/\partial t^2\right) \qquad (2.2)$$

where $\rho = \gamma/g$ is the density of the material of the bar. However,

$$\sigma = (\text{strain})(\text{Young's modulus}) = (\partial u/\partial x)E \qquad (2.3)$$

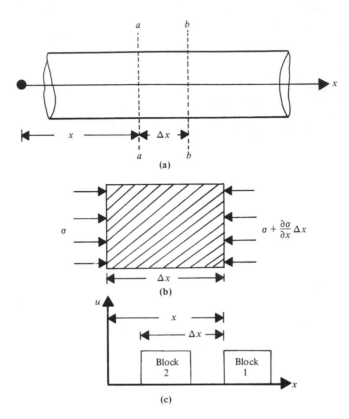

FIGURE 2.1 Longitudinal elastic waves in a bar.

Substitution of Eq. (2.3) into (2.2) yields

$$\partial^2 u / \partial t^2 = (E/\rho)(\partial^2 u / \partial x^2)$$

or

$$\partial^2 u / \partial t^2 = v_c^2(\partial^2 u / \partial x^2) \qquad (2.4)$$

where

$$v_c = \sqrt{E/\rho} \qquad (2.5)$$

The term v_c is the velocity of the *longitudinal* stress wave propagation. This fact can be demonstrated as follows. The solution to Eq. (2.4) can be written in the form

$$u = F(v_c t + x) + G(v_c t - x) \qquad (2.6)$$

where $F(v_c t + x)$ and $G(v_c t - x)$ represent some functions of $(v_c t + x)$ and $(v_c t - x)$, respectively. At a given time t, let the function $F(v_c t + x)$ be represented by block 1 in Figure 2.1c, and

$$u_t = F(v_c t + x)$$

At time $t + \Delta t$, the function will be represented by block 2 in Figure 2.1c. Thus

$$u_{t+\Delta t} = F\left[v_c(t + \Delta t) + (x - \Delta x)\right] \qquad (2.7)$$

If the block moves unchanged in shape from position 1 to position 2,

$$u_t = u_{t+\Delta t}$$

or

$$F(v_c t + x) = F\left[v_c(t + \Delta t) + (x - \Delta x)\right]$$

or

$$v_c \Delta t = \Delta x \qquad (2.8)$$

Thus the velocity of the longitudinal stress wave propagation is equal to $\Delta x / \Delta t = v_c$. In a similar manner, it can be shown that the function $G(v_c t - x)$ represents a wave traveling in the positive direction of x.

If the bar described above is confined, so that no lateral expansion is possible, then the above equation can be modified as

$$\frac{\partial^2 u}{\partial t^2} = v_c'^2 \frac{\partial^2 u}{\partial x^2} \qquad (2.9)$$

where

$$v_c' = \sqrt{M/\rho} \qquad (2.10)$$

and M is the constrained modulus $= E(1 - \mu)/[(1 - 2\mu)(1 + \mu)]$ where μ is Poisson's ratio.

Reflection of Elastic Stress Waves at the End of a Bar

Bars must terminate at some point. One needs to consider the case of what happens when one of these disturbances, $F(v_c t - x)$ or $G(v_c t + x)$, meets the end of the bar.

Figure 2.2a shows a compression wave moving along a bar in the positive x direction. Additionally, a tension wave of the same length is moving along the negative direction of x. When the two waves meet each other (at section $a-a$), the compression and tension cancel each other, resulting in zero stress; however, the particle velocity is doubled. This is because the particle velocity for a compression wave is in the direction of the motion and, in the tension wave, the particle velocity is opposed to the direction of motion. The section $a-a$ corresponds to having the stress condition that a free end of a bar would have. Thus it can be said that, at the free end of a bar, a *compression wave is reflected back as a tension wave* having the same magnitude and shape. In a similar manner, a tension wave is reflected back as a compression wave at the free end of a bar.

Figure 2.2b shows a bar in which two identical compression waves are traveling in opposite directions. When the two waves cross each other at

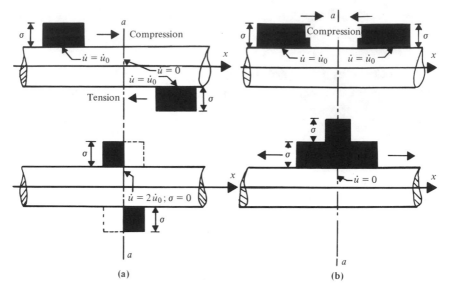

FIGURE 2.2 Reflection of stress waves: **(a)** at a free end of a bar; **(b)** at a fixed end of a bar.

section $a-a$, the magnitude of the stress will be doubled. However, the particle velocity $\dot{u}$ will be equal to zero, which is similar to the condition of the fixed end of a bar: A compression wave is reflected back as a compression wave of the same magnitude and shape, but the stress is doubled at the fixed end. In a similar manner, a tension wave is reflected back as a tension wave at the fixed end of a bar.

2.2 TORSIONAL WAVES IN A BAR

Figure 2.3 shows a rod to which a torque T is applied at a distance x, and the end at x will be rotated through an angle θ. The torque at the section located at a distance $x + \Delta x$ can be given by $T + (\partial T/\partial x)\Delta x$ and the corresponding rotation by $\theta + (\partial\theta/\partial x)\Delta x$. Applying Newton's second law of motion,

$$-T + \left(T + \frac{\partial T}{\partial x}\Delta x\right) = \rho J \Delta x \frac{\partial^2\theta}{\partial t^2} \qquad (2.11)$$

where J is the polar moment of inertia of the cross section of the bar.

However, torque T can be expressed by the relation

$$T = JG\frac{\partial\theta}{\partial x} \qquad (2.12)$$

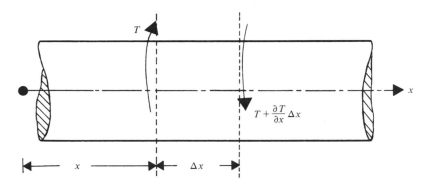

FIGURE 2.3 Torsional waves in a bar.

Substitution of Eq. (2.12) into Eq. (2.11) results in

$$\frac{\partial^2\theta}{\partial t^2} = \frac{G}{\rho}\frac{\partial^2\theta}{\partial x^2} \qquad (2.13)$$

or

$$\frac{\partial^2\theta}{\partial t^2} = v_s^2\frac{\partial^2\theta}{\partial x^2} \qquad (2.14)$$

where

$$v_s = \sqrt{G/\rho} \qquad (2.15)$$

is the velocity of torsional waves. Note that Eqs. (2.14) and (2.4) are of similar form.

2.3 LONGITUDINAL VIBRATION OF SHORT BARS

The solution to the wave equations for short bars vibrating in a natural mode can be written in the general form as

$$u(x,t) = U(x)(A_1\sin\omega_n t + A_2\cos\omega_n t) \qquad (2.16)$$

where A_1 and A_2 are constants, ω_n is the natural circular frequency of vibration, and $U(x)$ is the amplitude of displacement along the length of the rod and is independent of time.

For longitudinal vibration of uniform bars, if Eq. (2.16) is substituted into Eq. (2.4), it yields

$$\frac{\partial^2 u(x,t)}{\partial x^2} - \frac{\rho}{E}\frac{\partial^2 u(x,t)}{\partial t^2} = 0$$

or

$$\frac{\partial^2 U(x)}{\partial x^2} + \frac{\rho}{E}(\omega_n^2) \cdot U(x) = 0 \qquad (2.17)$$

The solution to Eq. (2.17) may be expressed in the form

$$U(x) = B_1 \sin(\omega_n x / v_c) + B_2 \cos(\omega_n x / v_c) \qquad (2.18)$$

where B_1 and B_2 are constants. These constants may be determined by the end condition to which a rod may be subjected.

End Condition: Free – Free

For this condition, the stress and thus the strain at the ends are zero. So at $x = 0$, $dU(x)/dx = 0$; and at $x = L$, $dU(x)/dx = 0$, where L is the length of the bar. Differentiating Eq. (2.18) with respect to x,

$$\frac{dU(x)}{dx} = \frac{B_1 \omega_n}{v_c}\cos\left(\frac{\omega_n x}{v_c}\right) - \frac{B_2 \omega_n}{v_c}\sin\left(\frac{\omega_n x}{v_c}\right) \qquad (2.19)$$

Substitution of the first boundary condition into Eq. (2.19) results in

$$0 = B_1\omega_n / v_c; \qquad \text{i.e.,} \quad B_1 = 0 \qquad (2.20)$$

Again, from the second boundary condition and Eq. (2.19),

$$0 = -(B_2\omega_n / v_c)\sin(\omega_n L / v_c)$$

Since B_2 is not equal to zero,

$$\omega_n L / v_c = n\pi \qquad (2.21)$$

or

$$\omega_n = n\pi v_c / L \qquad (2.22)$$

where $n = 1, 2, 3, \ldots$. Thus,

$$v_c = \frac{\omega_n L}{n\pi} \qquad (2.23)$$

The equation for the amplitude of displacement for this case can be given by combining Eqs. (2.18), (2.20), and (2.23), or

$$U(x) = B_2\cos(n\pi x / L) \qquad (2.24)$$

The variation of the nature of $U(x)$ for the first two harmonics (i.e., $n = 1$ and 2) is shown in Figure 2.4a. The equation for $u(x, t)$ for all modes of vibration can also be given by combining Eqs. (2.24) and (2.16).

End Condition: Fixed – Fixed

For a fixed–fixed end condition, at $x = 0$, $U(x) = 0$ (i.e., displacement is zero); and at $x = L$, $U(x) = 0$.

43

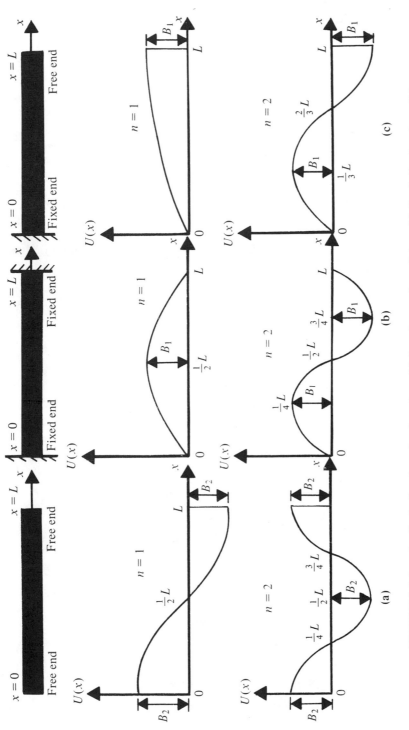

FIGURE 2.4 Longitudinal vibration of a short bar: (**a**) free–free end condition; (**b**) fixed–fixed end condition; (**c**) fixed–free end condition.

Substituting the first boundary condition into Eq. (2.18) results in

$$0 = B_2 \tag{2.25}$$

Again, combining the second boundary condition and Eq. (2.18),

$$0 = B_1 \sin(\omega_n L/v_c)$$

Since $B_1 \neq 0$,

$$\frac{\omega_n L}{v_c} = n\pi \tag{2.26}$$

where $n = 1, 2, 3, \ldots$; or

$$\omega_n = n\pi v_c/L \tag{2.27}$$

or

$$v_c = \omega_n L/(n\pi) \tag{2.28}$$

The displacement amplitude equation can now be given by combining Eqs. (2.18), (2.25), and (2.27) as

$$U(x) = B_1 \sin(n\pi x/L) \tag{2.29}$$

Figure 2.4b shows the variation of $U(x)$ for the first two harmonics ($n = 1$ and 2).

End Condition: Fixed – Free

The boundary conditions for this case can be given as follows:

at $x = 0$ (fixed end), $U(x) = 0$

at $x = L$ (free end), $dU(x)/dx = 0$

From the first boundary condition and Eq. (2.18),

$$U(x) = 0 = B_2 \tag{2.30}$$

Again, from the second boundary condition and Eq. (2.18),

$$\frac{dU(x)}{dx} = 0 = \frac{B_1 \omega_n}{v_c} \cos\left(\frac{\omega_n L}{v_c}\right) \tag{2.31}$$

or

$$\omega_n L/v_c = (2n-1)\tfrac{1}{2}\pi$$

where $n = 1, 2, 3, \ldots$; so

$$\omega_n = \tfrac{1}{2}(2n-1)\pi(v_c/L) \tag{2.32}$$

The displacement amplitude equation can now be written by combining Eqs. (2.18), (2.30), and (2.32) as

$$U(x) = B_1 \sin\left[\tfrac{1}{2}(2n-1)\pi x/L\right] \tag{2.33}$$

Figure 2.4c shows the variation of $U(x)$ for the first two harmonics.

2.4 TORSIONAL VIBRATION OF SHORT BARS

The torsional vibration of short bars can be treated in a manner similar to the longitudinal vibration given in Section 2.3 by writing the equation for natural modes of vibration as

$$\theta(x,t) = \Theta(x)(A_1 \sin \omega_n t + A_2 \cos \omega_n t) \tag{2.34}$$

where Θ = amplitude of angular distorsion, and A_1 and A_2 are constants.
   Solution of Eqs. (2.14) and (2.34) result in

$$\omega_n = n\pi v_s/L \tag{2.35}$$

for the free–free end and fixed–fixed end conditions, and

$$\omega_n = \tfrac{1}{2}(2n-1)\pi v_s/L \tag{2.36}$$

for the fixed–free end condition, where L is the length of the bar and $n = 1,2,3,\dots$.

2.5 EXPERIMENTAL DETERMINATION OF v_c AND v_s FOR SOILS

The laboratory experimental determination of v_c and v_s can be done by the *travel time* method or by using a *resonant column* device. Each of these methods is briefly presented here.

2.5.1 Travel Time Method

Using electronic equipment, the time t_c required for travel of elastic waves through a soil specimen of length L can be measured. For *longitudinal* waves,

$$v_c = L/t_c \tag{2.37}$$

The Young's modulus can then be calculated from Eq. (2.5) as

$$v_c = \sqrt{E/\rho}$$

or

$$E = \rho v_c^2 = \rho L^2/t_c^2 \tag{2.38}$$

   If the soil specimen is *confined laterally*, then the travel time will give the value of v_c' as shown in Eq. (2.10). Thus $v_c' = L/t_c'$, and

$$M = \rho L^2/t_c'^2 \tag{2.39}$$

where t_c' is the travel time of longitudinal waves in a laterally confined specimen.
   Similarly, if the travel time t_s for *torsional waves* through a soil of length L is determined, the velocity v_s can be given by $v_s = L/t_s$, and

$$G = \rho v_s^2 = \rho L^2/t_s^2 \tag{2.40}$$

46

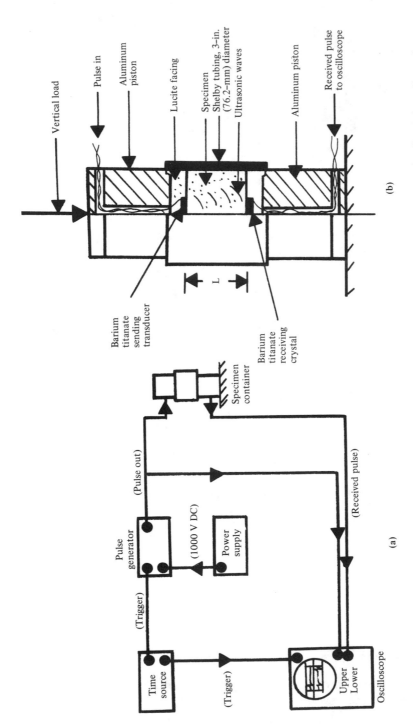

FIGURE 2.5 Travel time method: **(a)** schematic diagram of the laboratory setup and container for the laboratory setup. [Whitman, R. V., and Lawrence, F. V. (1963). "Discussion on Elastic Wave Velocities in Granular Soil," *Journal of the Soil Mechanics and Foundations Division, ASCE*, 89 (SM5), Figs. 21 and 22, p. 113 and 114.]

Vertical load

Pulse in

Aluminum piston

Lucite facing

Specimen
Shelby tubing, 3-in.
(76.2–mm) diameter

Ultrasonic waves

Aluminum piston

Received pulse
to oscilloscope

L

Barium
titanate
sending
transducer

Barium
titanate
receiving
crystal

(b)

Specimen container

(Pulse out)

Pulse generator

(1000 V DC)

Power supply

(Received pulse)

(Trigger)

Time source

(Trigger)

Upper
Lower

Oscilloscope

(a)

Whitman and Lawrence (1963) have given some test results for v_c' in 20–30 Ottawa sand. The schematic diagram of the apparatus for measuring v_c' is shown in Figure 2.5a. The soil specimen was confined in 3-in. (76.2-mm) diameter Shelby tube (Figure 2.5b). Vertical load was applied by an aluminum piston. In this system, a pulse was sent from one piezoelectric crystal and received by a second at the opposite end. The received signal was displayed on an oscilloscope, which allowed measurement of t_c'. It was found that the velocity v_c' increases with the increase of axial pressure.

2.5.2 Resonant Column Method

In this method, a soil column is excited to vibrate in one of its natural modes. Once the frequency at resonance is known, the wave velocity can easily be determined. The soil column in the resonant column device can be excited longitudinally or torsionally yielding velocities of v_c or v_s, respectively.

Several types of resonant column equipment have been used in the past. One of the earlier types of resonant column device was used by Wilson and Dietrich (1960) for testing of clay specimens; a schematic diagram of it is shown in Figure 2.6 (one for longitudinal vibration and one for torsional vibration). An audio frequency generator, whose output is amplified by a simple audio amplifier, supplied a variable frequency voltage to a driver unit adapted from a loudspeaker. For the longitudinal vibration (Figure 2.6a), the moving diaphragm of the driver was connected to a clamped-rim diaphragm by an aluminum rod. The clamped-rim diaphragm had a natural frequency several times larger than that of the soil specimen. The soil specimen to be tested was encased by a rubber membrane and first consolidated in a triaxial chamber by a desired pressure. It was then placed on a base attached to the oscillating diaphragm. A lightweight cap was attached to the top of the specimen; a phonograph crystal was suspended by a rubber thread from a tie rod; the needle of the crystal rested against the cap of the specimen. The output of the pickup was recorded on the screen of a cathode ray oscilloscope.

For the torsional vibration apparatus (Figure 2.6b) the moving diaphragm of the driver was connected by an aluminum rod to an aluminum clamp. This provided a torsional twist to the base of the specimen. The soil specimen encased in a rubber membrane was placed on a base (after proper consolidation in a triaxial chamber) along with a lightweight cap. A phonograph crystal picked up the output which was recorded on the screen of a cathode ray oscilloscope.

At the beginning of a test, the specimen was placed under a vacuum equal to the original chamber confining pressure. By gradually adjusting the oscillator and measuring the amplitude of vibration at the free end (top of the specimen), the resonant frequency of the soil specimen was determined.

48

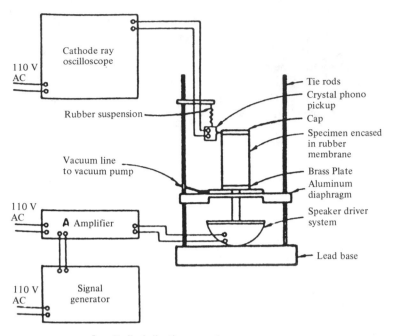

Longitudinal vibration apparatus

(a)

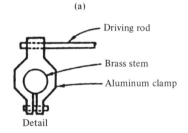

Detail

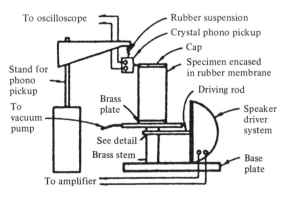

Torsional vibration apparatus

(b)

FIGURE 2.6 Schematic diagram of resonant column device. [Wilson, S. D., and Dietrich, D. J. (1960). "Effect of Consolidation Pressure on Elastic and Strength Properties of Clay," *Proceedings, Research Conference on Shear Strength of Cohesive Soils, ASCE*, Fig. 4, p. 423.]

The vibration of the specimens corresponded to a fixed–free condition (fixed at the lower end and free at the upper). The magnitude of v_c was determined from longitudinal vibration tests as follows. From Eq. (2.32), for $n = 1$ (first normal mode of vibration),

$$\omega_n = \tfrac{1}{2}\pi v_c / L$$

but $\omega_n = 2\pi f_n$, where f_n is the natural frequency; so

$$2\pi f_n = \tfrac{1}{2}\pi v_c / L$$

or

$$v_c = 4 f_n L \tag{2.41}$$

The Young's modulus can then be calculated as

$$v_c = 4 f_n L = \sqrt{E/\rho} \tag{2.42}$$

or

$$E = 16\rho f_n^2 L^2 \tag{2.43}$$

In a similar manner, the magnitude of v_s can be calculated from torsional vibration tests as follows:

$$\omega_n = 2\pi f_n = \tfrac{1}{2}\pi v_s / L \qquad \text{for} \quad n = 1 \tag{2.36}$$

or

$$v_s = 4 f_n L$$

or

$$G = 16\rho f_n^2 L^2 \tag{2.44}$$

Some of the results obtained by Wilson and Dietrich (1960) are presented in Table 2.1. Note that these results are for *low amplitudes of vibration*: strain amplitudes of the order of about 10^{-4} in./in. or less for longitudinal vibrations and about 10^{-4} rad for torsional vibrations. Also presented in Table 2.1 are the values of Poisson's ratio calculated by using the relation

$$\mu = E/2G - 1 \tag{2.45}$$

Hardin and Richart (1963) have reported the use of two types of resonant column device—one for longitudinal vibration and the other for torsional vibration. The specimens were free at each end (free–free end condition). A schematic diagram of the laboratory experimental setup is shown in Figure 2.7. The power supply and amplifier No. 1 were used to amplify the sinusoidal output signal of the oscillator which had a frequency range of

TABLE 2.1 Values of Young's Modulus, Shear Modulus, and Poisson's Ratio for Some Clays[a]

Name of clay	Liquid limit (%)	Plastic limit (%)	Water content	Undrained compression strength (lb/ft²)	(kN/m²)	E (lb/in.²)	(kN/m²)	G (lb/in.²)	(kN/m²)	μ
Cambridge clay	39	21	35.4	2160	103.5	9450	65,205			
Mexico City clay	400±		400±	1800	86.25	830	5727			
Mississippi gumbo	94	31	49.5	1040	49.83	4900	33,810			
Duwamish silt	61	37	60	2720	130.33	5700	39,330	1860	12,834	0.53
Birch Bay clay	34	17	22.8	2000	95.83	12,300	84,870			
	24	13	16.6	1720	82.42	10,800	74,520	3820	26,358	0.61
	33	16	19.4	3440	164.83	16,500				
	35	17	20.2	4840	231.92	23,600		8100		
Montana clay	50	18	15–18	12,000 av.	575	30,000–40,000	207,000–276,000	10,000 av.	69,000	
Whidbey Island clay	62	23	52.8	460	22.04	2910	20,079	900	6210	0.62
Idaho clay	21	16		7200	345	26,000	170,400			
	34	24		11,200	536.7	28,000	193,200			
	35	22		8800	421.7	40,000	276,000			
	45	25		11,200	536.7	42,000	289,800			

[a]Wilson, S. D., and Dietrich, D. J. (1960). "Effect of Consolidation Pressure on Elastic Strength Properties of Clay." *Proceedings, Research Conference on Shear Strength of Cohesive Soils*, ASCE, Table 1, p. 427.

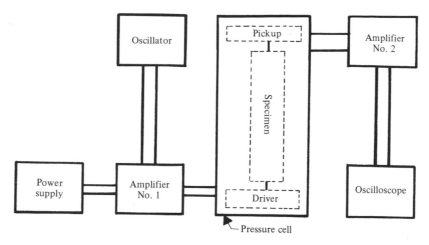

FIGURE 2.7 Schematic diagram of experimental setup for resonant column test. [Hardin, B. O., and Richart, F. E., Jr. (1963). "Elastic Wave Velocities in Granular Soils," *Journal of the Soil Mechanics and Foundations Division*, ASCE, 89 (SM1), Fig. 4, p. 43.]

5–600,000 cps. The amplified signals were fed into the driver producing desired vibrations. Figure 2.8a shows the schematic diagram of the driver for the torsional oscillation. Similarly, the schematic diagram of the driver for the longitudinal vibration is shown in Fig. 2.8b. These devices will give results for low amplitude vibration conditions. With free–free end condi-

FIGURE 2.8 Drawings for steady-state vibration drivers in the resonant column device with free–free end conditions: **(a)** for torsional vibration; **(b)** for longitudinal vibration. [Hardin, B. O., and Richart, F. E., Jr. (1963). "Elastic Wave Velocities in Granular Soils," *Journal of the Soil Mechanics and Foundations Division*, ASCE, 89 (SM1), Fig. 5, p. 46.]

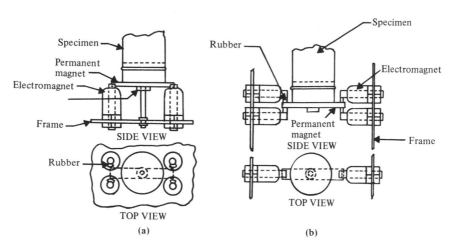

tions, for longitudinal vibrations at resonance,

$$v_c = \omega_n L / n\pi \tag{2.28}$$

For $n = 1$ (i.e., first normal mode of vibration),

$$v_c = \omega_n L / \pi = 2\pi f_n L / \pi = 2 f_n L$$

or

$$v_c = \sqrt{E/\rho} = 2 f_n L$$

or

$$E = 4 f_n^2 \rho L^2 \tag{2.46}$$

Similarly, for torsional vibration, at resonance (with $n = 1$),

$$v_s = 2 f_n L \tag{2.47}$$

or

$$v_s = \sqrt{G/\rho} = 2 f_n L$$

or

$$G = 4 f_n^2 \rho L^2 \tag{2.48}$$

Hall and Richart (1963) have used two other types of resonant column device (one for longitudinal vibration and the other for torsional vibration). The end conditions for these two types of device were *fixed – free*: fixed at the bottom and free at the top of the specimen. The general layouts of the laboratory setup for these equipments were almost the same as those shown in Figure 2.7 except for the fact that the driver and the pickup were located at the top of the specimen (Figure 2.9a, for torsional vibration, and Figure 2.9b, for longitudinal vibration). Since the driver and the pickup were located close together, a correction circuit was introduced to correct the inductive coupling between the driver and the pickup. The driver and pickup were attached to a common frame. The differences in construction and arrangement of the driver and the pickup produced either longitudinal or torsional vibration of the specimen.

Derivation of Expressions for v_c and E for Use in the Fixed – Free Type Resonant Column Test

An equation for the circular natural frequency for the longitudinal vibration of short rods with fixed–free end conditions was derived in Eq. (2.32) as

$$\omega_n = \frac{(2n-1)\pi}{2} \frac{v_c}{L}$$

However, in a fixed–free type resonant column test, the driving mechanism and the motion monitoring device have to be attached to the top of the specimen. This will, in effect, change the boundary conditions assumed in

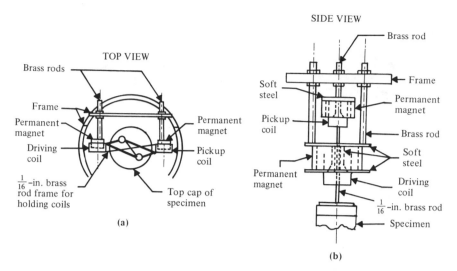

FIGURE 2.9 Driving and measuring components for a fixed–free resonant column device. [Hall, J. R., Jr., and Richart, F. E., Jr. (1963). "Dissipation of Elastic Wave Energy in Granular Soils," *Journal of the Soil Mechanics and Foundations Division*, *ASCE*, 89 (SM6), Fig. 6, p. 42.]

deriving Eq. (2.32), so a modified equation for the circular natural frequency has to be derived. This can be done as follows, with reference to Figure 2.10a.

Let the mass of the attachments placed on the specimen be equal to m. For the vibration of the soil column in a natural mode,

$$u(x,t) = U(x)(A_1\sin \omega_n t + A_2\cos \omega_n t) \qquad (2.16)$$

and

$$U(x) = B_1\sin(\omega_n x/v_c) + B_2\cos(\omega_n x/v_c) \qquad (2.18)$$

At $x = 0$, $U(x) = 0$. Thus, B_2 in Eq. (2.18) is zero and

$$U(x) = B_1\sin(\omega_n x/v_c) \qquad (2.49)$$

At $x = L$, the inertia force F of mass m is acting on the soil column and can be expressed as

$$F = -m\frac{\partial^2 u}{\partial t^2} \qquad (2.50)$$

Also, the strain

$$\frac{\partial u}{\partial x} = \frac{F}{AE} \qquad (2.51)$$

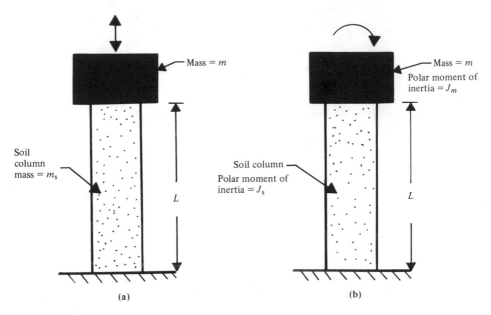

FIGURE 2.10 Derivation of: **(a)** Eq. (2.56); **(b)** Eq. (2.60).

where A is the cross-sectional area of the specimen and E is Young's modulus. Combining Eqs. (2.16), (2.49), and (2.51), we get

$$\frac{F}{AE} = \frac{\partial u}{\partial x} = \left(\frac{\partial U}{\partial x}\right)(A_1 \sin \omega_n t + A_2 \sin \omega_n t)$$

$$= \frac{\partial}{\partial x}\left[B_1 \sin\left(\frac{\omega_n x}{v_c}\right)\right](A_1 \sin \omega_n t + A_2 \sin \omega_n t)$$

$$= (B_1 \omega_n / v_c)\left[\cos(\omega_n x / v_c)\right](A_1 \sin \omega_n t + A_2 \sin \omega_n t) \quad (2.52)$$

Again, combining Eqs. (2.16), (2.49), and (2.50),

$$F = -m\left(\frac{\partial^2 u}{\partial t^2}\right) = -m\left[B_1 \sin\left(\frac{\omega_n x}{v_c}\right)\right]\left(\frac{\partial^2}{\partial t^2}\right)(A_1 \sin \omega_n t + A_2 \sin \omega_n t)$$

$$= m\omega_n^2 B_1 \sin(\omega_n x / v_c)(A_1 \sin \omega_n t + A_2 \sin \omega_n t) \quad (2.53)$$

Now, from Eqs. (2.52) and (2.53),

$$\frac{AE}{v_c}\cos\left(\frac{\omega_n x}{v_c}\right) = m\omega_n \sin\left(\frac{\omega_n x}{v_c}\right) \quad (2.54)$$

At $x = L$,

$$AE = m\omega_n v_c \tan(\omega_n L / v_c) \quad (2.55)$$

TABLE 2.2 Values of α and Corresponding $AL\gamma/W$ [Eq. (2.56)]

$AL\gamma/W$	α (rad)	$AL\gamma/W$	α (rad)
0.1	0.32	1	0.86
0.3	0.53	2	1.08
0.5	0.66	4	1.27
0.7	0.75	10	1.43

However $v_c = \sqrt{E/\rho}$, or $E = v_c^2\rho$. Substitution of this in Eq. (2.55) gives

$$Av_c^2\rho = m\omega_n v_c\tan(\omega_n L/v_c)$$

$$\frac{A\rho}{m} = \frac{\omega_n}{v_c}\tan\left(\frac{\omega_n L}{v_c}\right)$$

$$\frac{AL\rho}{m} = \frac{\omega_n L}{v_c}\tan\left(\frac{\omega_n L}{v_c}\right)$$

or

$$AL\gamma/W = \alpha\tan\alpha \tag{2.56}$$

where $\gamma = \rho g$ is the unit weight of soil, $W = mg$ is the weight of the attachments on top of the specimen, and

$$\alpha = \omega_n L/v_c \tag{2.57}$$

The values of α corresponding to some values of $AL\gamma/W$ [Eq. (2.56)] are given in Table 2.2.

In any resonant column test, the ratio $AL\gamma/W$ will be known. With a known value of $AL\gamma/W$, the value of α can be determined and the natural frequency of vibration can be obtained from the test. Thus,

$$\alpha = \omega_n L/v_c = 2\pi f_n L/v_c$$

or

$$v_c = 2\pi f_n L/\alpha \tag{2.58}$$

The Young's modulus of the soil can then be obtained as

$$E = \rho v_c^2 = \rho(2\pi f_n L/\alpha)^2 = 39.48(f_n^2 L^2/\alpha^2)\cdot\rho \tag{2.59}$$

Derivation of Expressions for v_s and G for Use in the Fixed–Free Type Resonant Column Test

In the resonant column tests where soils specimens are subjected to torsional oscillations with fixed–free end conditions, the mass of the driving and motion monitoring devices (Figure 2.10b) can also be taken into

account. For this condition, an equation similar to Eq. (2.56) can be derived
which is of the form

$$\frac{J_s}{J_m} = \frac{\omega_n L}{v_s} \tan \frac{\omega_n L}{v_s} = \alpha \tan \alpha \tag{2.60}$$

where J_s is the mass polar moment of inertia of the soil specimen and J_m is
the mass polar moment of inertia of the attachments with mass m. Thus,

$$v_s = \omega_n L / \alpha = 2\pi f_n L / \alpha \tag{2.61}$$

and

$$G = \rho v_s^2 = 39.48 \left(f_n^2 L^2 / \alpha^2 \right) \cdot \rho \tag{2.62}$$

Typical Laboratory Test Results from Free – Free and Fixed – Free Types of Resonant Column Test

Typical results of v_c and v_s for No. 20–30 Ottawa Sand compacted at a void
ratio of about 0.55 are shown in Chapter 3 (Figure 3.6, p. 93). These tests
were conducted by using the free–free and fixed–free type resonant column
devices developed by Hardin and Richart (1963) and Hall and Richart
(1963). Note that these results are for tests conducted with *low amplitudes of
vibration*. Based on the results given in Figure 3.6, the following general
conclusions can be drawn:

1. The values of v_c and v_s in soils increase with the increase of the
   effective average confining pressure $\bar{\sigma}_0$.
2. The values of v_c and v_s for saturated soils are slightly lower than those
   for dry soils. This can be accounted for by the increase of the unit
   weight of soil due to the presence of water in the void spaces.

2.6 CORRELATIONS FOR SHEAR MODULUS OF GRANULAR SOILS: LOW AMPLITUDE VIBRATION

Hardin and Richart (1963) have reported the results of several resonant
column tests conducted in dry Ottawa sands. The shear wave velocities
determined from these tests are shown in Fig. 2.11. The peak-to-peak shear
strain amplitude for these tests was 10^{-3} rad. From Fig. 2.11 it may be seen
that the values of v_s are independent of the gradation, grain-size distribution, and also the relative density of compaction. However, v_s is dependent
on the void ratio and the effective confining pressure and can be expressed
by the following empirical relations:

$$v_s = (170 - 78.2e)\bar{\sigma}_0^{1/4} \qquad \text{for} \quad \bar{\sigma}_0 \geq 2000 \text{ lb/ft}^2 \tag{2.63}$$

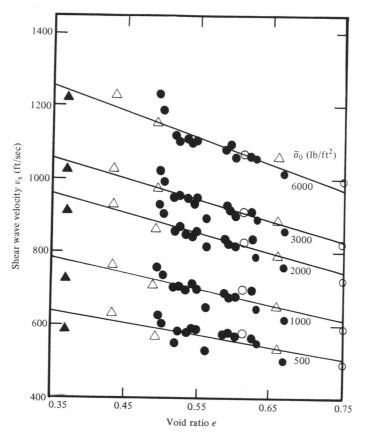

FIGURE 2.11 Variation of shear wave velocity with $\bar{\sigma}_0$ for round-grained dry Ottawa sand: ● No. 20–No. 30; ○ No. 80–no. 140; ▲ 74.8% of No. 20–No. 30 and 25.2% of No. 80–No. 140; △ No. 20–No. 140, well graded. Note: 1 m/sec = 3.28 ft/sec; 1 lb/ft^2 = 47.88 N/m^2. [Hardin, B. O., and Richart, F. E., Jr. (1963). "Elastic Wave Velocities in Granular Soils," *Journal of the Soil Mechanics and Foundations Division*, *ASCE*, 89 (SM1), Fig. 15, p. 59.]

and

$$v_s = (119 - 56e)\bar{\sigma}_0^{0.3} \qquad \text{for} \quad \bar{\sigma}_0 < 2000 \text{ lb/ft}^2 \qquad (2.64)$$

where e = void ratio and v_s and $\bar{\sigma}_0$ are in ft/sec and lb/ft^2, respectively.

In SI units, these relations may be expressed as

$$v_s = (19.7 - 9.06e)\bar{\sigma}_0^{1/4} \qquad \text{for} \quad \bar{\sigma}_0 \geqslant 95.8 \text{ kN/m}^2 \qquad (2.65)$$

and

$$v_s = (11.36 - 5.35e)\bar{\sigma}_0^{0.3} \qquad \text{for} \quad \bar{\sigma}_0 < 95.8 \text{ kN/m}^2 \qquad (2.66)$$

In Eqs. (2.65) and (2.66), the units of v_s and $\bar{\sigma}_0$ are in m/sec and N/m^2, respectively.

Several experimental results for shear wave velocity in extremely *angular crushed quartz sands* have also been reported by Hardin and Richart (1963). Based on these results, the value of v_s for angular sands can be expressed by the empirical relation

$$v_s = (159 - 53.5e)\ \bar{\sigma}_0^{1/4} \tag{2.67}$$
$$\text{(ft/sec)} \qquad\qquad\qquad \text{(lb/ft}^2\text{)}$$

In SI units,

$$v_s = (18.43 - 6.2e)\ \bar{\sigma}_0^{1/4} \tag{2.68}$$
$$\text{(m/sec)} \qquad\qquad\qquad \text{(N/m}^2\text{)}$$

Based on the shear wave velocity relations presented above, the shear modulus of sands for *low amplitudes of vibration* can be given by the following relations (Hardin and Black, 1968):

$$G = \frac{2630(2.17 - e)^2}{1 + e}\bar{\sigma}_0^{1/2} \qquad \text{for round grained sands} \tag{2.69}$$

and

$$G = \frac{1230(2.97 - e)^2}{1 + e}\bar{\sigma}_0^{1/2} \qquad \text{for angular grained sands} \tag{2.70}$$

The units of G and $\bar{\sigma}_0$ in Eqs. (2.69) and (2.70) are in lb/in.^{2}. In SI units (kN/m^2), these relations can be expressed as

$$G = \frac{6908(2.17 - e)^2}{1 + e}\bar{\sigma}_0^{1/2} \qquad \text{round grained} \tag{2.71}$$

and

$$G = \frac{3230(2.97 - e)^2}{1 + e}\bar{\sigma}_0^{1/2} \qquad \text{angular grained} \tag{2.72}$$

At this time, it needs to be pointed out that for a soil specimen subjected to a stress condition such that $\bar{\sigma}_1 \neq \bar{\sigma}_2 \neq \bar{\sigma}_3$ (where $\bar{\sigma}_1$, $\bar{\sigma}_2$, and $\bar{\sigma}_3$ are the major, intermediate, and minor effective principal stresses, respectively), the average effective confining pressure is

$$\bar{\sigma}_0 = \tfrac{1}{3}(\bar{\sigma}_1 + \bar{\sigma}_2 + \bar{\sigma}_3) = \text{effective octahedral stress}$$

This value of $\bar{\sigma}_0$ can be used in Eqs. (2.63)–(2.72).

2.7 CORRELATION FOR SHEAR MODULUS OF COHESIVE SOILS: LOW AMPLITUDE VIBRATION

Lawrence (1965), Hardin and Black (1968), and Humphries and Wahls (1968) have reported results of laboratory tests for shear modulus of cohesive soils. The experimental studies of Hardin and Black and those of Humphries and Wahls were conducted by resonant column devices. Lawrence conducted his tests on cylindrical specimens and measured the travel time for high-frequency shear pulses to determine v_s. These tests indicate that, for *normally consolidated clays* of modest sensitivity, Eq. (2.70) [or (2.72)] can be used to predict the shear modulus reasonably. This fact is demonstrated in Figure 2.12. The experimental values of G shown in this figure have been calculated by Hardin and Black (1968) from the shear wave velocities of kaolinite and Boston Blue clay determined by Lawrence.

FIGURE 2.12 Experimental values of G for some normally consolidated clays: □ dispersed kaolinite; ● tap water kaolinite; △ salt flocculated kaolinite; ▽ flocculated Boston Blue clay; ○ dispersed Boston Blue clay; — — — Eq. (2.70) [or (2.72)]. [Hardin, B. O., and Black, W. L. (1968). "Vibration Modulus of Normally Consolidated Clay," *Journal of the Soil Mechanics and Foundations Division*, *ASCE*, 94 (SM2), Fig. 4, p. 361.]

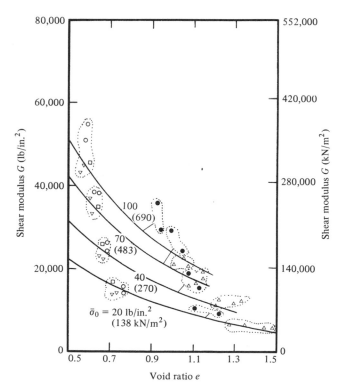

TABLE 2.3 Soil Properties for Undisturbed Specimens Reported in Figure 2.13[a]

Soil No.	Symbol in Figure 2.13	Name & Remarks	ASTM Classification
1	○	Lick Creek Silt Loam: Normally consolidated with wood particles, 8.5–10.5 ft (2.58–3.2 m), Kentucky	ML
2	●	Lick Creek Silt: Depth 25 ft (7.62 m), Kentucky	ML
3	■	Leda Clay: Sensitive Canadian clay	CH
4	△	San Francisco Bay Mud: Depth 25–28 ft (7.62–8.54 m)	CL–CH
5	▲	San Francisco Bay Mud: Depth 94 ft (28.66m)	CH
6	▽	Rhodes Creek Silty Clay: Depth 8.5–11 ft (2.59–3.35 m), Kentucky	CL
7	▼	Little River Silt Loam: Depth 8.5–10.5 ft (2.59–3.2 m), Kentucky	CL
8	◆	Virginia Clay: Depth 43–50 ft (13.1– 15.24 m), Block sample	CH
9	✕	San Francisco Sand: Depth 60–65 ft (18.29–19.82 m)	SM
10	◇	Floyd Brown Loam: Depth 8–10.5 ft (2.44– 3.2 m), Kentucky	CL

[a]Hardin, B. O., and Black, W. L. (1969). "Closure on Vibration Modulus of Normally Consolidated Clay," *Journal of the Soil Mechanics and Foundations Division*, *ASCE*, 95 (SM6), Table 3, p. 1532.

Hardin and Black (1969) also reported the shear moduli for some undisturbed clay specimens collected from the field. The properties of these specimens are given in Table 2.3. For *normally consolidated clays*, Eq. (2.70) [or (2.72)] can be written in a general form as

$$G = C_1 F(e) \bar{\sigma}_0^{1/2} \tag{2.73}$$

where $C_1 = $ const and

$$F(e) = (2.97 - e)^2 / (1 + e) \tag{2.74}$$

TABLE 2.3 (*continued*)

Natural Moisture Content w_n	Natural Void Ratio e_n	Liquid Limit	Plastic Limit	Activity	Specific Gravity of Soil Solids
23	0.61	22	20	0.29	2.67
36	0.95	34	27	0.81	2.73
71	1.98	59	23	0.47	2.74
52	1.42	49	25	1.11	2.75
44	1.2	79	27	1.16	2.75
36	0.98	42	22	0.68	2.73
19	0.59	29	15	0.79	2.70
33	0.9	54	28		
18	0.51	NP	NP		2.80
25	0.66	34	22	0.64	2.65

or

$$G = CF(e) \qquad (2.75)$$

where

$$C = C_1 \bar{\sigma}_0^{1/2} \qquad (2.76)$$

If the value of $\bar{\sigma}_0$ is constant, C will be a constant; and, hence, $G \propto F(e)$. This fact is demonstrated in Figure 2.13 for the normally consolidated undisturbed clays described in Table 2.3.

For highly sensitive clay, the special structure tends to increase the value of the shear modulus as compared to that for a normal clay at similar void

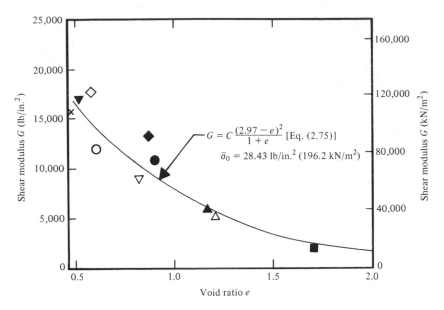

FIGURE 2.13 Effect of void ratio on the shear modulus of undisturbed normally consolidated clays. See Table 2.3 for explanation of symbols. [Hardin, B. O., and Black, W. L. (1969). "Closure on Vibration Modulus of Normally Consolidated Clay," *Journal of the Soil Mechanics and Foundations Division*, ASCE, 95 (SM6), Fig. 14, p. 1533.]

ratio and similar value of $\bar{\sigma}_0$. This can be seen for the case of a Nevada clay (activity = 2.13) in Figure 2.14.

Shear Modulus of Overconsolidated Clays

If a soil specimen is first consolidated under a hydrostatic stress of $\bar{\sigma}_c$ ($\approx \sigma_c$) and then tested under an effective confining pressure of $\bar{\sigma}_0$ ($\bar{\sigma}_0 < \bar{\sigma}_c$), the soil will be preconsolidated. The overconsolidation ratio (OCR) can be given by

$$OCR = \bar{\sigma}_c / \bar{\sigma}_0 \qquad (2.77)$$

The shear modulus of an overconsolidated clay of moderate sensitivity can be expressed by an empirical relation (Hardin and Drnevich, 1972)

$$G = \frac{1230(2.97 - e)^2}{1 + e} (OCR)^K \bar{\sigma}_0^{1/2} \qquad (lb/in.^2) \qquad (2.78)$$

or

$$G = \frac{3230(2.97 - e)^2}{1 + e} (OCR)^K \bar{\sigma}_0^{1/2} \qquad (kN/m^2) \qquad (2.79)$$

The term K in Eqs. (2.78) and (2.79) is dependent on the plasticity index of soils (Table 2.4).

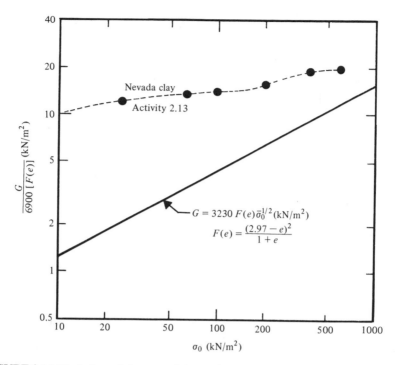

FIGURE 2.14 Variation of shear modulus with $\bar{\sigma}_0$ for a sensitive Nevada clay. Note: 1 lb/in.^{2} = 6.9 kN/m^2. [Hardin, B. O., and Black, W. L. (1969). "Closure on Vibration Modulus of Normally Consolidated Clay," *Journal of the Soil Mechanics and Foundations Division*, ASCE, 95 (SM6), Fig. 20, p. 1536.]

Effect of Secondary Consolidation on the Shear Modulus of Clay

If a saturated clay specimen is subjected to a hydrostatic confining pressure σ_0, the pore water pressure in the specimen will immediately increase. If drainage from the specimen is allowed, its volume will gradually decrease (i.e., a *decrease* of void ratio) accompanied by an *increase* of effective

TABLE 2.4 Values[a] of K

Plasticity index	K
0	0
20	0.18
40	0.30
60	0.41
80	0.48
≥ 100	0.5

[a]Hardin, B. O., and Drnevich, V. P. (1972). "Shear Modulus and Damping Soils: Design Equation and Curves," *Journal of the Soil Mechanics and Foundations Division*, ASCE, 98 (SM7), Table 1, p. 672.

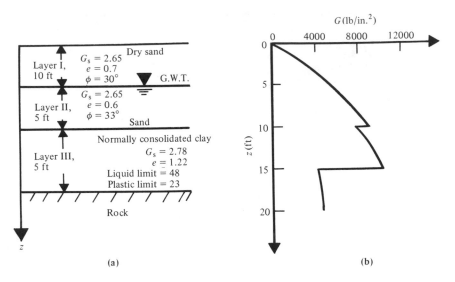

FIGURE 2.15

confining pressure. If this specimen is tested for determination of shear modulus, it will be noticed that G is increasing with the progress of primary consolidation following the general relation given by Eq (2.70) [or Eq. (2.72)]. However it has been shown by Hardin and Black (1968) and Humphries and Wahls (1968) that the shear modulus continues to increase during the period of secondary consolidation. Note that during secondary consolidation, although there is a decrease of the void ratio of the specimen, the *effective* confining pressure $\bar{\sigma}_0$ is equal to the total confining pressure σ_0. The total increase of G at any time of the secondary consolidation of the specimen cannot be fully accounted for by the change of void ratio alone. Hence, a portion of the increase of the shear modulus may be due to the change in the soil structure. This may, however, by destroyed by changes in effective stress.

Example 2.1 A soil profile is shown in Figure 2.15a. Calculate and plot the variation of shear modulus with depth (for low amplitude of vibration).

Solution: At any depth z,

$$\bar{\sigma}_0 = \tfrac{1}{3}(\bar{\sigma}_1 + \bar{\sigma}_2 + \bar{\sigma}_3) \qquad \bar{\sigma}_2 = \bar{\sigma}_3 = K_0\bar{\sigma}_1$$

where K_0 is the coefficient of earth pressure at rest and $\bar{\sigma}_1$ is the vertical effective pressure.

For sands,

$$K_0 = 1 - \sin\phi \tag{E2.1a}$$

For normally consolidated clays (Brooker and Ireland, 1965),

$$K_0 = 0.4 + 0.007(\text{PI}) \qquad \text{for} \quad 0 \leqslant \text{PI} \leqslant 40 \tag{E2.1b}$$

where PI is the plasticity index. Also,

$$K_0 = 0.68 + 0.001(PI - 40) \qquad \text{for} \quad 40 \leqslant PI \leqslant 80 \qquad (E2.1c)$$

Calculation of Effective Unit Weights

z = 0–10 ft

$$\gamma_{dry} = \gamma_{eff} = G_s\gamma_w/(1+e) = 2.65(62.4)/1.7 = 97.27 \text{ lb/ft}^3$$

z = 10–15 ft

$$\gamma_{eff} = \gamma_{sat} - \gamma_w = (G_s + e)\gamma_w/(1+e) - \gamma_w$$
$$= (G_s - 1)\gamma_w/(1+e) = (2.65 - 1)62.4/1.6$$
$$= 64.35 \text{ lb/ft}^3$$

z = 15–20 ft

$$\gamma_{eff} = (G_s - 1)\gamma_w/(1+e) = (2.78 - 1)62.4/2.22 = 50.03 \text{ lb/ft}^3$$

Table E2.1 can now be prepared.

TABLE E2.1

Depth z (ft)	$\bar{\sigma}_1$ (lb/ft^2)	$\bar{\sigma}_2 = \bar{\sigma}_3 = K_0\bar{\sigma}_1$ (lb/ft^2)^a	$\bar{\sigma}_0$ (lb/ft^2)	e	G [Eq. (2.70)]^b (lb/in^2)
0	0	0	0	0.7	0
5	(97.27)(5) = 486.35	243.18^c	324.24	0.7	5595
10 (In Layer I)	(97.27)(10) = 972.7	486.35^c	648.47	0.7	7912
10 (In Layer II)	972.7	442.58^d	619.29	0.6	8955
15	972.7 + (64.35)(5) = 1294.45	588.97^d	824.04	0.6	10329
15 (In Layer III)	1294.45	744.30^e	927.69	1.22	4306
20	1294.45 + (50.03)(5) = 1544.6	888.15^e	1106.96	1.22	4704

^{a}*Note*: The coefficients of earth pressure at rest given in Eqs. (E2.1b, c) are for normally consolidated clay. Relations for estimation of K_0 for overconsolidated clays have also been provided by Brooker and Ireland (1965). For peats, the coefficient of earth pressure at rest varied from 0.53 to 0.3 (Edil and Dhowian, 1981).

^{b}Shear modulus $G = 1230(2.97 - e)^2/1 + e\bar{\sigma}_0^{1/2}$. The variation of G with depth is shown in Figure 2.15b.

$^c K_0 = 1 - \sin\phi = 1 - \sin 30° = 0.5$.

$^d K_0 = 1 - \sin\phi = 1 - \sin 33° = 0.45$.

$^e PI = LL - PL = 48 - 23 = 25$; $K_0 = 0.4 + 0.007(PI) = 0.4 + 0.007(25) = 0.575$.

2.8 SHEAR MODULUS FOR LARGE STRAIN AMPLITUDES

For solid cylindrical specimens torsionally excited by resonant column devices (as described in Section 2.5), the shear strain varies from *zero at the center* to a *maximum at the periphery*, and it is difficult to evaluate a representative strain. For that reason, Drnevich et al. (1967) used hollow cylindrical soil specimens in resonant column device to determine the shear modulus of sand at large strain amplitudes. The inside and outside diameters of the hollow specimens were 40 and 50 mm, respectively. The variation

FIGURE 2.16 Effect of strain amplitude on shear modulus. Note: 1 lb/in.^{2} = 6.9. kN/m^2. [(a) Drnevich, V. P., Hall, J. R., Jr., and Richart, F. E., Jr. (1967). "Effect of the Amplitude of Vibration on the Shear Modulus of Sand," *Proceedings, International Symposium on Wave Propagation and Dynamic Properties of Earth Materials.* University of New Mexico Press, Albuquerque, Fig. 9, p. 198.]

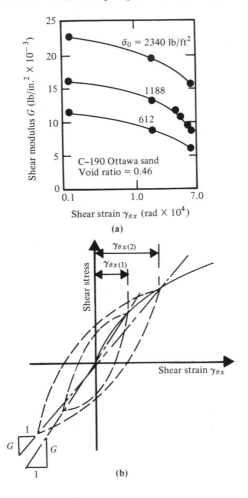

of the shear modulus of dense C-190 Ottawa sand with the shear strain amplitude ($\gamma_{\theta x}$) is shown in Figure 2.16a. Note that the value of G decreases with $\gamma_{\theta x}$—but more rapidly for $\gamma_{\theta x} > 10^{-4}$. This is true for all soils. The reason for this can be explained by the use of Figure 2.16b, which is a stress–strain diagram for a soil. The stress–strain relationships of soils are curvilinear. The shear modulus that is determined experimentally is the secant modulus determined by joining the extreme points on the hysteresis loop. Note that, when the amplitude of strain is small [i.e., $\gamma_{\theta x} = \gamma_{\theta x(1)}$ (Figure 2.16b)], the value of G is larger as compared to that for the larger strain level (i.e., $\gamma_{\theta x} = \gamma_{\theta x(2)}$. A more detailed description of shear modulus of soils at large strain amplitudes is given in Chapter 8.

2.9 EFFECT OF PRESTRAINING ON THE SHEAR MODULUS OF SOILS

The effect of shear modulus of soils due to prestraining has also been reported by Drnevich et al. (1967). These tests were conducted by using C-190 Ottawa sand specimens. The specimens were first vibrated at a large

FIGURE 2.17 Effect of number of cycles of high amplitude vibration on shear modulus determined at low amplitude. C-190 Ottawa sand, void ratio = 0.46. [Drnevich, V. P., Hall, J. R., Jr., and Richart, F. E., Jr. (1967). "Effect of the Amplitude of Vibration on the Shear Modulus of Sand," *Proceedings, International Symposium on Wave Propagation and Dynamic Properties of Earth Materials,* University of New Mexico Press, Albuquerque, Fig. 7, p. 197.]

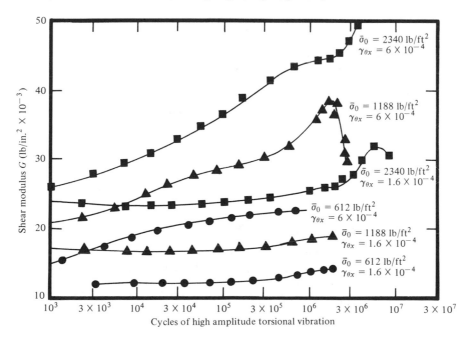

amplitude for a certain number of cycles under a constant effective confining pressure ($\bar{\sigma}_0$). After that the shear modulii were determined by torsionally vibrating the specimens at small amplitudes (shearing strain $< 10^{-5}$). Figure 2.17 shows the results of six series of this type of test for dense sand (void ratio = 0.46). In general, the value of G increased with the increase of prestrain cycles. More studies are needed in this area.

2.10 DETERMINATION OF INTERNAL DAMPING FROM RESONANT COLUMN TESTS

2.10.1 General Theory

In Chapter 1, the derivation of the expression for the logarithmic decrement was given as

$$\delta = \ln(X_n / X_{n+1}) = 2\pi D / \sqrt{1 - D^2} \qquad (1.70)$$

where δ is the logarithmic decrement and D is the damping ratio. The preceding equation is for the case of free vibration of a mass–spring–dashpot system. In the above equation, the notation for the amplitude of vibration has been taken as X instead of Z as given in Chapter 1. This is for the reason of consistency with the notation used in Section 2.3 where x is taken as the distance. The damping ratio is given by the expression

$$D = c/c_c = c/2\sqrt{km_s} \qquad (1.47b)$$

Note that the term m_s is the mass of the soil specimen whereas in Chapter 1, m was the mass of the foundation. For soils, the value of D is small and Eq. (1.70) can be approximated as

$$\delta = \ln(X_n / X_{n+1}) = 2\pi D \qquad (2.80)$$

Now, combining Eqs. (1.47b) and (2.80)

$$\delta = \pi c / \sqrt{km_s} \qquad (2.81)$$

2.10.2 Experimental Determination

The logarithmic decrement of a soil specimen (and hence the damping ratio D) can be easily measured by using a fixed–free type of resonant column device. The soil specimen is first set into steady-state forced vibration. The driving power is then shut off and the decay of the amplitude of vibration

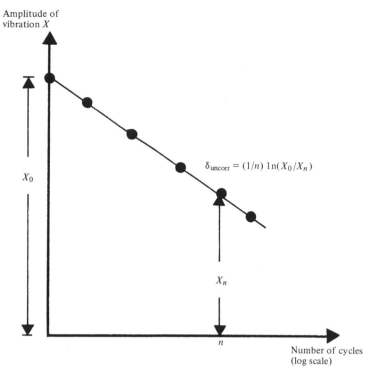

FIGURE 2.18 Plot of the amplitude of vibration against the corresponding number of cycles for determination of logarithmic decrement.

can be plotted against the corresponding number of cycles. This will plot as a straight line on a semilog graph paper as shown in Figure 2.18. The logarithmic decrement can then be evaluated as

$$\delta_{\text{uncorr}} = (1/n)\ln(X_0/X_n) \tag{2.82}$$

However, in a fixed–free type of resonant column device, the driving and the motion monitoring equipments are placed on the top of the specimen. Hence, for determination of the true logarithmic decrement of the soil specimen, a correction to Eq. (2.82) is necessary. This has been discussed by Hall and Richart (1963). Consider the case of longitudinal vibration of a soil column as shown in Figure 2.19, in which m is the mass of the attachments on the top of the soil specimen and m_s is that of the soil specimen. With the addition of mass m, Eq. (2.81) can be rewritten as

$$\delta_{\text{uncorr}} = \pi c / \sqrt{k(m_s + m)} \tag{2.83}$$

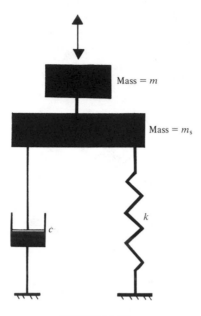

FIGURE 2.19

From Eqs. (2.81) and (2.83),

$$\delta/\delta_{\text{uncorr}} = \sqrt{(m_s + m)/m_s} = \sqrt{1 + m/m_s} \qquad (2.84)$$

In order to use Eq. (2.84), it will be required to convert the mass m_s into an equivalent concentrated mass; this can be shown to be equal to $0.405m_s$. Thus, replacing m_s in Eq. (2.84) by $0.405m_s$,

$$\delta = \delta_{\text{uncorr}}\sqrt{1 + m/(0.405m_s)} \qquad (2.85)$$

A similar correction may be used for specimens subjected to torsional vibration and is of the form

$$\delta = \delta_{\text{uncorr}}\sqrt{1 + \frac{J_m}{0.405J_s}} \qquad (2.86)$$

where J_s and J_m are the polar moments of inertia of the soil specimen and the attachments, respectively.

2.10.3 Laboratory Experimental Results

Using resonant column devices, Hall and Richart (1963) conducted several tests to determine the logarithmic decrement of Ottawa sand (No. 20–No. 30) and novaculite No. 1250, a fine quartz powder that is considered to be a silt. Equations (2.85) and (2.86) were used for calculation of logarithmic

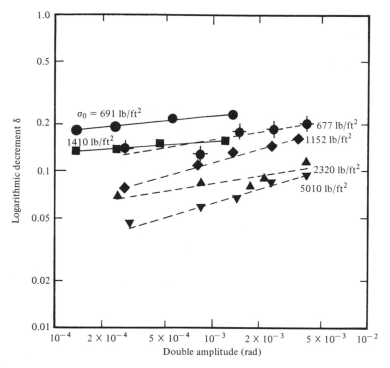

FIGURE 2.20 Variation of logarithmic decrement with amplitude for dry ($e = 0.67$, broken line) and saturated ($e = 0.64$, solid line) Ottawa sand in torsional oscillation. Note: 1 lb/in.^{2} = 6.9 kN/m^2. [Hall, J. R., Jr., and Richart, F. E., Jr. (1963). "Dissipation of Elastic Wave Energy in Granular Soils," *Journal of the Soil Mechanics and Foundations Division, ASCE*, 89 (SM6), Fig. 9, p. 45.]

decrement for cases of longitudinal vibrations and torsional vibrations, respectively. Figure 2.20 shows the nature of variation of δ with the amplitude of *torsional oscillations* for a dry and a saturated Ottawa sand at various effective confining pressures ($\bar{\sigma}_0$). The amplitude of oscillation plotted in the abscissa of Figure 2.20 corresponds to the steady-state amplitude at which a specimen vibrated before the power was shut off. Based on the works of Hall and Richart, the following general conclusions regarding the variation of δ in sand can be drawn:

1. For a given $\bar{\sigma}_0$, the value of δ in *dry* sand generally decreases with the decrease of the amplitude of vibration. For the first normal mode of vibration (longitudinal and torsional), δ is approximately proportional to the 0.25 power of the amplitude.

2. For a given $\bar{\sigma}_0$, in saturated sands, the value of the logarithmic decrement in torsional oscillation varies between the 0.0 and 0.13 power of the amplitude.

3. For a given $\bar{\sigma}_0$, the value of δ for saturated sands in longitudinal oscillation does not vary considerably with the amplitude.

4. For a given amplitude of vibration, the logarithmic decrement decreases with the increase of the effective confining pressure (for longitudinal and torsional oscillations).

Hardin (1965) suggested a relation for δ of *dry* sand in torsional oscillation for low amplitude vibration as

$$\delta = 9\pi\left(\gamma_{\theta x}\right)^{0.2}\left(\bar{\sigma}_0\right)^{-0.5} \tag{2.87}$$

Equation (2.87) is valid for $\gamma_{\theta x} = 10^{-6}$–$10^{-4}$ and $\bar{\sigma}_0 = 500$–3000 lb/ft^2 (24–144 kN/m^2). Also note that $\bar{\sigma}_0$ is in lb/ft^2.

Figure 2.21 shows the results of a series of logarithmic decrement tests conducted by Hall and Richart (1963) in saturated novaculite No. 1250. It may be seen that the variation of the logarithmic decrement with amplitude of vibration does not plot as a straight line on a log–log paper as in the case

FIGURE 2.21 Variation of logarithmic decrement with amplitude for Novaculite No. 1250 after rebounding from 7250 lb/ft^2 (347 kN/m^2) to 4130 lb/ft^2 (198 kN/m^2) in torsional oscillation. [Hall, J. R., Jr., and Richart, F. E., Jr. (1963). "Dissipation of Elastic Wave Energy in Granular Soils," *Journal of the Soil Mechanics and Foundations Division, ASCE*, 89 (SM6), Fig. 18, p. 51.]

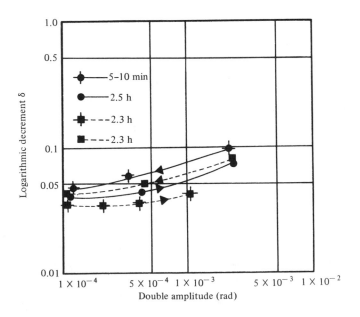

of sand. Also, the value of δ is dependent on the stress history and the time for which it has been applied.

More discussion regarding the evaluation of damping in soils is found in Chapter 8.

PROBLEMS

2.1. A uniformly graded dry sand specimen was tested in a resonant column device. The shear wave velocity v_s determined by torsional vibration of the specimen was 760 ft/sec. The longitudinal wave velocity determined by using a similar specimen was 1271 ft/sec. Determine:
   a. Poisson's ratio
   b. Young's modulus and shear modulus if the void ratio and the specific gravity of soil solids of the specimen were 0.5 and 2.65, respectively.

2.2. A clayey soil specimen was tested in a resonant column device (torsional vibration, free–free end condition) for determination of shear modulus. Given a specimen length of 90 mm, diameter of 35.6 mm, mass of 170 g, and a frequency at normal mode of vibration ($n = 1$) of 790 cps, determine the shear modulus of the specimen in kN/m^2.

2.3. The Poisson's ratio for the clay specimen described in Problem 2.2 is 0.52. If a similar specimen is vibrated longitudinally in a resonant column device (free–free end condition), what would be its frequency at normal mode of vibration ($n = 1$)?

2.4. An angular-grained sand has maximum and minimum void ratios of 1.1 and 0.55, respectively. Using Eq. (2.72), determine and plot the variation of shear modulus vs relative density ($D_r = 0\%–100\%$) for mean confining pressures of 50, 100, 150, 200, and 300 kN/m^2.

2.5. A 20-m-thick sand layer in the field is underlain by rock. The ground water table is located at a depth of 5 m measured from the ground surface. Determine the shear modulus of this sand at a depth of 10 m below the ground surface, given a void ratio of 0.6, a specific gravity of soil solids of 2.68, and a 36° angle of friction of sand. Assume the sand to be round grained.

2.6. A remolded clay specimen was consolidated by a hydrostatic pressure of 30 $lb/in.^2$. The specimen was then allowed to swell under a hydrostatic pressure of 15 $lb/in.^2$. The void ratio at the end of swelling was 0.8. If this clay were subjected to a torsional vibration in a resonant column test, what would be its shear modulus? The liquid and plastic limits of the clay are 58 and 28, respectively.

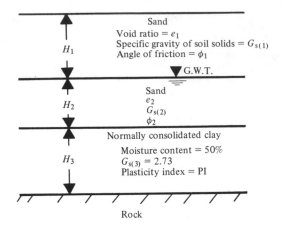

FIGURE P2.7

2.7. Refer to Figure P2.7. Given

$H_1 = 2$ m	$e_1 = 0.6$	$\phi_1 = 35°$
$H_2 = 8$ m	$e_2 = 0.7$	$\phi_2 = 30°$
$H_3 = 3$ m	$G_{s(1)} = 2.68$	PI of clay $= 32$
	$G_{s(2)} = 2.65$	

Estimate and plot the variation of shear modulus with depth for the soil profile.

2.8. Repeat Problem 2.7. Given

$H_1 = H_2 = H_3 = 20$ ft	$\phi_1 = 28°$
$e_1 = 0.8$	$\phi_2 = 32°$
$e_2 = 0.68$	$G_{s(1)} = G_{s(2)} = 2.66$
	PI of clay $= 20$

REFERENCES

Brooker, E. W., and Ireland, H. O. (1965). "Earth Pressure at Rest Related to Stress History," *Canadian Geotechnical Journal* 2 (1), 1–15.

Drnevich, V. P., Hall, J. R., Jr., and Richart, F. E., Jr. (1967). "Effects of the Amplitude of Vibration on the Shear Modulus of Sand," *Proceedings, International Symposium on Wave Propagation and Dynamic Properties of Earth Materials*, University of New Mexico Press, pp. 189–199.

Edil, T. B., and Dhowian, A. W. (1981). "At-Rest Pressure of Peat Soils," *Journal of the Geotechnical Engineering Division* 107 (GT2), 201–217.

Hall, J. R., Jr., and Richart, F. E., Jr. (1963). "Dissipation of Elastic Wave Energy in Granular Soils," *Journal of the Soil Mechanics and Foundations Division, ASCE,* 89 (SM6), 27–55.

Hardin, B. O. (1965), "The Nature of Damping in Sands," *Journal of the Soil Mechanics and Foundations Division, ASCE* 91 (SM1), 63–97.

Hardin, B. O., and Black, W. L. (1968). "Vibration Modulus of Normally Consolidated Clay," *Journal of the Soil Mechanics and Foundations Division, ASCE* 94 (SM2), 353–369.

Hardin, B. O., and Black, W. L. (1969). "Closure of Vibration Modulus of Normally Consolidated Clay," *Journal of the Soil Mechanics and Foundations Division, ASCE* 95 (SM6), 1531–1537.

Hardin, B. O., and Drnevich, V. P. (1972). "Shear Modulus and Damping in Soils: Design Equations and Curves," *Journal of the Soil Mechanics and Foundations Division, ASCE* 98 (SM7), 667–692.

Hardin, B. O., and Richart, F. E., Jr. (1963). "Elastic Wave Velocities in Granular Soils," *Journal of the Soil Mechanics and Foundations Division, ASCE* 89 (SM1), 33–65.

Humphries, W. K., and Wahls, H. E. (1968). "Stress History Effects on Dynamic Modulus of Clay," *Journal of the Soil Mechanics and Foundations Division, ASCE* 94 (SM2), 371–389.

Lawrence, F. V., Jr. (1965). "Ultrasonic Shear Wave Velocities in Sand and Clay," *Response of Soils to Dynamic Loading* , Report No. 23, MIT, Cambridge, Massachusetts.

Whitman, R. V., and Lawrence, F. V. (1963). "Discussion on Elastic Wave Velocities in Granular Soils," *Journal of the Soil Mechanics and Foundations Division, ASCE* 89 (SM5), 112–118.

Wilson, S. D., and Dietrich, D. J. (1960). "Effect of Consolidation Pressure on Elastic and Strength Properties of Clay," *Proceedings, Research Conference on Shear Strength of Cohesive Soils, ASCE,* pp. 419–435.

3
STRESS WAVES
IN THREE DIMENSIONS

In this chapter, the propagation of stress waves in three dimensions is presented. In order to derive the mathematical equations for stress waves, one needs to be acquainted with the notations and sign conventions for stresses in three dimensions.

Figure 3.1 shows a soil element whose sides measure dx, dy, and dz. The normal stresses acting on the planes normal to the x, y and z axes are σ_x, σ_y, and σ_z, respectively. The sign conventions for the normal stresses are *positive* when they are directed into the surface. The shear stresses are τ_{xy}, τ_{yx}, τ_{yz}, τ_{zy}, τ_{zx}, and γ_{xz}. The notations for the shear stresses are as follows:

If τ_{ij} is a shear stress, it means that it is acting on a plane normal to the i axis and its direction is parallel to the j axis. A shear stress is considered *positive* if it is directed in the negative j direction while acting on a plane whose outward normal is in the positive i direction. For example, all shear stresses are positive in Figure 3.1. Note, for equilibrium,

$$\tau_{xy} = \tau_{yx} \tag{3.1}$$

$$\tau_{xz} = \tau_{zx} \tag{3.2}$$

$$\tau_{yz} = \tau_{zy} \tag{3.3}$$

3.1 EQUATION OF MOTION IN AN ELASTIC MEDIUM

Figure 3.2 shows the stresses acting on a soil element with sides measuring dx, dy, and dz. For obtaining the differential equations of motion, one needs to sum the forces in the x, y, and z directions. Along the x direction,

$$\sum F_x = \left[\sigma_x - \left(\sigma_x + \frac{\partial \sigma_x}{\partial x}dx\right)\right](dy)(dz) + \left[\tau_{zx} - \left(\tau_{zx} + \frac{\partial \tau_{zx}}{\partial z}dz\right)\right](dx)(dy)$$

$$+ \left[\tau_{yx} - \left(\tau_{yx} + \frac{\partial \tau_{yx}}{\partial y}dy\right)\right](dx)(dz) + \rho(dx)(dy)(dz) \cdot \frac{\partial^2 u}{\partial t^2} = 0$$

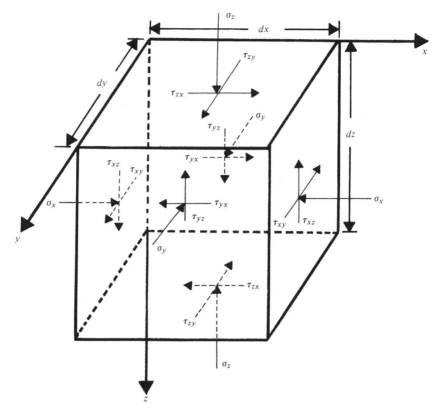

FIGURE 3.1 Notations for normal and shear stresses in cartesian coordinate system.

where ρ is the soil density and u is the displacement component along the x direction. Alternatively,

$$\frac{\partial \sigma_x}{\partial x} + \frac{\partial \tau_{yx}}{\partial y} + \frac{\partial \tau_{zx}}{\partial z} = \rho \frac{\partial^2 u}{\partial t^2} \tag{3.4}$$

Similarly, summing forces on the soil element in the y and z directions

$$\frac{\partial \sigma_y}{\partial y} + \frac{\partial \tau_{xy}}{\partial x} + \frac{\partial \tau_{zy}}{\partial z} = \rho \frac{\partial^2 v}{\partial t^2} \tag{3.5}$$

and

$$\frac{\partial \sigma_z}{\partial z} + \frac{\partial \tau_{xz}}{\partial x} + \frac{\partial \tau_{yz}}{\partial y} = \rho \frac{\partial^2 w}{\partial t^2} \tag{3.6}$$

where v and w are the components of displacement in the y and z directions, respectively.

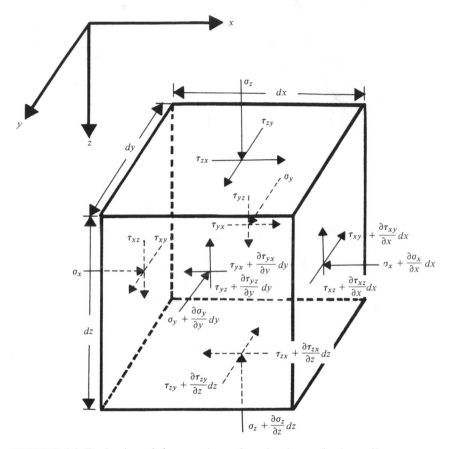

FIGURE 3.2 Derivation of the equations of motion in an elastic medium.

3.2 STRAIN

In Section 3.1, u, v, and w were taken to be the displacements in the x, y, and z directions, respectively. The equations for strains and rotations of elastic and isotropic materials in terms of displacements are as follow:

$$\epsilon_x = \frac{\partial u}{\partial x} \tag{3.7}$$

$$\epsilon_y = \frac{\partial v}{\partial y} \tag{3.8}$$

$$\epsilon_z = \frac{\partial w}{\partial z} \tag{3.9}$$

$$\gamma_{xy} = \frac{\partial v}{\partial x} + \frac{\partial u}{\partial y} \tag{3.10}$$

$$\gamma_{yz} = \frac{\partial w}{\partial y} + \frac{\partial v}{\partial z} \tag{3.11}$$

$$\gamma_{zx} = \frac{\partial u}{\partial z} + \frac{\partial w}{\partial x} \tag{3.12}$$

$$\bar{\omega}_x = \frac{1}{2}\left(\frac{\partial w}{\partial y} - \frac{\partial v}{\partial z}\right) \tag{3.13}$$

$$\bar{\omega}_y = \frac{1}{2}\left(\frac{\partial u}{\partial z} - \frac{\partial w}{\partial x}\right) \tag{3.14}$$

$$\bar{\omega}_z = \frac{1}{2}\left(\frac{\partial v}{\partial x} - \frac{\partial u}{\partial y}\right) \tag{3.15}$$

where

ϵ_x, ϵ_y, and ϵ_z = normal strains in the direction of x, y, and z, respectively

γ_{xy} = shearing strain between the planes xz and yz

γ_{yz} = shearing strain between the planes yx and zx

γ_{zx} = shearing strain between the planes zy and xy

$\bar{\omega}_x, \bar{\omega}_y$, and $\bar{\omega}_z$ = the components of rotation about the x, y, and z axes.

These derivations are given in most of the textbooks on the theory of elasticity (e.g., Timoshenko and Goodier, 1970). Hence, this is not covered here in detail.

3.3 HOOKE'S LAW

For an elastic, isotropic material, the normal strains and normal stresses can be related by the following equations:

$$\epsilon_x = (1/E)\left[\sigma_x - \mu(\sigma_y + \sigma_z)\right] \tag{3.16}$$

$$\epsilon_y = (1/E)\left[\sigma_y - \mu(\sigma_x + \sigma_z)\right] \tag{3.17}$$

$$\epsilon_z = (1/E)\left[\sigma_z - \mu(\sigma_x + \sigma_y)\right] \tag{3.18}$$

where ϵ_x, ϵ_y, and ϵ_z are the respective normal strains in the directions of x, y, and z, E is Young's modulus, and μ is Poisson's ratio.

The shear stresses and the shear strains can be related by the following equations:

$$\tau_{xy} = G\gamma_{xy} \tag{3.19}$$

$$\tau_{yz} = G\gamma_{yz} \tag{3.20}$$

$$\tau_{zx} = G\gamma_{zx} \tag{3.21}$$

where the shear modulus

$$G = \tfrac{1}{2}E/(1+\mu) \tag{3.22}$$

and γ_{xy}, γ_{yz}, and γ_{zx} are the shear strains.

Equations (3.16)–(3.18) can be solved to express normal stresses in terms of normal strains as

$$\sigma_x = \lambda\bar{\epsilon} + 2G\epsilon_x \tag{3.23}$$

$$\sigma_y = \lambda\bar{\epsilon} + 2G\epsilon_y \tag{3.24}$$

$$\sigma_z = \lambda\bar{\epsilon} + 2G\epsilon_z \tag{3.25}$$

where

$$\lambda = \mu E/\big[(1+\mu)(1-2\mu)\big] \tag{3.26}$$

$$\bar{\epsilon} = \epsilon_x + \epsilon_y + \epsilon_z \tag{3.27}$$

From Eqs. (3.22) and (3.26), it is easy to see that

$$\mu = \tfrac{1}{2}\lambda/(\lambda + G) \tag{3.28}$$

3.4 EQUATIONS FOR STRESS WAVES IN AN INFINITE ELASTIC MEDIUM

3.4.1 Compression Waves

Equations (3.4)–(3.6) give the equations of motion in terms of stresses. Now, considering Eq. (3.4) and noting that $\tau_{xy} = \tau_{yx}$ and $\tau_{xz} = \tau_{zx}$,

$$\rho\frac{\partial^2 u}{\partial t^2} = \frac{\partial\sigma_x}{\partial x} + \frac{\partial\tau_{xy}}{\partial y} + \frac{\partial\tau_{xz}}{\partial z} \tag{3.4}$$

Substitution of Eqs. (3.19), (3.21), and (3.23) into Eq. (3.4) yields

$$\rho\frac{\partial^2 u}{\partial t^2} = \frac{\partial}{\partial x}\big(\lambda\bar{\epsilon} + 2G\epsilon_x\big) + \frac{\partial}{\partial y}\big(G\gamma_{xy}\big) + \frac{\partial}{\partial z}\big(G\gamma_{xz}\big)$$

Again, substitution of Eqs. (3.10) and (3.12) into the above expression will yield

$$\rho\frac{\partial^2 u}{\partial t^2} = \frac{\partial}{\partial x}\big(\lambda\bar{\epsilon} + 2G\epsilon_x\big) + G\frac{\partial}{\partial y}\left(\frac{\partial v}{\partial x} + \frac{\partial u}{\partial y}\right) + G\frac{\partial}{\partial z}\left(\frac{\partial u}{\partial z} + \frac{\partial w}{\partial x}\right)$$

or

$$\rho\frac{\partial^2 u}{\partial t^2} = \lambda\frac{\partial\bar{\epsilon}}{\partial x} + G\left(\frac{\partial^2 u}{\partial x^2} + \frac{\partial^2 v}{\partial x\,\partial y} + \frac{\partial^2 w}{\partial x\,\partial z} + \frac{\partial^2 u}{\partial x^2} + \frac{\partial^2 u}{\partial y^2} + \frac{\partial^2 u}{\partial z^2}\right) \tag{3.29}$$

But

$$\frac{\partial^2 u}{\partial x^2} + \frac{\partial^2 v}{\partial x\,\partial y} + \frac{\partial^2 w}{\partial x\,\partial z} = \frac{\partial\bar{\epsilon}}{\partial x} \tag{3.30}$$

So

$$\rho \frac{\partial^2 u}{\partial t^2} = (\lambda + G) \frac{\partial \bar{\epsilon}}{\partial x} + G \nabla^2 u \qquad (3.31)$$

where

$$\nabla^2 = \frac{\partial^2}{\partial x^2} + \frac{\partial^2}{\partial y^2} + \frac{\partial^2}{\partial z^2} \qquad (3.32)$$

Similarly, by proper substitution in Eqs. (3.5) and (3.6), the following relations can be obtained:

$$\rho \frac{\partial^2 v}{\partial t^2} = (\lambda + G) \frac{\partial \bar{\epsilon}}{\partial y} + G \nabla^2 v \qquad (3.33)$$

and

$$\rho \frac{\partial^2 w}{\partial t^2} = (\lambda + G) \frac{\partial \bar{\epsilon}}{\partial z} + G \nabla^2 w \qquad (3.34)$$

Now, differentiating Eqs. (3.31)–(3.34) with respect to x, y, and z, respectively, and adding

$$\rho \frac{\partial^2}{\partial t^2} \left(\frac{\partial u}{\partial x} + \frac{\partial v}{\partial y} + \frac{\partial w}{\partial z} \right) = (\lambda + G) \left(\frac{\partial^2 \bar{\epsilon}}{\partial x^2} + \frac{\partial^2 \bar{\epsilon}}{\partial y^2} + \frac{\partial^2 \bar{\epsilon}}{\partial z^2} \right)$$

$$+ G \nabla^2 \left(\frac{\partial u}{\partial x} + \frac{\partial v}{\partial y} + \frac{\partial w}{\partial z} \right)$$

or

$$\rho \frac{\partial^2 \bar{\epsilon}}{\partial t^2} = (\lambda + G)(\nabla^2 \bar{\epsilon}) + G(\nabla^2 \bar{\epsilon}) = (\lambda + 2G) \nabla^2 \bar{\epsilon} \qquad (3.35)$$

Therefore

$$\frac{\partial^2 \bar{\epsilon}}{\partial t^2} = \frac{(\lambda + 2G)}{\rho} \nabla^2 \bar{\epsilon} = v_p^2 \nabla^2 \bar{\epsilon} \qquad (3.36)$$

where

$$v_p = \sqrt{(\lambda + 2G)/\rho} \qquad (3.37)$$

Equation (3.37) is in the same form as the wave equation given in Eq. (2.4). Also note that $\bar{\epsilon}$ is the volumetric strain and v_p is the *velocity of the dilatational waves*. This is also referred to as the *primary wave* or *P-wave* or *compression wave*. Also another fact that needs to be pointed out here is that the expression for v_c was given in Chapter 2 as $v_c = \sqrt{E/\rho}$. Comparing the expressions for v_c and v_p, one can see that the velocity of compression waves is faster than v_c.

3.4.2 Distortional Waves or Shear Waves

Differentiating Eq. (3.33) with respect to z and Eq. (3.34) with respect to y,

$$\rho \frac{\partial^2}{\partial t^2}\left(\frac{\partial v}{\partial z}\right) = (\lambda + G)\frac{\partial \bar{\epsilon}}{(\partial y)(\partial z)} + G \nabla^2 \frac{\partial v}{\partial z} \qquad (3.38)$$

and

$$\rho \frac{\partial^2}{\partial t^2}\left(\frac{\partial w}{\partial y}\right) = (\lambda + G)\frac{\partial \bar{\epsilon}}{(\partial y)(\partial z)} + G \nabla^2 \frac{\partial w}{\partial y} \qquad (3.39)$$

Subtracting Eq. (3.38) from (3.39) yields

$$\rho \frac{\partial}{\partial t^2}\left(\frac{\partial w}{\partial y} - \frac{\partial v}{\partial z}\right) = G \nabla^2 \left(\frac{\partial w}{\partial y} - \frac{\partial v}{\partial z}\right)$$

However, $\partial w/\partial y - \partial v/\partial z = 2\bar{\omega}_x$ [Eq. (3.13)]; thus,

$$\rho \frac{\partial^2 \bar{\omega}_x}{\partial t^2} = G \nabla^2 \bar{\omega}_x \qquad (3.40)$$

or

$$\frac{\partial^2 \bar{\omega}_x}{\partial t^2} = \frac{G}{\rho} \nabla^2 \bar{\omega}_x = v_s^2 \nabla^2 \bar{\omega}_x \qquad (3.41)$$

where $v_s = \sqrt{G/\rho}$.

Equation (3.41) represents the equation for distortional waves and the *velocity* of propagation is v_s. This is also referred to as the *shear wave* or *S-wave*. Comparison of the shear wave velocity given above with that in a rod [Eq. (2.14)] shows that they are the same. Using the process of similar manipulation, one can also obtain two more equations similar to Eq. (3.41):

$$\frac{\partial^2 \bar{\omega}_y}{\partial t^2} = v_s^2 \nabla^2 \bar{\omega}_y \qquad (3.42)$$

and

$$\frac{\partial^2 \bar{\omega}_z}{\partial t^2} = v_s^2 \nabla^2 \bar{\omega}_z \qquad (3.43)$$

3.5 RAYLEIGH WAVES

Equations derived in Section 3.4 are for stress waves in the body of an infinite, elastic, and isotropic medium. Another type of wave, called a *Rayleigh wave*, also exists near the boundary of an elastic half-space. This type of wave was first investigated by Lord Rayleigh (1885). In order to study this, consider a plane wave through an elastic medium with a plane boundary as shown in Figure 3.3. Note the plane $x - y$ is the boundary of the elastic half-space and z is positive downwards. Let u and w represent the

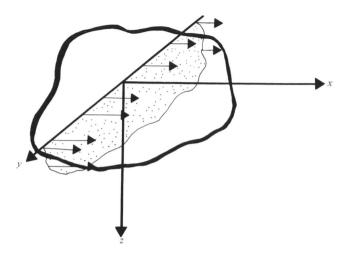

FIGURE 3.3 Plane wave through an elastic medium with a plane boundary.

displacements in the directions x and z, respectively, and be independent of y. Therefore,

$$u = \frac{\partial \phi}{\partial x} + \frac{\partial \psi}{\partial z} \tag{3.44}$$

and

$$w = \frac{\partial \phi}{\partial z} - \frac{\partial \psi}{\partial x} \tag{3.45}$$

where ϕ and ψ are two potential functions. The dilation $\bar{\epsilon}$ can be defined as

$$\bar{\epsilon} = \epsilon_x + \epsilon_y + \epsilon_z = \frac{\partial u}{\partial x} + \frac{\partial v}{\partial y} + \frac{\partial w}{\partial z}$$

$$= \left(\frac{\partial^2 \phi}{\partial x^2} + \frac{\partial^2 \psi}{\partial x \partial z} \right) + (0) + \left(\frac{\partial^2 \phi}{\partial z^2} - \frac{\partial^2 \psi}{\partial x \partial z} \right) = \frac{\partial^2 \phi}{\partial x^2} + \frac{\partial^2 \phi}{\partial z^2} = \nabla^2 \phi \tag{3.46}$$

Similarly, the rotation in the $x-z$ plane can be given by

$$2\bar{\omega}_y = \frac{\partial u}{\partial z} - \frac{\partial w}{\partial x} = \frac{\partial^2 \psi}{\partial x^2} + \frac{\partial^2 \psi}{\partial z^2} = \nabla^2 \psi \tag{3.47}$$

Substituting Eqs. (3.44) and (3.46) into Eq. (3.31) yields

$$\rho \frac{\partial^2}{\partial t^2} \left(\frac{\partial \phi}{\partial x} + \frac{\partial \psi}{\partial z} \right) = (\lambda + G) \frac{\partial}{\partial x} \left(\nabla^2 \phi \right) + G \nabla^2 \left(\frac{\partial \phi}{\partial x} + \frac{\partial \psi}{\partial z} \right)$$

or

$$\rho \frac{\partial}{\partial x}\left(\frac{\partial^2 \phi}{\partial t^2} \right) + \rho \frac{\partial}{\partial z}\left(\frac{\partial^2 \psi}{\partial t^2} \right) = (\lambda + 2G)\frac{\partial}{\partial x}(\nabla^2 \phi) + G\frac{\partial}{\partial z}(\nabla^2 \psi) \quad (3.48)$$

In a similar manner, substituting Eqs. (3.45) and (3.46) into Eq. (3.34), we get

$$\rho \frac{\partial}{\partial z}\left(\frac{\partial^2 \phi}{\partial t^2} \right) - \rho \frac{\partial}{\partial x}\left(\frac{\partial^2 \psi}{\partial t^2} \right) = (\lambda + 2G)\frac{\partial}{\partial z}(\nabla^2 \phi) - G\frac{\partial}{\partial x}(\nabla^2 \psi) \quad (3.49)$$

Equations (3.48) and (3.49) will be satisfied if (1) $\rho(\partial^2 \phi / \partial t^2) = (\lambda + 2G)\nabla^2 \phi$ or

$$\frac{\partial^2 \phi}{\partial t^2} = \left(\frac{\lambda + 2G}{\rho} \right) \nabla^2 \phi = v_p^2 \nabla^2 \phi \quad (3.50)$$

and (2) $\rho(\partial^2 \psi / \partial t^2) = G\nabla^2 \psi$ or

$$\frac{\partial^2 \psi}{\partial t^2} = \frac{G}{\rho} \nabla^2 \psi = v_s^2 \nabla^2 \psi \quad (3.51)$$

Now, consider a sinusoidal wave traveling in the positive x direction. Let the solutions of ϕ and ψ be expressed as

$$\phi = F(z)\exp[i(\omega t - fx)] \quad (3.52)$$

and

$$\psi = G(z)\exp[i(\omega t - fx)] \quad (3.53)$$

where $F(z)$ and $G(z)$ are functions of depth

$$f = 2\pi / \text{wavelength} \quad (3.54)$$

$$i = \sqrt{-1} \quad (3.55)$$

Substituting Eq. (3.52) into Eq. (3.50), we get

$$(\partial^2 / \partial t^2)\{F(z)\exp[i(\omega t - fx)]\} = v_p^2 \nabla^2\{F(z)\exp[i(\omega t - fx)]\}$$

or

$$-\omega^2 F(z) = v_p^2 [F''(z) - f^2 F(z)] \quad (3.56)$$

Similarly, substituting Eq. (3.53) into Eq. (3.51) results in

$$-\omega^2 G(z) = v_s^2 [G''(z) - f^2 G(z)] \quad (3.57)$$

where

$$F''(z) = \partial^2 F(z)/\partial z^2 \quad (3.58)$$

$$G''(z) = \partial^2 G(z)/\partial z^2 \quad (3.59)$$

Equations (3.56) and (3.57) can be rearranged to the form

$$F''(z) - q^2 F(z) = 0 \quad (3.60)$$

and

$$G''(z) - s^2 G(z) = 0 \qquad (3.61)$$

where

$$q^2 = f^2 - \omega^2/v_p^2 \qquad (3.62)$$

$$s^2 = f^2 - \omega^2/v_s^2 \qquad (3.63)$$

Solutions to Eqs. (3.60) and (3.61) can be given as

$$F(z) = A_1 e^{-qz} + A_2 e^{qz} \qquad (3.64)$$

and

$$G(z) = B_1 e^{-sz} + B_2 e^{sz} \qquad (3.65)$$

where A_1, A_2, B_1, and B_2 are constants.

From Eqs. (3.64) and (3.65), it can be seen that A_2 and B_2 must be equal to zero; otherwise $F(z)$ and $G(z)$ will approach infinity with depth which is not the type of wave that is considered here. With A_2 and B_2 equal to zero,

$$F(z) = A_1 e^{-qz} \qquad (3.66)$$

$$G(z) = B_1 e^{-sz} \qquad (3.67)$$

Combining Eqs. (3.52) and (3.66) and Eqs. (3.53) and (3.67)

$$\phi = \left(A_1 e^{-qz} \right) \left[e^{i(\omega t - fx)} \right] \qquad (3.68)$$

and

$$\psi = \left(B_1 e^{-sz} \right) \left[e^{i(\omega t - fx)} \right] \qquad (3.69)$$

The boundary conditions for the two preceding equations are at $z = 0$, $\sigma_z = 0$, $\tau_{zx} = 0$, and $\tau_{zy} = 0$. From Eq. (3.25),

$$\sigma_{z(z=0)} = \lambda \bar{\epsilon} + 2G\epsilon_z = \lambda \bar{\epsilon} + 2G(\partial w/\partial z) = 0 \qquad (3.70)$$

Combining Eqs. (3.45), (3.46), and (3.68)–(3.70), one obtains

$$A_1 \left[(\lambda + 2G)q^2 - \lambda f^2 \right] - 2iB_1 Gfs = 0 \qquad (3.71)$$

or

$$A_1/B_1 = 2iGfs / \left[(\lambda + 2G)q^2 - \lambda f^2 \right] \qquad (3.72)$$

Similarly,

$$\tau_{zx(z=0)} = G\gamma_{zx} = G(\partial w/\partial x + \partial u/\partial z) = 0 \qquad (3.73)$$

Again, combining Eqs. (3.44), (3.45), (3.68), (3.69), and (3.73),

$$2iA_1 fq + \left(s^2 + f^2 \right) B_1 = 0$$

or

$$A_1/B_1 = -\left(s^2 + f^2 \right)/2ifq \qquad (3.74)$$

Equating the right-hand sides of Eqs. (3.72) and (3.74),

$$2iGfs/[(\lambda+2G)q^2-\lambda f^2] = -(s^2+f^2)/2ifq$$
$$4Gf^2sq = (s^2+f^2)[(\lambda+2G)q^2-\lambda f^2]$$

or

$$16G^2f^4s^2q^2 = (s^2+f^2)^2[(\lambda+2G)q^2-\lambda f^2]^2 \qquad (3.75)$$

Substituting for q and s and then dividing both sides of Eq. (3.75) by G^2f^8, we get

$$16\left(1-\frac{\omega^2}{v_p^2f^2}\right)\left(1-\frac{\omega^2}{v_s^2f^2}\right) = \left[2-\left(\frac{\lambda+2G}{G}\right)\frac{\omega^2}{v_p^2f^2}\right]^2\left(2-\frac{\omega^2}{v_s^2f^2}\right)^2 \qquad (3.76)$$

From Eq. (3.54)

$$\text{wavelength} = 2\pi/f \qquad (3.77)$$

However,

$$\text{wavelength} = \frac{\text{velocity of wave}}{(\omega/2\pi)} = \frac{v_r}{(\omega/2\pi)} \qquad (3.78)$$

where v_r is the *Rayleigh* wave velocity. Thus, from Eqs. (3.77) and (3.78), $2\pi/f = 2\pi v_r/\omega$ or

$$f = \omega/v_r \qquad (3.79)$$

So,

$$\frac{\omega^2}{v_p^2f^2} = \frac{\omega^2}{v_p^2(\omega^2/v_r^2)} = \frac{v_r^2}{v_p^2} = \alpha^2 V^2 \qquad (3.80)$$

Similarly,

$$\frac{\omega^2}{v_s^2f^2} = \frac{\omega^2}{v_s^2(\omega^2/v_r^2)} = \frac{v_r^2}{v_s^2} = V^2 \qquad (3.81)$$

where

$$\alpha^2 = v_s^2/v_p^2 \qquad (3.82)$$

However $v_p^2 = (\lambda+2G)/\rho$ and $v_s^2 = G/\rho$. Thus

$$\alpha^2 = v_s^2/v_p^2 = G/(\lambda+2G) \qquad (3.83)$$

The term α^2 can also be expressed in terms of Poisson's ratio. From the relations given in Eq. (3.28),

$$\lambda = 2\mu G/(1-2\mu) \qquad (3.84)$$

Substitution of the above relation in Eq. (3.83) yields

$$\alpha^2 = \frac{G}{2\mu G/(1-2\mu)+2G} = \frac{(1-2\mu)G}{2\mu G+2G-4\mu G} = \frac{(1-2\mu)}{(2-2\mu)} \qquad (3.85)$$

TABLE 3.1 Values of V [Eq. (3.86)]

μ	$V = v_r/v_s$
0.25	0.919
0.29	0.926
0.33	0.933
0.4	0.943
0.5	0.955

Again, substituting Eqs. (3.80), (3.81), and (3.83) into Eq. (3.76),

$$16(1 - \alpha^2 V^2)(1 - V^2) = (2 - V^2)^2(2 - V^2)^2$$

or

$$V^6 - 8V^4 - (16\alpha^2 - 24)V^2 - 16(1 - \alpha^2) = 0 \qquad (3.86)$$

The above equation is a cubic equation in V^2. For a given value of Poisson's ratio, the proper value of V^2 can be found and, hence, so can the value of v_r in terms of v_p or v_s. An example of this is shown in Example 3.1. Table 3.1 gives some values of v_r/v_s ($=V$) for various values of Poisson's ratio.

Displacement of Rayleigh Waves

From Eqs. (3.44) and (3.45),

$$u = \frac{\partial \phi}{\partial x} + \frac{\partial \psi}{\partial z} \qquad (3.44)$$

and

$$w = \frac{\partial \phi}{\partial z} - \frac{\partial \psi}{\partial x} \qquad (3.45)$$

Substituting the relations developed for ϕ and ψ [Eqs. (3.68), (3.69)] in the above equations, one obtains

$$u = -(if A_1 e^{-qz} + B_1 s e^{-sz})[e^{i(\omega t - fx)}] \qquad (3.87)$$

$$w = -(A_1 q e^{-qz} - B_1 i f e^{-sz})[e^{i(\omega t - fx)}] \qquad (3.88)$$

However, from Eq. (3.74), $B_1 = -2iA_1 fq/(s^2 + f^2)$. Substituting this relation in Eqs. (3.87) and (3.88) gives

$$u = A_1 f i \left(-e^{-qz} + \frac{2qs}{s^2 + f^2} e^{-sz} \right) [e^{i(\omega t - fx)}] \qquad (3.89)$$

and

$$w = A_1 q \left(-e^{-qz} + \frac{2f^2}{s^2 + f^2} e^{-sz} \right) [e^{i(\omega t - fx)}] \qquad (3.90)$$

From the above two equations, it is obvious that the rate of attenuation of the displacement along the x direction with depth z will depend on the factor U, where

$$U = -e^{-qz} + \frac{2qs}{s^2 + f^2}e^{-sz} = -e^{-(q/f)(fz)} + \left[\frac{2(q/f)(s/f)}{s^2/f^2 + 1}\right]e^{-(s/f)(fz)}$$

(3.91)

Similarly, the rate of attenuation of the displacement along the z direction with depth will depend on

$$W = -e^{-qz} + \frac{2f^2}{s^2 + f^2}e^{-sz} = -e^{-(q/f)(fz)} + \frac{2}{s^2/f^2 + 1}e^{-(s/f)(zf)}$$

(3.92)

However,

$$q^2 = f^2 - \omega^2/v_p^2 \tag{3.62}$$

or

$$\frac{q^2}{f^2} = 1 - \frac{\omega^2}{f^2 v_p^2} = 1 - \frac{v_r^2}{v_p^2} = 1 - \alpha^2 V^2 \tag{3.93}$$

Also,

$$s^2 = f^2 - \omega^2/v_s^2 \tag{3.63}$$

$$\frac{s^2}{f^2} = 1 - \frac{\omega^2}{f^2 v_s^2} = 1 - \frac{v_r^2}{v_s^2} = 1 - V^2 \tag{3.94}$$

If the Poisson's ratio is known, one can determine the value of V from Eq. (3.86). Substituting the above determined values of V in Eqs. (3.93) and (3.94), q/f and s/f can be determined; hence, U and W are determinable as functions of z and f. From Example 3.1, it can be seen that for $\mu = 0.25$, $V = 0.9194$. Thus,

$$\frac{q^2}{f^2} = 1 - \alpha^2 V^2 = 1 - \left(\frac{1-2\mu}{2-2\mu}\right)V^2 = 1 - \left(\frac{1-0.5}{2-0.5}\right)(0.9194)^2 = 0.7182$$

or

$$q/f = 0.8475$$

$$s^2/f^2 = 1 - V^2 = 1 - (0.9194)^2 = 0.1547$$

or

$$s/f = 0.3933$$

Substituting these values of q/f and s/f into Eqs. (3.91) and (3.92),

$$U_{(\mu=0.25)} = -\exp(-0.8475fz) + 0.5773\exp(-0.3933fz) \tag{3.95}$$

$$W_{(\mu=0.25)} = -\exp(-0.8475fz) + 1.7321\exp(-0.3933fz) \tag{3.96}$$

Based on Eqs. (3.95) and (3.96), the following observations can be made:

1. The magnitude of U decreases rapidly with increasing value of fz. At $fz = 1.21$, U becomes equal to zero; so, at $z = 1.21/f$, there is no motion parallel to the surface. It has been shown in Eq. (3.54) that $f = 2\pi/(\text{wavelength})$. Thus, at $z = 1.21/f = 1.21(\text{wavelength})/2\pi = 0.1926(\text{wavelength})$, the value of U is zero. At greater depths, U becomes finite; however it is of the opposite sign, so the vibration takes place in opposite phase.
2. The magnitude of W first increases with fz, reaches a maximum value at $z = 0.076(\text{wavelength})$ (i.e., $fz = 0.4775$), and then decreases with depth.

Figure 3.4 shows a nondimensional plot of the variation of amplitude of vertical and horizontal components of Rayleigh waves with depth for

FIGURE 3.4 Variation of the amplitude of vibration of the horizontal and vertical components of Rayleigh waves with depth ($\mu = 0.25$).

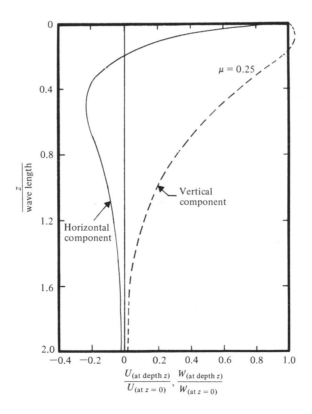

$\mu = 0.25$. Equations (3.95) and (3.96) show that the path of a particle in the medium is an *ellipse* with its *major axis normal to the surface*.

Example 3.1 Given $\mu = 0.25$, determine the value of the Rayleigh wave velocity in terms of v_s.

   Solution: From Eq. (3.86),

$$V^6 - 8V^4 - (16\alpha^2 - 24)V^2 - 16(1 - \alpha^2) = 0$$

For $\mu = 0.25$,

$$\alpha^2 = \frac{1 - 2\mu}{2 - 2\mu} = \frac{1 - 0.5}{2 - 0.5} = \frac{1}{3}$$

$$V^6 - 8V^4 - \left(\tfrac{16}{3} - 24\right)V^2 - 16\left(1 - \tfrac{1}{3}\right) = 0$$

$$3V^6 - 24V^4 + 56V^2 - 32 = 0$$

$$(V^2 - 4)(3V^4 - 12V^2 + 8) = 0$$

Therefore,

$$V^2 = 4, \quad 2 + 2/\sqrt{3}, \quad 2 - 2/\sqrt{3}$$

If $V^2 = 4$,

$$s^2/f^2 = 1 - V^2 = 1 - 4 = -3$$

and s/f is imaginary. This is also the case for $V^2 = 2 + 2/\sqrt{3}$.

   Keeping Eqs. (3.89), (3.91), and (3.90), (3.92) in mind, one can see that when q/f and s/f are imaginary, it does not yield the type of wave that is being discussed here. Thus,

$$V^2 = 2 - 2/\sqrt{3} \qquad V = v_r/v_s = 0.9194$$

or

$$v_r = 0.9194 v_s$$

3.6 ATTENUATION OF THE AMPLITUDE OF ELASTIC WAVES WITH DISTANCE

If an impulse of short duration is created at the surface of an elastic half space, the body waves travel into the medium with hemispherical wave fronts as shown in Figure 3.5a. The *Rayleigh waves* will propagate radially outwards along a *cylindrical wave front*. At some distance from the point of disturbance, the vertical displacement of the ground will be of the nature shown in Figure 3.5b. Since *P*-waves are the fastest, they will arrive first followed by *S*-waves and then the Rayleigh waves. As may be seen from Figure 3.5b, the vertical ground displacement due to the Rayleigh wave arrival is much greater than that for *P*- and *S*-waves. The amplitude of disturbance gradually decreases with distance.

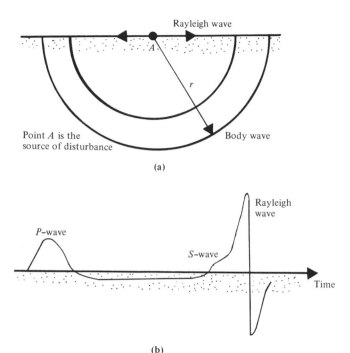

FIGURE 3.5 (a) Propagation of body waves and Rayleigh waves; **(b)** vertical disturbance at a point on the surface due to an impulse of short duration.

When *body waves* spread out along a hemispherical wave front, the energy is distributed over an area that increases with the square of the radius:

$$E' \propto 1/r^2 \qquad (3.97)$$

where E' is the energy per unit area and r is the radius. However, the amplitude is proportional to the square root of the energy per unit area:

$$\text{Amplitude} \propto \sqrt{E'} \propto \sqrt{1/r^2}$$

or

$$\text{Amplitude} \propto 1/r \qquad (3.98)$$

Along the surface of the half space only, the amplitude of the body waves are proportional to $1/r^2$.

Similarly, the amplitude of the Rayleigh waves, which spread out in a *cylindrical* wave front, are proportional to $1/\sqrt{r}$. Thus the attenuation of the amplitude of the *Rayleigh* waves is *slower* than that for the body waves.

The loss of the amplitude of waves due to spreading out is called *geometrical damping*. In addition to the above damping, there is another type of loss—that from *absorption* in real earth material. This is called

TABLE 3.2 Values of β [Eq. (3.99)] [a]

No.	Soil	Absorption coefficient [b] β	
		$(m)^{-1}$	$(ft)^{-1}$
1	Yellow water saturated fine grained sand	0.10	0.0305
2	Yellow water saturated fine grained sand in a frozen state	0.06	0.0183
3	Gray water saturated sand with laminae of peat and organic silt	0.04	0.0122
4	Clayey sands with laminae of more clayey sands and of clays with some sand and silt, above ground water level	0.04	0.0122
5	Heavy water saturated brown clays with some sand silt	0.04– 0.12	0.0122– 0.0366
6	Marly chalk	0.10	0.0305
7	Loess and loessial soil	0.1	0.0305

[a] From *Dynamics of Bases and Foundations* by D. D. Barkan. Copyright © 1962 McGraw-Hill. Used with the permission of McGraw-Hill Book Company.
[b] 1 m = 3.28 ft.

material damping. Thus, accounting for both types of damping, the vertical amplitude of Rayleigh waves can be given by the relation

$$\bar{w}_n = \bar{w}_1 \sqrt{\frac{r_1}{r_n}}\, \exp\left[-\beta(r_n - r_1)\right] \tag{3.99}$$

where $\bar{w}_n$ and $\tilde{w}_1$ are vertical amplitudes at distances r_n and r_1, and β is the absorption coefficient.

The above equation as given by Bornitz (1931). (See also Hall and Richart 1963.) The magnitude of β depends on the type of soil. Table 3.2 gives some representative values of β determined experimentally.

3.7 COMPRESSION AND SHEAR WAVES IN SATURATED SOILS

Wave propagation through saturated soils involves the soil skeleton and water in the void spaces. A comprehensive theoretical study of this problem is given by Biot (1956). This study shows that there are two compressive waves and one shear wave through the saturated medium. Some investigators have referred to the two compressive waves as the *fluid wave* (transmitted through the fluid) and the *frame wave* (transmitted through the soil structure), although there is coupled motion of the fluid and the frame waves. As far as the shear wave is concerned, the pore water has no rigidity to shear. Hence, the shear wave in the soil is dependent only on the properties of the soil skeleton.

Figure 3.6a shows the theoretical variation of the *compressive frame wave* velocities in dry and saturated sands, based on Biot's theory, using the values of the constants representative for a quartz sand (Hardin and Richart, 1963). Along with that, for comparison purposes, are shown the

FIGURE 3.6 **(a)** Comparison of experimental and theoretical results for compressive frame wave velocities in dry ■ and saturated ● Ottawa sand; **(b)** variation of shear wave velocity with confining pressure for Ottawa sand: ● dry, ■ drained, ▲ saturated. Note: 1 m = 3.28 ft, 1 lb/ft^2 = 47.9 N/m^2. [Hardin, B. O., and Richart, F. E., Jr. (1963). "Elastic Wave Velocities in Granular Solids," *Journal of the Soil Mechanics and Foundations Division*, ASCE, 89 (SM6), Fig. 3, p. 40, and Fig. 7, p. 47.]

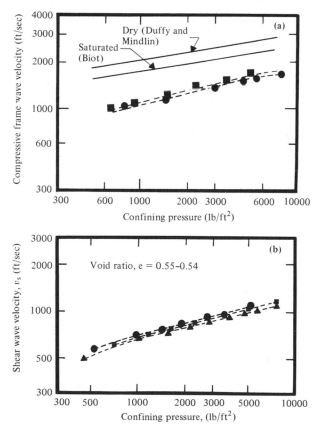

experimental *longitudinal wave* velocities (v_c from Chapter 2) for dry and saturated Ottawa sands. For a given confining pressure, the difference of wave velocities between dry and saturated specimens is negligible and may be accounted for by the difference in the unit weight of the soil.

The velocity of *compression* waves (v_w) through water can be expressed as

$$v_w = \sqrt{B_w/\rho_w} \qquad (3.100)$$

where B_w is the bulk modulus of water and ρ_w is the density of water. Usually the value of v_w is of the order of 4800 ft/sec (1463 m/sec).

Figure 3.6b shows the variation of the experimental shear wave velocity for dry, drained, and saturated Ottawa sand. It may be noted that for a given confining pressure the range of variation of v_s is very small.

3.8 REFLECTION AND REFRACTION OF ELASTIC BODY WAVES

When an elastic stress wave impinges on the boundary of two layers, the wave is reflected and refracted. As has already been shown, there are two types of body wave: compression waves (*P*-waves) and shear waves (*S*-waves). In the case of *P*-waves, the direction of the movement of the particles coincides with the direction of propagation. This is shown by the arrows in Figure 3.7a. The shear waves can be separated into two components:

1. *SH-waves*, in which the motion of the particles is in the plane of propagation as shown by the arrows in Figure 3.7b, and
2. *SV-waves*, in which the motion of the particles is perpendicular to the plane of propagation as shown by a dark dot in Figure 3.7c.

If a *P*-wave impinges on the boundary between two layers as shown in Figure 3.8a, there are two reflected waves and two refracted waves. The reflected waves consist of: a *P*-wave and an *SV*-wave, respectively shown as P_1 and SV_1 in layer 1. Likewise, the refracted waves also consist of a *P*-wave

FIGURE 3.7 *P*-wave, *SV*-wave, and *SH*-wave.

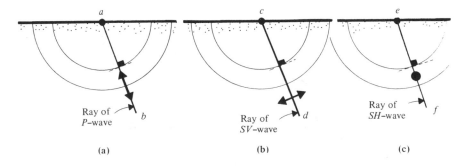

| (a) | (b) | (c) |

95

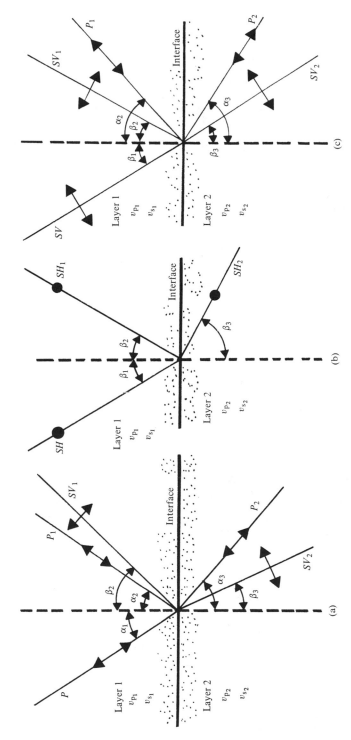

FIGURE 3.8 Reflection and refraction for **(a)** an incident *P*-ray; **(b)** an incident *SH*-ray; and **(c)** an incident *SV*-ray.

and an SV-wave, shown as P_2 and SV_2 in layer 2. Referring to the angles in Figure 3.8a, it can be shown that

$$\alpha_1 = \alpha_2 \tag{3.101}$$

and

$$\frac{\sin \alpha_1}{v_{p_1}} = \frac{\sin \alpha_2}{v_{p_1}} = \frac{\sin \beta_2}{v_{s_1}} = \frac{\sin \alpha_3}{v_{p_2}} = \frac{\sin \beta_3}{v_{s_2}} \tag{3.102}$$

where v_{p_1} and v_{p_2} are the velocities of the P-wave front in layers 1 and 2, respectively, and v_{s_1} and v_{s_2} are the velocities of the corresponding shear wave front.

If an SH-wave impinges the boundary between two layers as shown in Figure 3.8b, there is one reflected SH-wave (SH_1) and one refracted SH-wave (SH_2). For this case,

$$\beta_1 = \beta_2 \tag{3.103}$$

and

$$\sin \beta_1 / v_{s_1} = \sin \beta_3 / v_{s_2} \tag{3.104}$$

Lastly, an SV-wave impinging the boundary between two layers, as shown in Figure 3.8c, results in two reflected waves (P_1, SV_1) (and two refracted waves (P_2, SV_2). For this case, $\beta_1 = \beta_2$ and

$$\frac{\sin \beta_1}{v_{s_1}} = \frac{\sin \alpha_2}{v_{p_1}} = \frac{\sin \beta_2}{v_{s_1}} = \frac{\sin \beta_3}{v_{s_2}} = \frac{\sin \alpha_3}{v_{p_2}} \tag{3.105}$$

For details of the mathematical derivations of the above facts, the reader is referred to Kolsky (1963, pp. 24–38).

3.9 PRINCIPLES OF SEISMIC REFRACTION SURVEY (HORIZONTAL LAYERING)

Sometimes, seismic refraction survey is used to determine wave propagation velocities through various soil layers in the field and to obtain thicknesses of each layer. Consider the case where there are two layers of soil as shown in Figure 3.9a. Let the velocities of P-waves in layers 1 and 2 be v_{p_1} and v_{p_2}, respectively, and let $v_{p_1} < v_{p_2}$. Point A is a source of impulsive energy. If seismic waves are generated at A, the energy from that point travels in hemispherical waves fronts. Consider the case of P-waves since they are the fastest. If a detecting device is placed at point B, which is located at a *small* distance x from A, the P-wave that travels through the upper medium will reach it before any other wave. The travel time for this first arrival may be given by

$$t = x/v_{p_1} \tag{3.106}$$

where $\overline{AB} = x$. Now consider the first arrival time of a P-wave at a point G

located at a larger distance. A spherical P-wave front that originates at A strikes the interface of the two layers. At some point C, the refracted P-wave front in the lower medium is such that the tangent to the sphere is perpendicular to the interface. In that case, the refracted P-ray (shown as P_2 in Figure 3.9a) is parallel to the boundary and travels with a velocity v_{p_2}. Note since $v_{p_1} < v_{p_2}$, this wave front travels faster that those above. From Eq. (3.102),

$$\sin\alpha_1/v_{p_1} = \sin\alpha_3/v_{p_2}$$

Since $\alpha_3 = 90°$, $\sin\alpha_3 = 1$ and

$$\alpha_1 = \sin^{-1}\left(v_{p_1}/v_{p_2}\right) = \alpha_c \qquad (3.107)$$

where α_c is the critical angle of incidence.

The wave front described above traveling with a velocity v_{p_2} creates oscillating stresses at the interface, and this generates wave fronts which spread out into the upper medium. These P-waves spread with a velocity of v_{p_1}. The spherical wave front traveling downwards from D (in layer 2) has a radius equal to DE after a time Δt. At the same time Δt, the spherical wave front traveling upwards from the point D has a radius equal to DF. The

FIGURE 3.9 Seismic refraction survey: horizontal layering.

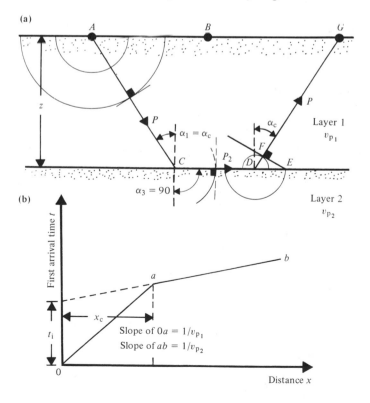

resultant wave front in the upper layer follows a line EF. It can be seen from the diagram that

$$\frac{v_{p_1}\Delta t}{v_{p_2}\Delta t} = \frac{DF}{DE} = \sin i_c \tag{3.108}$$

The ray DFG thus forms an angle i_c with the vertical. It can be mathematically shown that, for x greater than a *critical value* x_c, the P-wave that travels the path $ACDG$ is the *first to arrive* at point G. Let the time of travel for the P-wave along the path $ACDG$ be equal to t. Thus, $t = t_{AC} + t_{CD} + t_{DG}$; or,

$$t = \left(\frac{z}{\cos i_c}\right)\frac{1}{v_{p_1}} + \frac{x - 2z \tan i_c}{v_{p_2}} + \left(\frac{z}{\cos i_c}\right)\frac{1}{v_{p_1}}$$

$$= \frac{x}{v_{p_2}} - \frac{2z \sin i_c}{v_{p_2}\cos i_c} + \frac{2z}{v_{p_1}\cos i_c}$$

where $x = \overline{AG}$. However, $v_{p_2} = v_{p_1}/\sin i_c$ [from Eq. (3.108)]; thus,

$$t = \frac{x}{v_{p_2}} - \frac{2z \sin^2 i_c}{v_{p_1}\cos i_c} + \frac{2z}{v_{p_1}\cos i_c} = \frac{x}{v_{p_2}} + \frac{2z}{v_{p_1}}\left(\frac{1 - \sin^2 i_c}{\cos i_c}\right)$$

$$= \frac{x}{v_{p_2}} + \frac{2z}{v_{p_1}}\cos i_c \tag{3.109}$$

Since $\sin i_c = v_{p_1}/v_{p_2}$,

$$\cos i_c = \sqrt{1 - \sin^2 i_c} = \sqrt{1 - \left(v_{p_1}/v_{p_2}\right)^2} \tag{3.110}$$

Substituting Eq. (3.110) into Eq. (3.109), we get

$$t = \frac{x}{v_{p_2}} + 2z\sqrt{v_{p_2}^2 - v_{p_1}^2}\Big/\left(v_{p_1}v_{p_2}\right) \tag{3.111}$$

When detecting instruments are placed at various distances from the source of disturbance to obtain first arrival time and the results are plotted, a graph like that shown in Figure 3.9b is obtained. The line $0a$ represents the data that follows Eq. (3.106). The slope of this line is $1/v_{p_1}$. The line ab represents that data which follows Eq. (3.111) and has a slope of $1/v_{p_2}$. Thus the velocities v_{p_1} and v_{p_2} can now be obtained.

If line ab is projected back to $x = 0$, one obtains

$$t = t_i = 2z\sqrt{v_{p_2}^2 - v_{p_1}^2}\Big/\left(v_{p_1}v_{p_2}\right)$$

or

$$z = \tfrac{1}{2}(t_i)v_{p_1}v_{p_2}\Big/\sqrt{v_{p_2}^2 - v_{p_1}^2} = \tfrac{1}{2}t_i v_{p_1}/\cos i_c \tag{3.112}$$

where t_i is the intercept time.

The thickness of layer 1 can now easily be obtained. The *critical distance* x_c (Figure 3.9b), beyond which the wave refracted at the interface arrives at the detector before the direct wave, can be obtained by equating the right-hand sides of Eqs. (3.106) and (3.111):

$$\frac{x_c}{v_{p_1}} = \frac{x_c}{v_{p_2}} + \frac{2z\sqrt{v_{p_2}^2 - v_{p_1}^2}}{v_{p_1}v_{p_2}}$$

or,

$$x_c = 2z\frac{\sqrt{v_{p_2}^2 - v_{p_1}^2}}{v_{p_1}v_{p_2}} \cdot \frac{v_{p_1}v_{p_2}}{v_{p_2} - v_{p_1}}$$

$$= 2z\sqrt{\frac{v_{p_2} + v_{p_1}}{v_{p_2} - v_{p_1}}} \tag{3.113}$$

The depth of the first layer can also be calculated from Eq. (3.113) as

$$z = \tfrac{1}{2}x_c\sqrt{(v_{p_2} - v_{p_1})/(v_{p_2} + v_{p_1})} \tag{3.114}$$

Refraction Survey in a Three-Layer Soil Medium

Figure 3.10 illustrates the case of refraction survey through a three-layered soil medium. Let $v_{p_1} < v_{p_2} < v_{p_3}$ be the P-wave velocities in layers 1, 2, and 3, respectively, as shown in Figure 3.10a. If A in Figure 3.10a is a source of disturbance, the P-wave traveling through layer 1 will arrive first at B, which is located a small distance away from A. The travel time for this can be given by Eq. (3.106) as $t = x/v_{p_1}$. At a greater distance x, the first arrival will correspond to the wave taking the path $ACDE$. The travel time for this can be given by Eq. (3.111) as

$$t = x/v_{p_2} + 2z_1\sqrt{v_{p_2}^2 - v_{p_1}^2}/(v_{p_1}v_{p_2})$$

where z_1 is the thickness of the top layer.

At a still larger distance, the first arrival corresponds to the path $AGHIJK$. Note that the refracted ray HI will travel with a velocity of v_{p_3}. The angle i_{c2} is the critical angle for layer 3,

$$i_{c2} = \sin^{-1}(v_{p_2}/v_{p_3}) \tag{3.115}$$

For this path ($AGHIJK$) the total travel time can be derived as

$$t = \frac{x}{v_{p_3}} + \frac{2z_1\sqrt{v_{p_3}^2 - v_{p_1}^2}}{v_{p_3}v_{p_1}} + \frac{2z_2\sqrt{v_{p_3}^2 - v_{p_2}^2}}{v_{p_3}v_{p_2}} \tag{3.116}$$

where z_2 is the thickness of layer 2.

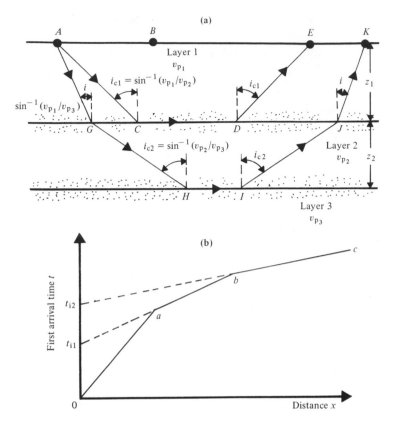

FIGURE 3.10 Refraction survey in a three-layer soil; **(a)** ray paths; **(b)** graph of t vs x. Slopes: $0a = 1/v_{p_1}$, $ab = 1/v_{p_2}$, $bc = 1/v_{p_3}$.

When detecting instruments are placed at various distances from the source of disturbance to obtain the first arrival times, these can be plotted in a t vs x graph, as shown in Figure 3.10b. The line $0a$ corresponds to Eq. (3.106), ab to Eq. (3.111), and bc to Eq. (3.116). The slopes of $0a$, ab, and bc are $1/v_{p_1}$, $1/v_{p_2}$, and $1/v_{p_3}$, respectively. The thickness of the first layer z_1 can be determined from the intercept time t_{i1} in a manner similar to that shown in Eq. (3.112):

$$z_1 = \tfrac{1}{2} t_{i1} v_{p_1} v_{p_2} \Big/ \sqrt{v_{p_2}^2 - v_{p_1}^2} \tag{3.112}$$

The thickness of the second layer can be obtained from Eq. (3.116). Referring to Figure 3.10b, the expression for the intercept time t_{i2} can be evaluated by substituting $x = 0$ into Eq. (3.116):

$$t = t_{i2} = \frac{2z_1 \sqrt{v_{p_3}^2 - v_{p_1}^2}}{v_{p_3} v_{p_1}} + \frac{2z_2 \sqrt{v_{p_3}^2 - v_{p_2}^2}}{v_{p_3} v_{p_2}}$$

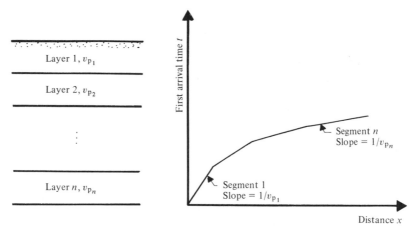

FIGURE 3.11 Refraction survey for multilayer soil.

or

$$z_2 = \frac{1}{2}\left(t_{i2} - \frac{2z_1\sqrt{v_{p_3}^2 - v_{p_1}^2}}{v_{p_3}v_{p_1}}\right)\frac{v_{p_3}v_{p_2}}{\sqrt{v_{p_3}^2 - v_{p_2}^2}} \qquad (3.117)$$

Refraction Survey for Multilayer Soil

In general, if there are n number of layers, the first arrival time at various distances from the source of disturbance will plot as shown in Figure 3.11. Note that there will be n segments in the t vs x plot. The slope of the nth segment will give the value of $1/v_{p_n}$ ($n = 1, 2, \dots$).

Typical Values of P-Wave Velocity in Soil

The value of *P*-wave velocity in a natural deposit of soil will depend on several factors such as confining pressure, moisture content, and void ratio.

TABLE 3.3 Typical Value Ranges of v_p at Shallow Depths

	Velocity[a] v_p	
Material	(ft/sec)	(m/sec)
Dry sand	600–4,000	183–1,220
Clay and wet soils	2,000–6,000	610–1,830
Loam	2,000–5,000	610–1,525
Sandstone	5,000–15,000	1,525–4,573
Shale	5,000–14,000	1,525–4,268

[a] 1 msec = 3.28 ft/sec.

Some typical values of v_p are given in Table 3.3 for soil and rock layers at the surface or shallow depths.

Example 3.2 Following are the results of a refraction survey (horizontal layering of soil) showing the distance x and time of first arrival t. Determine the P-wave velocities of the soil layers and their thicknesses.

x (m)	t (msec)	x (m)	t (msec)
2.5	5.5	35	38.2
5	11.1	45	46.1
7.5	16.1	55	51.3
15	24.0	60	52.8
25	30.8		

Solution: The time–distance plot is given in Figure 3.12. From the graph,

$$v_{p_1} = 5/(10.6 \times 10^{-3}) = 472 \text{ m/sec}$$

$$v_{p_2} = 10/(7.2 \times 10^{-3}) = 1389 \text{ m/sec}$$

$$v_{p_3} = 10/(3 \times 10^{-3}) = 3333 \text{ m/sec}$$

$$t_{i1} = 13.5 \times 10^{-3} \text{ sec}; \qquad t_{i2} = 35.6 \times 10^{-3} \text{ sec}.$$

FIGURE 3.12

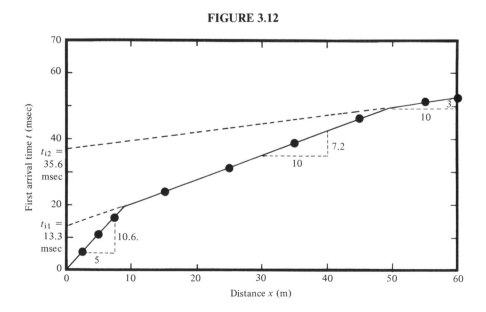

From Eq. (3.112),

$$z_1 = \frac{t_{i1}v_{p_1}v_{p_2}}{2\sqrt{v_{p_2}^2 - v_{p_1}^2}} = \frac{(13.5\times 10^{-3})(472)(1389)}{2\sqrt{(1389)^2 - (472)^2}} = 3.39 \text{ m}$$

From Eq. (3.117),

$$z_2 = \frac{1}{2}\left[t_{i2} - \frac{2z_1\sqrt{v_{p_3}^2 - v_{p_1}^2}}{v_{p_3}v_{p_1}}\right]\frac{v_{p_3}v_{p_2}}{\sqrt{v_{p_3}^2 - v_{p_2}^2}}$$

$$= \frac{1}{2}\left[35.6\times 10^{-3} - \frac{(2)(3.39)\sqrt{(3333)^2 - (472)^2}}{(3333)(472)}\right]\frac{(3333)(1389)}{\sqrt{(3333)^2 - (1389)^2}}$$

$$= \tfrac{1}{2}(0.02138)(1528) = 16.33 \text{ m}$$

3.10 REFRACTION SURVEY IN SOILS WITH INCLINED LAYERING

Figure 3.13a shows two soil layers. The interface of soil layers 1 and 2 is inclined at an angle β with respect to the horizontal. Let the P-wave velocities in layers 1 and 2 be v_{p_1} and v_{p_2}, respectively ($v_{p_1} < v_{p_2}$). If a disturbance is created at A, a detector at point B, which is a small distance away from A, first receives the P-wave traveling through layer 1. The time for its arrival may be given by

$$t_d = x/v_{p_1} \tag{3.118}$$

However, at a larger distance, the first arrival is for the P-wave following the path $ACDE$, which consists of three segments. The time taken can thus be written as

$$t_d = t_{AC} + t_{CD} + t_{DE} \tag{3.119}$$

Referring to Figure 3.13a,

$$t_{AC} = z'/(v_{p_1}\cos i_c) \tag{3.120}$$

$$t_{CD} = CD/v_{p_2} = (AA_4 - AA_1 - A_2A_3 - A_3A_4)/v_{p_2}$$
$$= (x\cos\beta - z'\tan i_c - z'\tan i_c - x\sin\beta\tan i_c)/v_{p_2} \tag{3.121}$$

$$t_{DE} = \frac{DA_3 + A_3E}{v_{p_1}} = \frac{(z'/\cos i_c) + x\sin\beta/\cos i_c}{v_{p_1}} \tag{3.122}$$

Substitution of Eqs. (3.120)–(3.122) into Eq. (3.119) and simplification yields

$$t_d = (2z'\cos i_c)/v_{p_1} + (x/v_{p_1})\sin(i_c + \beta) \tag{3.123}$$

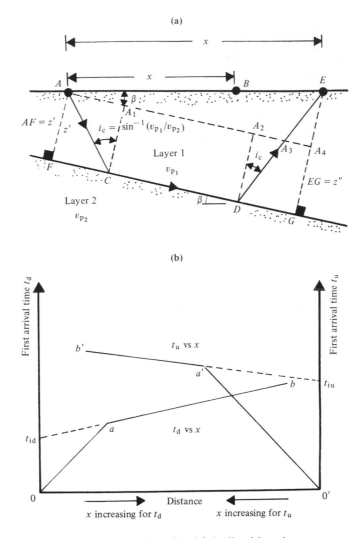

FIGURE 3.13 Refraction survey in soils with inclined layering.

If the source of disturbance is E and the detector is placed at A, the first arrival time along the refracted ray path can be given by

$$t_{\mathrm{u}} = (2z''\cos i_{\mathrm{c}})/v_{\mathrm{p}_1} + (x/v_{\mathrm{p}_1})\sin(i_{\mathrm{c}} - \beta) \qquad (3.124)$$

In the actual survey, one can have (1) a source of disturbance such as A and observe the first arrival time at several points to its right, and also have (2) a source of disturbance such as E and observe the first arrival time at several points to its left. These results can be plotted in a graphical form as shown in Figure 3.13b (t vs x). From Figure 3.13b, note that the slopes of $0a$ and $0'a'$ are both $1/v_{p_1}$. The slope of branch ab is $\sin(i_c + \beta)/v_{p_1}$, as can be seen from Eq. (3.123). Similarly, the slope of branch $a'b'$ is $\sin(i_c - \beta)/v_{p_1}$ [see Eq. (3.124)]. Let

$$m_d = \sin(i_c + \beta)/v_{p_1} \tag{3.125}$$

and

$$m_u = \sin(i_c - \beta)/v_{p_1} \tag{3.126}$$

From Eq. (3.125),

$$i_c = \sin^{-1}(m_d v_{p_1}) - \beta \tag{3.127}$$

Similarly, from Eq. (3.126),

$$i_c = \sin^{-1}(m_u v_{p_1}) + \beta \tag{3.128}$$

Solving the two preceding equations,

$$i_c = \tfrac{1}{2}\left[\sin^{-1}(v_{p_1} m_d) + \sin^{-1}(v_{p_1} m_u)\right] \tag{3.129}$$

and

$$\beta = \tfrac{1}{2}\left[\sin^{-1}(v_{p_1} m_d) - \sin^{-1}(v_{p_1} m_u)\right] \tag{3.130}$$

Once i_c is determined, the value of v_{p_2} can be obtained as

$$v_{p_2} = v_{p_1}/\sin i_c \tag{3.131}$$

Again, referring to Figure 3.13b, if ab and $a'b'$ branches are projected back, they will intercept the time axes at t_{id} and t_{iu}, respectively. From Eqs. (3.123) and (3.124), it can be seen that

$$t_{id} = 2z' \cos i_c/v_{p_1}$$

or

$$z' = \tfrac{1}{2} t_{id} v_{p_1}/\cos i_c \tag{3.132}$$

and

$$t_{iu} = 2z'' \cos i_c/v_{p_1}$$

or

$$z'' = \tfrac{1}{2} t_{iu} v_{p_1}/\cos i_c \tag{3.133}$$

Since i_c and v_{p_1} are known, and t_{id} and t_{iu} can be determined from a graph, one can obtain the values of z' and z''.

Example 3.3 Referring to Figure 3.13a, the results of a refraction survey are as tabulated. The distance between A and E is 60 m.

Point of disturbance A		Point of disturbance E	
Distance from A (m)	Time of first arrival (msec)	Distance from E (m)	Time of first arrival (msec)
5	12.1	5	11.5
10	25.2	10	22.8
15	35.3	15	34.5
20	48.0	20	44.8
30	60.2	30	69.1
40	68.5	40	78.1
50	76.8	50	82.8
60	85.1	60	87.5

Determine

a. v_{p_1} and v_{p_2},
b. z' and z'', and
c. β.

Solution: The time–distance records are plotted in Figure 3.14.

a. Determination of v_{p_1} and v_{p_2}. From the branch $0a$,

$$v_{p_1} = 10/(25 \times 10^{-3}) = 400 \text{ m/sec}$$

From the branch $0'a'$,

$$v_{p_1} = 10/(22 \times 10^{-3}) = 455 \text{ m/sec}$$

The average value of $v_{p_1} = 427.5$ m/sec.
From the slope of the branch ab,

$$m_d = 8.8 \times 10^{-3}/10 = 0.88 \times 10^{-3}$$

Again, from the slope of branch $a'b'$,

$$m_u = 5 \times 10^{-3}/10 = 0.5 \times 10^{-3}$$

$$i_c = \tfrac{1}{2}\left[\sin^{-1}(v_{p_1} m_d) + \sin^{-1}(v_{p_1} m_u)\right] \qquad (3.129)$$

$$\sin^{-1}(v_{p_1} m_d) = \sin^{-1}\left[(427.5)(0.88 \times 10^{-3})\right] = 22.07°$$

$$\sin^{-1}(v_{p_1} m_u) = \sin^{-1}\left[(427.5)(0.5 \times 10^{-3})\right] = 12.33°$$

Hence, $i_c = \tfrac{1}{2}(22.07 + 12.33) = 17.2°$. Using Eq. (3.131),

$$v_{p_2} = v_{p_1}/\sin i_c = 427.5/\sin(17.2) = 1444 \text{ m/sec}$$

b. Determination of z' and z''. From Eq. (3.132),

$$z' = \tfrac{1}{2} t_{id} v_{p_1}/\cos i_c$$

$$t_{id} = 35.9 \times 10^{-3} \text{ sec (Figure 3.14)}$$

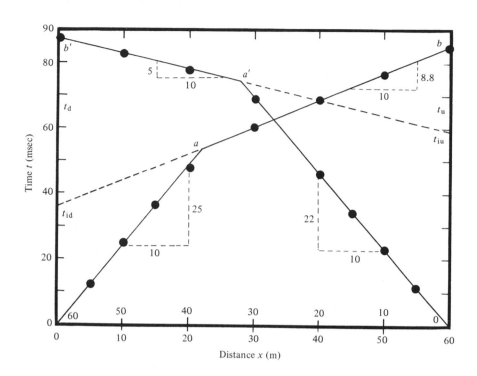

FIGURE 3.14

so

$$z' = \tfrac{1}{2}(35.9 \times 10^{-3})(427)/\cos(17.2) = 8.03 \text{ m}$$

Again, from Eq. (3.133),

$$z'' = \tfrac{1}{2}t_{iu}v_{p_1}/\cos i_c$$

From Figure 3.14, $t_{iu} = 59.8 \times 10^{-3}$ sec:

$$z'' = \tfrac{1}{2}(59.8 \times 10^{-3})(427)/\cos(17.2) = 13.37 \text{ m}$$

c. Determination of β. From Eq. (3.130),

$$\beta = \tfrac{1}{2}\left[\sin^{-1}(v_{p_1}m_d) - \sin^{-1}(v_{p_1}m_u)\right]$$
$$= \tfrac{1}{2}(22.07° - 12.33°) = 4.87°$$

3.11 REFLECTION SURVEY IN SOIL

3.11.1 Horizontal Layering

Reflection surveys can also be conducted to obtain information about the soil layers. Figure 3.15a shows a two-layered soil system with point of disturbance A. If a recorder is placed at C, a distance x away from A, the

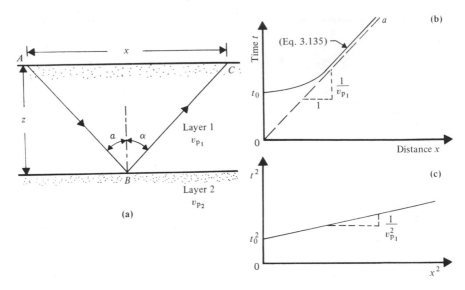

FIGURE 3.15 Reflection survey in soil: horizontal layering.

travel time for the reflected P-wave can be given by

$$t = (AB + BC)/v_{p_1} = (2/v_{p_1})\sqrt{z^2 + \left(\tfrac{1}{2}x\right)^2} \qquad (3.134)$$

where t is the total travel time for the ray path ABC.

From the above equation, the thickness of layer 1 can be obtained as

$$z = \tfrac{1}{2}\sqrt{\left(v_{p_1}t\right)^2 - x^2} \qquad (3.135)$$

If the travel times t for the reflected P-waves are obtained at various distances x, they can be plotted in a graphical form as shown in Figure 3.15b. Note that the time–distance curve obtained from Eq. (3.134) is a hyperbola. The line $0a$ shown in Figure 3.15b is the time–distance plot for the direct P-waves traveling through layer 1 (compare the line $0a$ in Figure 3.9b). The slope of this line is $1/v_{p_1}$.

If the time–distance curve obtained from the reflection data is extended back, it intersects the time axis at a distance t_0. From Eq. (3.134), it can be seen that at $x = 0$,

$$t_0 = 2z/v_{p_1}$$

or

$$z = \tfrac{1}{2}t_0 v_{p_1} \qquad (3.136)$$

With v_{p_1} and t_0 known, the thickness z of the top layer can be calculated.

Another convenient way to interpret the reflection survey records is to plot a graph of t^2 vs x^2. From Eq. (3.134),

$$t^2 = \left(4/v_{p_1}^2\right)\left[z^2 + \left(\tfrac{1}{2}x\right)^2\right] = \left(1/v_{p_1}^2\right)\left(4z^2 + x^2\right) \qquad (3.137)$$

The above relationship shows that the plot of t^2 vs x^2 is a straight line as shown in Figure 3.15c. The slope of this line is $1/v_{p_1}^2$ and the intercept on the t^2 axis is equal to t_0^2.

$$t_0^2 = 4z^2/v_{p_1}^2$$

so

$$z^2 = \tfrac{1}{4}t_0^2 v_{p_1}^2 \qquad (3.138)$$

With t_0^2 and $v_{p_1}^2$ known, the thickness of the top layer can now be calculated.

Example 3.4 The results of a reflection survey on a relatively flat area (shale underlain by granite) are given below. Determine the velocity of P-waves in the shale.

Distance from point of disturbance (ft)	Time for first reflection (sec)
100	1.0
300	1.002
500	1.003
700	1.007
900	1.011
1100	1.017
1300	1.023

Solution: Using the time–distance records, the following table can be prepared.

x (ft)	x^2 (ft^2)	t (sec)	t^2 (sec^2)
100	10,000	1.0	1.0
300	90,000	1.002	1.004
500	250,000	1.003	1.006
700	490,000	1.007	1.014
900	810,000	1.011	1.022
1100	1,210,000	1.017	1.034
1300	1,690,000	1.023	1.046

A plot of t^2 vs x^2 is shown in Figure 3.16. From the plot,

$$v_{p_1} = \sqrt{(\Delta x)^2/(\Delta t)^2} = \sqrt{800,000/0.023} = 5898 \text{ ft/sec}$$

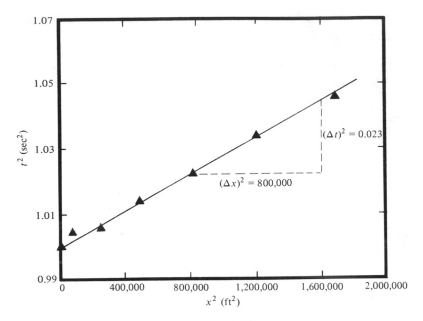

FIGURE 3.16

3.11.2 Inclined Layering

Figure 3.17 considers the case of reflection survey where the reflecting boundary is inclined at an angle β with respect to the horizontal and A is the point for the source of disturbance. The reflected P-wave reaching point C takes the path ABC. Referring to Figure 3.17,

$$AB + BC = A'B + BC = A'C$$

but

$$(A'C)^2 = (A'A_2)^2 + (A_2C)^2 \qquad (3.139)$$

$$A'A_2 = AA' \cos \beta = 2z' \cos \beta \qquad (3.140)$$

$$A_2C = A_2A + AC = 2z' \sin \beta + x_C \qquad (3.141)$$

Substituting Eqs. (3.140) and (3.141) into Eq. (3.139),

$$A'C = \sqrt{(2z' \cos \beta)^2 + (2z' \sin \beta + x_C)^2}$$

$$= \sqrt{4z'^2 + x_C^2 + 4z'x_C \sin \beta}$$

Thus, the travel time for the reflected P-wave along the path ABC is

$$t_C = A'C/v_{p_1} = \frac{1}{v_{p_1}} \cdot \sqrt{4z'^2 + x_C^2 + 4z'x_C \sin \beta} \qquad (3.142)$$

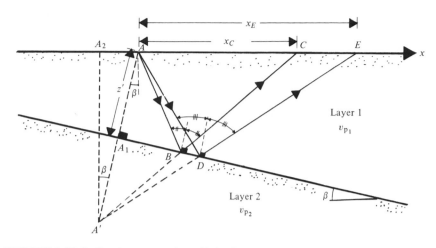

FIGURE 3.17 Reflection survey in soil: inclined layering.

In a similar manner, the time of arrival for the reflected P-waves received at point E can be given by

$$t_E = \frac{1}{v_{p_1}} \cdot \sqrt{4z'^2 + x_E^2 + 4z'x_E \sin \beta} \qquad (3.143)$$

Combining Eqs. (3.142) and (3.143),

$$\sin \beta = \frac{v_{p_1}^2 \left(t_E^2 - t_C^2 \right)}{4z'(x_E - x_C)} - \frac{x_E + x_C}{4z'} \qquad (3.144)$$

Now, let $\bar{t} = \frac{1}{2}(t_E + t_C)$ and $\Delta t = t_E - t_C$. Substitution of this relation in Eq. (3.144) gives

$$\sin \beta = \frac{v_{p_1}^2 \bar{t}(\Delta t)}{2z'(x_E - x_C)} - \frac{x_E + x_C}{4z'} \qquad (3.145)$$

If x_C is equal to zero, the above equation becomes

$$\sin \beta = v_{p_1}^2 \bar{t}(\Delta t)/(2z'x_E) - (x_E/4z') \qquad (3.146)$$

If $x_0 = 0$ and $\beta = 0$ (i.e., the reflecting layer is horizontal), from Eq. (3.146),

$$\Delta t = x_E^2/2v_{p_1}^2 \bar{t} \qquad (3.147)$$

If $x_C = 0$ and $\Delta t > x_E^2/(2v_{p_1}^2 \bar{t})$, the reflecting layer is sloping down in the direction of positive x as shown in Figure 3.17. If $x_C = 0$ and $\Delta t < x_E^2/(2v_{p_1}^2 \bar{t})$, the reflecting layer is sloping down in the direction of negative x (i.e., opposite to what is shown in Figure 3.17).

In actual practice, the source of disturbance A (Figure 3.18) is generally placed midway between the two detectors, so $x_E = -x_C = x$. Thus, from

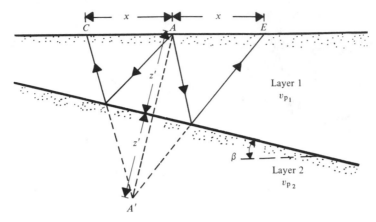

FIGURE 3.18

Eq. (3.145),

$$\sin \beta = v_{p_1}^2 \bar{t}(\Delta t)/(4z'x) \qquad (3.148)$$

Referring to Figure 3.18, $AA' = 2z' = \frac{1}{2}(A'C + A'E)$; so

$$2z'/v_{p_1} = \frac{1}{2}\left(A'C/v_{p_1} + A'E/v_{p_1}\right) = \frac{1}{2}(t_C + t_E) = \bar{t} \qquad (3.149)$$

Combining Eqs. (3.148) and (3.149),

$$\sin \beta = v_{p_1}(\Delta t)/2x \qquad (3.150)$$

Example 3.5 Refer to Figure 3.18. Given $x = 280$ ft, $t_C = 0.026$ sec, and $t_E = 0.038$ sec, determine β and z'. (The value of v_{p_1} has previously been determined to be 1350 ft/sec.)

Solution

$$\bar{t} = \frac{1}{2}(t_C + t_E) = \frac{1}{2}(0.026 + 0.038) = 0.032 \text{ sec}$$
$$\Delta t = t_E - t_C = 0.038 - 0.026 = 0.012 \text{ sec}$$

From Eq. (3.150),

$$\beta = \sin^{-1}\left[v_{p_1}(\Delta t)/2x\right] = \sin^{-1}\left[\frac{1}{2}(1350 \times 0.012)/(280)\right] = 1.66°$$

From Eq. (3.149),

$$2z'/v_{p_1} = \bar{t} \quad \text{or} \quad z' = \frac{1}{2}\left(\bar{t}v_{p_1}\right) = \frac{1}{2}(0.032)(1350) = 21.6 \text{ ft}$$

3.12 SUBSOIL EXPLORATION BY STEADY-STATE VIBRATION TECHNIQUE

In this technique, a circular plate placed on the ground surface is vibrated vertically by a sinusoidal load (Figure 3.19a). This vibration sends out Rayleigh waves, and the vertical motion of the ground surface is predomi-

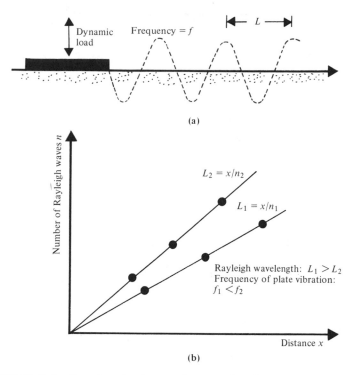

FIGURE 3.19 Subsoil exploration by steady-state vibration.

nantly due to these waves. This can be picked up by motion transducers. The velocity of the Rayleigh waves can be given by

$$v_r = fL \tag{3.151}$$

where f is the frequency of the vibration of the plate and L is the wavelength.

If the wavelength L can be measured, the velocity of the Rayleigh waves can be calculated. The wavelength is generally determined by the number of waves occurring at a given distance x. For a given frequency f_1 the wavelength can be given by

$$L_1 = x/n_1 \tag{3.152}$$

where n_1 is the number of waves at a distance x for frequency f_1. This is shown in Figure 3.19b.

As has been shown in Table 3.1 (p. 87), the Rayleigh wave velocity is approximately equal to the shear wave velocity,

$$v_r \approx v_s \tag{3.153}$$

It has been previously mentioned that, for all practical purposes, the Rayleigh wave travels through the soil within a depth of one wavelength. Hence, for a given frequency f, if the wavelength L is known, then the value

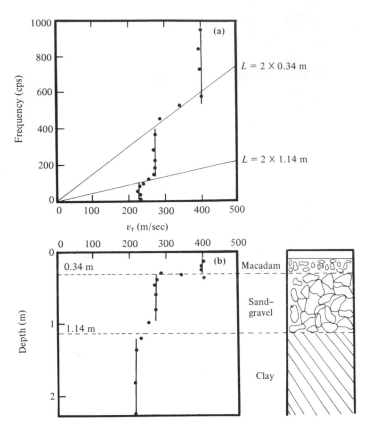

FIGURE 3.20 Velocity v_r as a function of (a) frequency and (b) depth determined by steady-state vibration technique. Note: 1 m = 3.28 ft. [Heukelom, W., and Foster, C. R. (1960). "Dynamic Testing of Pavements," *Journal of the Soil Mechanics and Foundations Division*, *ASCE*, 86 (SM1), Part I, Fig. 13, p. 18.]

of v_s determined by the above technique will represent the soil conditions at an average depth of $\frac{1}{2}L$. Thus, for *a large value of f*, the value of v_s is representative of soil conditions at *a smaller depth*; and, for *a small value of f*, the v_s obtained is representative of the soil conditions at *a larger depth*.

Figure 3.20 shows the results of wave propagation on a stratified pavement system obtained by using this technique (Heukelom and Foster, 1960).

3.13 SOIL EXPLORATION BY UP-, DOWN-, AND CROSS-HOLE SHOOTING

Shooting Up the Hole

In this technique, a hole is drilled into the ground and a detector is placed at the ground surface. Charges are exploded at various depths in the whole and the direct travel time of body waves (P or S) along the boundary of the hole is measured. Thus, the values of v_p and v_s of various soil layers can be easily obtained. There is a definite *advantage* in this technique because it determines the *shear wave velocities* of various layers of soil; the refraction and reflection techniques only give the *P*-wave velocity. However, below the ground water table, the compression waves will travel through water. The first arrival for points below the water table will usually be for this type of wave and is generally higher than the compression waves for soils. On the other hand, shear waves cannot travel through water, and the shear wave velocity measured above or below the water table will give the measured wave velocity of the soil.

Shooting Down the Hole

Shear wave velocity determination of various soil layers by shooting down the hole has been described by Schwarz and Musser (1972), Beeston and McEvilly (1977), and Larkin and Taylor (1979). Figure 3.21 shows a schematic diagram for the down-hole method of seismic wave testing as presented by Larkin and Taylor, which relies on measuring the time interval of *SH*-waves traveling between the ground surface and the subsurface points. A bidirectional impulsive source for the propagation of *SH*-waves is placed on the surface adjacent to a borehole. A horizontal sensitive transducer is located at a depth in the borehole; the depth of the transducer is varied throughout the length of the borehole. The shear wave velocity can

FIGURE 3.21 The down-hole method of seismic wave testing as conducted by Larkin and Taylor (1979).

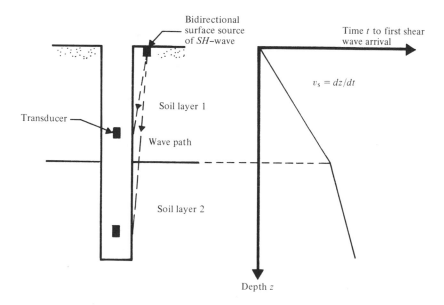

then be obtained as

$$v_s = \Delta z / \Delta t \qquad (3.154)$$

where z is the depth below the ground surface and t is the time of travel of shear wave from the surface impulsive source to the transducer.

The results of a down-hole shooting test as conducted by Larkin and Taylor (1979) have been compiled in Table 3.4. The shear wave velocities have been determined by using Eq. (3.154). Since the P-wave arrival times down the hole are recorded by the transducer, they are also given in Table 3.4. The velocity v_p for the entire 50-m depth is constant and of the order of 1500 m/sec (4921 ft/sec). This value corresponds to the compressive wave velocity in water. (Note that although the water table is located at a depth of about 10 m, the soil above that is fully saturated.)

During the process of field investigation, Larkin and Taylor determined that the shear strains at depths of 3 and 50 m were about 1×10^{-6} and 0.3×10^{-6}, respectively. In order to compare the field and laboratory values of v_s, some undisturbed samples from various depths were collected. The shear wave velocity of various specimens at a shear strain of 1×10^{-6} were determined. A comparison of laboratory and field test results shows that, for similar soils, the value of $v_{s(lab)}$ is considerably lower than that obtained in the field. For *the range of soil tested*,

$$v_{s(lab)} \approx 0.25 v_{s(field)} + 83 \qquad (in \ m/sec)$$

TABLE 3.4 Results of a Down-Hole Shooting Test[a]

Depth below the ground surface (m)	Description of soil	Shear wave velocity v_s (m/sec)	P-wave velocity,[b] v_p (m/sec)
0–3	Silty clay	135	1500
3–5.8	Silty sand	135	1500
5.8–8.9	Silty clay	258	1500
8.9–19.7	Sandy clay grading into clay	143	1500
19.7–24.4	Firm silty clay	334	1500
24.4–28.1	Silty sand	334	1500
28.1–29.1	Firm silty clay	334	1500
29.1–44.2	Very firm silty sand	334	1500
44.2–50	Sandstone and siltstone	1000	1500

[a] Compiled from Larkin and Taylor (1979).
[b] Water table located at a depth of about 10 m below the ground surface.

Larkin and Taylor also defined a quantity called a sample disturbance factor D_F,

$$D_F = \left(v_{s(\text{field})} / v_{s(\text{lab})} \right)^2 = G_{\text{field}} / G_{\text{lab}} \tag{3.155}$$

The average value of D_F for this investigation varied from about 1 for $v_{s(\text{field})} = 140$ m/sec to about 4 for $v_{s(\text{field})} = 400$ m/sec. This shows that somewhat small disturbances in sampling could introduce large errors in the evaluation of representative shear moduli of soils.

Cross-Hole Shooting

This technique of seismic surveying relies on the measurement of SV-wave velocity. In this procedure, two vertical holes are bored into the ground a given distance apart (Figure 3.22). Shear waves are generated by a vertical impact at the bottom of one bore hole. The arrival of the body waves is recorded by a vertically sensitive transducer placed at the bottom of the

FIGURE 3.22 Schematic diagram of cross-hole seismic survey technique. [Stokoe, K. H., and Woods, R. D. (1972). "*In Situ* Wave Velocity by Cross-Hole Method," *Journal of the Soil Mechanics and Foundations Division*, *ASCE*, 98 (SM5), Fig. 1, p. 445.]

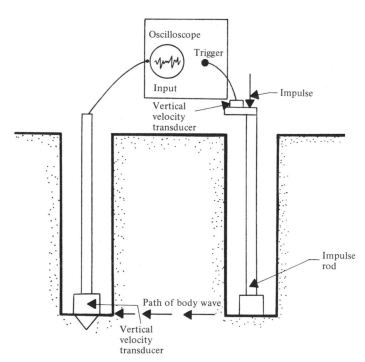

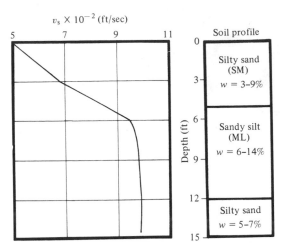

FIGURE 3.23 Shear wave velocity vs depth from cross-hole seismic survey; $w =$ moisture content. Note: 1 m = 3.28 ft. [Stokoe, K. H., and Woods, R. D. (1972). "*In Situ* Wave Velocity by Cross-Hole Method," *Journal of the Soil Mechanics and Foundations Division, ASCE*, 98 (SM5), Fig. 5, p. 449.]

other bore hole at the same depth. Thus,

$$v_s = L/t \qquad (3.156)$$

where t is the travel time for the shear wave and L is the length between the two bore holes.

Figure 3.23 shows the plot of shear wave velocity against depth for a test site obtained using the cross-hole shooting technique of seismic surveying (Stokoe and Woods, 1972).

PROBLEMS

3.1. For a Poisson's ratio of 0.33, solve Eq. (3.86) and determine the value of v_r/v_s.

3.2. Use the results of v_r/v_s (for Poisson's ratio of 0.33) obtained from Problem 3.1 in Eqs. (3.91) and (3.92) and express U and W as a function of fz.

3.3. **a.** With the results of Problem 3.2, plot a graph showing the variation of z/wavelength vs $U_{(\text{depth } z)}/U_{(\text{depth}=0)}$.
   b. Repeat Problem 3.3a for the vertical component of Rayleigh waves (plot z/wavelength vs $W_{(\text{depth } z)}/W_{(\text{depth}=0)}$.

3.4. Following are the results of a refraction survey. Assuming that the soil layers are horizontal, determine the P-wave velocities in the soil layers and the thickness of the top layer.

Distance (ft)	Time of first arrival (msec)	Distance (ft)	Time of first arrival (msec)
25	49.08	150	174.2
50	81.96	200	202.8
75	122.8	250	228.6
100	148.2	300	256.7

3.5. Repeat Problem 3.4 for the following:

Distance (m)	Time of first arrival (msec)	Distance (m)	Time of first arrival (msec)
10	19.23	100	125.82
20	38.40	150	138.72
30	57.71	200	152.61
40	76.90	250	166.81
60	115.40	300	178.31
80	120.71		

Comment about the material encountered in the second layer.

3.6. Repeat Problem 3.4 for the following, also determining the thickness of the second layer of soil encountered.

Distance (m)	Time of first arrival (msec)	Distance (m)	Time of first arrival (msec)
10	41.66	60	119.21
15	62.51	70	128.11
20	83.37	80	136.22
30	91.82	90	141.00
40	101.22	100	143.81
50	110.16	120	152.00

3.7. Refer to Figure 3.13a for the results of the following refraction survey:

Distance from point of disturbance A (ft)	Time of first arrival (msec)	Distance from point of disturbance E (ft)	Time of first arrival (msec)
0	0	0	0
20	20	20	20
40	40	40	40.1
60	60	60	59.8
80	78.2	80	79.7
120	92.8	120	121.0
200	122.2	200	167.2
280	149.8	280	175.1
360 (Point E)	177.9	360 (Point A)	180.2

Determine:
a. The *P*-wave velocities in the two layers
b. z' and z''
c. The angle β

3.8. The results of a reflection survey are given below. Determine the velocity of *P*-waves in the top layer and its thickness.

Distance from the shot point (m)	Time for first reflection (msec)
10	32.5
20	39.05
30	48.02
40	58.3
60	80.78
100	128.55

3.9. For a reflection survey, refer to Figure 3.18, in which *A* is the shot point. Distances $AC = AE = 180$ m. The times for arrival of the first reflected wave at points *C* and *E* are 45.0 msec and 64.1 msec, respectively. If the *P*-wave velocity in Layer 1 is 280 m/sec, determine β and z'.

3.10. The results of a subsoil exploration by steady-state vibration technique are given below (Section 3.12).

Distance from the plate vibrated x (m)	No. of waves n_1	Frequency of vibration of the plate f (cps)
10	41	900
10	18	400
10	9	200
10	4.55	100
10	2.65	90
10	2.3	75
10	1.77	60
10	1.47	50

Make necessary calculations and plot the variation of the wave velocity with depth.

REFERENCES

Barkan, D. D. (1962). *Dynamics of Bases and Foundations*, McGraw-Hill, New York.

Beeston, H. E., and McEvilly, T. V. (1977). "Shear Wave Velocities from Down Hole Measurements," *Journal of the International Association for Earthquake Engineering* 5 (2), 181–190.

Biot, M. A. (1956). "Theory of Propagation of Elastic Waves in a Fluid Saturated Solid," *Journal of the Acoustical Society of America* (28), 168–178.

Bornitz, G. (1931). *Über die Ausbreitung der von Graszkolbenmaschinen erzeugten Bodenschwingungen in die Tiefe*, J. Springer, Berlin.

Hall, J. R., Jr. and Richart, F. E., Jr. (1963). "Dissipation of Elastic Wave Energy in Granular Soils," *Journal of the Soil Mechanics and Foundations Division, ASCE* 89 (SM6), 27–56.

Hardin, B. O., and Richart, F. E., Jr. (1963). "Elastic Wave Velocities in Granular Soils," *Journal of the Soil Mechanics and Foundations Division, ASCE* 89 (SM1), 33–65.

Heukelom, W., and Foster, C. R. (1960). "Dynamic Testing of Pavements," *Journal of the Soil Mechanics and Foundations Division, ASCE* 86 (SM1, Part 1), 1–28.

Kolsky, H. (1963). *Stress Waves in Solids*, Dover, New York.

Larkin, T. J., and Taylor, P. W. (1979). "Comparison of Down Hole and Laboratory Shear Wave Velocities," *Canadian Geotechnical Journal* 16 (1), 152–162.

Rayleigh, Lord (1885). "On Waves Propagated Along the Plane Surface of an Elastic Solid," *Proceedings, London Mathematical Society* 17, 4–11.

Richart, F. E., Jr., Hall, J. R., and Woods, R. D. (1970). *Vibration of Soils and Foundations*, Prentice-Hall, Englewood Cliffs, NJ.

Schwarz, S. D., and Musser, J. (1972). "Various Techniques for Making *In Situ* Shear Wave Velocity Measurements: A Description and Evaluation," *Proceedings, Microzonation Conference, Seattle, Washington* 2, 593.

Stokoe, K. H., and Woods, R. D. (1972). "*In Situ* Shear Wave Velocity by Cross-Hole Method," *Journal of the Soil Mechanics and Foundations Division, ASCE* 98 (SM5), 443–460.

Timoshenko, S. P., and Goodier, J. N. (1970). *Theory of Elasticity*, McGraw-Hill, New York.

4

AIR BLAST LOADING ON GROUND

The displacement of the surface of the earth's crust due to a nuclear explosion is caused by cratering and also due to the *air blast loading*. Immediately below the center of explosion, a crater is formed due to the conversion of thermonuclear energy into mechanical energy. The center of the explosion is called *ground zero*. At large distances from ground zero, displacement of ground is caused by the high air pressure initiated by the explosion. A knowledge of the stresses and displacements caused by the air blast loading at large distances from ground zero is required for the design of underground protective structures. This is the subject of discussion in this chapter.

4.1 OVERPRESSURE

The nature of variation with distance of the high air pressure caused by a nuclear explosion is shown in Figure 4.1. The front of the air pressure moves radially outwards like a ring load with a near vertical front with a peak value of p_0 and decays in an exponential manner. The pressure front moves with a velocity V as shown in Figure 4.1. Under certain conditions, the overpressure becomes negative after some time. Figure 4.2 shows the variation of p_0 for various yields with distance.

Where p_0 is in the range 200–25 lb/in.^{2} (1380–172.5 kN/m^2), it has been observed that there is a rise time t_r for the overpressure front as shown in Figure 4.3a. The value of t_r for a 1 kiloton (4.184×10^{12} J) yield is 3–5 msec.

As mentioned before, at a given point on the ground surface, the overpressure becomes negative after some time as shown in Figure 4.3b. The duration of the positive phase of the overpressure is generally referred to as t_+. The variation of the duration of positive phase of overpressure with p_0 for a 1-megaton yield (4.184×10^{15} J) is shown in Figure 4.4. The value of t_+

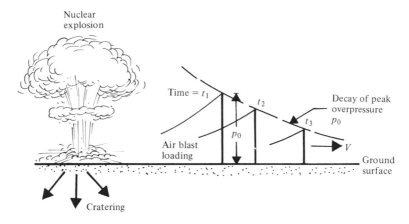

FIGURE 4.1 Air blast loading.

FIGURE 4.2 Variation of pear overpressure with yield W and distance from ground zero. Note: 1 lb/in.^{2} = 6.9 kN/m^2; 1 m = 3.2 ft; 1 megaton = 4.184×10^{15} J. [U.S. Air Force (1962), Fig. 3.1, p. 3-37.]

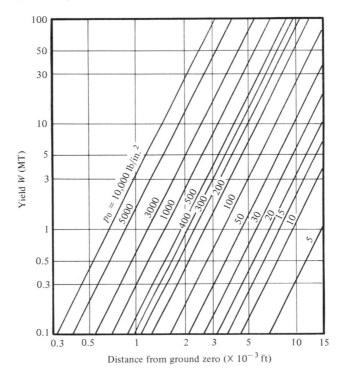

124

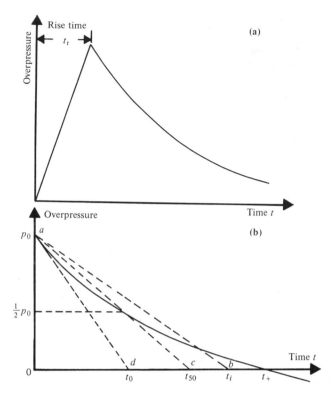

FIGURE 4.3 Definition of **(a)** rise time of overpressure at a point; **(b)** duration of positive phase of overpressure.

FIGURE 4.4 Variation of t_+ with peak overpressure p_0 for yield $W = 1$ megaton $(4.184 \times 10^{15}$ J). [U.S. Air Force (1962), Fig. 3.5, p. 3-41.]

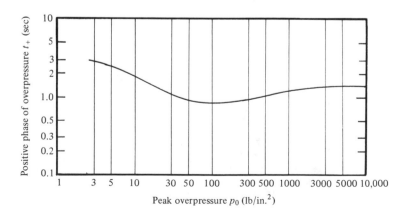

for a given p_0 due to a weapon yield W can be estimated by the equation

$$t_{+(\text{yield}=W)} = t_{+(W=1)}W^{1/3} \tag{4.1}$$

In some cases, for simplified calculation purposes, the positive phase of the overpressure–time relationship is approximated as a single triangle pulse having a duration t_i as shown in Figure 4.3b (triangle $0ab$). This is useful for determination of the maximum response of structures after the passage of the pressure pulse. In this case, the total impulse due to the triangular pulse $0ab$ is equal to that of the actual overpressure–time curve. Similarly, for the analysis of the response of structures in the early part of the overpressure history, a triangular pulse such as $0ad$ (Figure 4.3b) can be used which has a duration of t_0. Note that the line ad is *tangent* to the initial part of the actual overpressure–time curve. Analysis of response to conditions intermediate to the pulses represented by $0ab$ and $0ad$ can be made by a simplified triangular pulse represented by $0ac$. The line ac intersects the actual overpressure–time curve at a point which has an ordinate equal to $\frac{1}{2}p_0$. The duration of the pulse is equal to t_{50}. Empirical relations between the weapon yield, peak overpressure, t_i, t_0, and t_{50} can be given as follows (U.S. Air Force, 1962):

$$t_i = 0.37(100/p_0)^{1/2}W^{1/3} \qquad \text{for} \quad 2 \leqslant p_0 \leqslant 10{,}000 \text{ lb/in.}^2 \tag{4.2}$$

$$t_{50} = 1.0(10/p_0)^{1/2}W^{1/3} \qquad \text{for} \quad 2 \leqslant p_0 \leqslant 30 \text{ lb/in.}^2 \tag{4.3}$$

$$t_{50} = 0.2(100/p_0)^{7/8}W^{1/3} \qquad \text{for} \ 30 \leqslant p_0 \leqslant 10{,}000 \text{ lb/in.}^2 \tag{4.4}$$

$$t_0 = 0.87(10/p_0)^{1/2}W^{1/3} \qquad \text{for} \quad 2 \leqslant p_0 \leqslant 30 \text{ lb/in.}^2 \tag{4.5}$$

$$t_0 = 0.15(100/p_0)W^{1/3} \qquad \text{for} \ 30 \leqslant p_0 \leqslant 10{,}000 \text{ lb/in.}^2 \tag{4.6}$$

where t_i, t_{50}, t_0 are in seconds, p_0 is in lb/in.^{2}, and weapon yield W is in MT.

Decay of Overpressure at a Given Point

The decay of overpressure with time at a given point has been given by Anderson (1960) as

$$p = p_0(1 - t/t_+)e^{-t/t_+} \tag{4.7}$$

where p is the overpressure at time t.

Borde (1964) has compiled a summary of air blasts and given an equation for overpressure decay in the form

$$p = p_0(Ae^{-\theta\bar{t}} + Be^{-\phi\bar{t}} + Ce^{-\psi\bar{t}})(1 - \bar{t}) \tag{4.8}$$

where

$$\bar{t} = t/t_+ \tag{4.9}$$

TABLE 4.1 Values of A, B, C, θ, ϕ, and ψ [Eq. (4.8)][a]

p_0		A	B	C	θ	ϕ	ψ
lb/in.^{2}	kN/m^2						
100	690	0.59	0.41	0	1.9	11.3	—
200	1380	0.43	0.435	0.135	2.5	19.0	81
300	2070	0.34	0.42	0.24	2.5	19.5	90
500	3450	0.26	0.37	0.37	2.6	20.5	103
1000	6900	0.15	0.30	0.55	2.9	21.5	130

[a]Hendron, A. J., and Auld, H. E. (1967). "The Effect of the Soil Properties on the Attenuation of Air-Blast Induced Ground Motions," *Proceedings*, *International Symposium on Wave Propagation and Dynamic Properties of Earth Materials*, University of New Mexico Press, Albuquerque, Table 1, p. 35.

The values of the nondimensional terms A, B, C, θ, ϕ, and ψ are given in Table 4.1 (Hendron and Auld, 1970).

Example 4.1 For a surface of yield 42.3×10^{15} J, determine

a. the peak overpressure at a distance of 1.5 km from ground zero
b. the duration positive phase
c. the overpressure at a distance of 1.50 km from ground zero after 1 sec of the arrival of the overpressure front

Solution

a. Peak overpressure. From Figure 4.2, for $W = 42.3 \times 10^{15}$ J (10.11 megaton) and distance equal to 1.50 km (4920 ft),

$$p_0 \approx 2070 \text{ kN/m}^2 \quad (300 \text{ lb/in.}^2)$$

b. Duration. From Figure 4.4, for $p_0 = 2070$ kN/m^2 and $W = 4.184 \times 10^{15}$ J, $t_+ = 1$ sec. For $W = 42.3 \times 10^{15}$ J, referring to Eq. (4.1),

$$t_{+(W=42.3 \times 10^{15} \text{ J})} / t_{+(W=4.184 \times 10^{15} \text{ J})} = (42.3/4.184)^{1/3}$$

or

$$t_{+(W=42.3 \times 10^{15} \text{ J})} = 1 \text{ sec}(42.3/4.184)^{1/3} = 2.16 \text{ sec}$$

c. Overpressure. From Eq. (4.8),

$$p = p_0 \left(Ae^{-\theta \bar{t}} + Be^{-\phi \bar{t}} + Ce^{-\psi \bar{t}} \right)(1 - \bar{t})$$

$$p_0 = 2070 \text{ kN/m}^2$$

$$t = 1 \text{ sec}$$

$$\bar{t} = \frac{t}{t_+} = 1/2.16 = 0.463$$

From Table 4.1, for $p_0 = 2070$ kN/m^2,

$$A = 0.34, \qquad B = 0.42, \qquad C = 0.24$$
$$\theta = 2.5, \qquad \phi = 19.5, \qquad \psi = 90$$

so

$$p_0 = 2070[0.34e^{-(2.5)(0.463)} + 0.42e^{(-19.5)(0.463)} + 0.24e^{-(90)(0.463)}](1 - 0.463)$$
$$= 118.8 \text{ kN/m}^2$$

Example 4.2 For Example 4.1, find t_i, t_{50}, and t_0.

Solution: Peak overpressure $p_0 = 300$ lb/in.^{2}, $W = 10.11$ megaton.

$$t_i = 0.37(100/p_0)^{1/2}W^{1/3} \tag{4.2}$$

so

$$t_i = 0.37(100/300)^{1/2}(10.11)^{1/3} = 0.462 \text{ sec}$$
$$t_{50} = 0.2(100/p_0)^{7/8}W^{1/3} \tag{4.3}$$

so

$$t_{50} = 0.2(100/300)^{7/8}(10.11)^{1/3} = 0.165 \text{ sec}$$
$$t_0 = 0.15(100/p_0)W^{1/3} \tag{4.6}$$

so

$$t_0 = 0.15(100/300)(10.11)^{1/3} = 0.108 \text{ sec}$$

4.2 GROUND MOTION DUE TO OVERPRESSURE

When the blast pressure reaches some point on the ground surface, it is propagated into the ground at seismic velocities. There are three possible cases for this.

1. At smaller distances from ground zero, overpressure front velocity V is large. If V is greater than the *dilatational seismic* velocity v_p it is referred to as a *superseismic* case. This is shown in Figure 4.5a. At a time $t = t_1$, the overpressure is at point A on the ground surface. The arrival of the blast front at A initiates the dilatational and shear waves which travel into the ground. After a time Δt (i.e., at time $t_1 + \Delta t$), the blast front reaches point B and seismic waves start to propagate. Thus BP and BS are the dilatational and shear wave fronts. The slope of the dilatational wave front with the horizontal ground surface can be given by

$$\alpha_1 = \sin^{-1}\left(\frac{AA_2}{AB}\right) = \sin^{-1}\left(\frac{v_p \Delta t}{V \Delta t}\right) = \sin^{-1}\left(\frac{v_p}{V}\right) \tag{4.10}$$

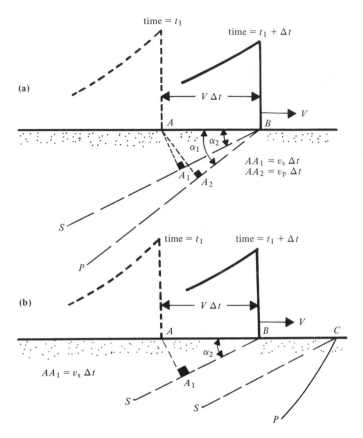

FIGURE 4.5 (a) Superseismic case; **(b)** transeismic case.

Similarly, the slope of the shear wave front with the horizontal is

$$\alpha_2 = \sin^{-1}(AA_1/AB) = \sin^{-1}(v_s/V) \qquad (4.11)$$

2. As the overpressure front moves forward, its velocity V gradually decreases. At some time, a condition is reached where $v_s < V < v_p$; This is called the *transeismic* case (Figure 4.5b). When the blast front is at point A at time t_1, it generates dilatational and shear waves. At time $t_1 + \Delta t$, the blast front is at B. The shear wave front BS forms an angle α_2 with the horizontal that can be given by Eq. (4.11). However, the dilatational wave front CP moves faster, since $v_p > V$. At the ground surface, the dilatational waves generate a second shear wave front given by CS.

3. At larger distances, when the blast front velocity V becomes less than v_s (i.e., $v_p > v_s > V$), the dilatational and shear wave fronts move ahead of the blast front. This is called *outrunning*.

4.3 CALCULATION OF GROUND MOTION DUE TO A BLAST FRONT WAVE

Figure 4.6 shows a plot of overpressure–time relationship for consideration of the steps involved in the calculation of ground movement with time at a given distance from *ground zero* for a *superseismic case*. It is also assumed that the propagation of the overstress into the ground is *one dimensional*.

The actual overpressure–time relationships can be approximated by a finite number of step functions, as shown by the broken lines in Figure 4.6. There is thus an initial stress pulse of magnitude p_1 which moves downward into the soil. This is followed by an unloading pulse of magnitude $\Delta p_1 = p_1 - p_2$. The velocity of movement of the initial stress pulse into the ground can be given by [see Eq. (2.10)],

$$v'_{c(L)} = \sqrt{M_L/\rho} \qquad (4.12)$$

where ρ is the soil density and M_L is the constrained secant modulus of soil for the loading branch (see Figure 4.11, p. 138).

The unloading pulse Δp moves with a velocity of

$$v'_{c(U)} = \sqrt{M_U/\rho} \qquad (4.13)$$

where M_U is the constrained secant modulus for the unloading branch of soil (see Figure 4.11).

FIGURE 4.6 Ground motion due to a blast wave front.

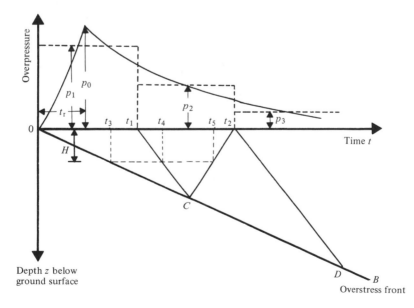

In Figure 4.6, OB is the initial overstress front due to the stress pulse p_1, and t_1C is the stress front due to the unloading pulse Δp. Note that the slope of the front t_1C is larger than the slope of the initial overstress front OB. This is because of the fact that $v'_{c(U)}$ is larger than $v'_{c(L)}$ since $M_U > M_L$. At point C, the wave front due to the unloading pulse meets the initial overstress front and is reflected back with a velocity equal to $v'_{c(U)} = \sqrt{M_U/\rho}$. In a similar manner, an unloading pulse equal to $p_3 - p_2$ starts traveling into the soil at time t_2 with a velocity of $\sqrt{M_U/\rho}$. Line t_2D represents the stress front for this unloading pulse.

It can be seen now that a region of soil acted on by a step change in pressure encounters a change in particle velocity. The change of particle velocity can be derived by the method of impulses (Heierli, 1962). This is explained below.

Change in Particle Velocity Due to Step Change in Pressure

Referring to Figure 4.7, the step change in stress is $\Delta\sigma$. Let it move from position z to $z + \Delta z$ in time Δt. Also let the velocities of particles be $\dot{w}$ and $\dot{w} + \Delta\dot{w}$ before and after the step change. The equation of continuity can be

FIGURE 4.7 Derivation of Eq. (4.20).

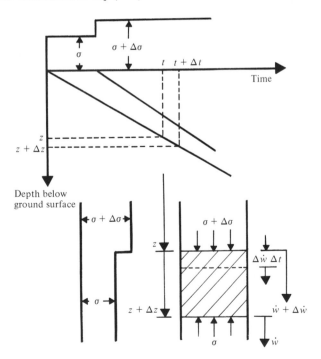

given by

$$\Delta\epsilon\,\Delta z = \Delta\dot{w}\Delta t \tag{4.14}$$

where $\Delta\epsilon$ is the change in strain. From Newton's second law of motion,

$$\rho\Delta\dot{w}\Delta z = \Delta\sigma\Delta t \tag{4.15}$$

where ρ is the soil density. The constrained secant modulus can be given by the equation,

$$M_{\rm L} = \frac{\Delta\sigma}{\Delta\epsilon} \tag{4.16}$$

or

$$\Delta\epsilon\,M_{\rm L} = \Delta\sigma \tag{4.17}$$

Combining Eqs. (4.14) and (4.17)

$$\frac{\Delta\sigma}{M_{\rm L}}\Delta z = \Delta\dot{w}\Delta t, \qquad \frac{\Delta\sigma}{M_{\rm L}}\frac{\Delta z}{\Delta t} = \Delta\dot{w}$$

or

$$\frac{\Delta\sigma}{M_{\rm L}}v'_{\rm c(L)} = \Delta\dot{w} \tag{4.18}$$

From Eq. (4.15),

$$\frac{\Delta z}{\Delta t} = v'_{\rm c(L)} - \frac{\Delta\sigma}{\rho\Delta\dot{w}} \tag{4.19}$$

Combining Eqs. (4.18) and (4.19), we get

$$\frac{\Delta\sigma}{M_{\rm L}}\left(\frac{\Delta\sigma}{\rho\Delta\dot{w}}\right) = \Delta\dot{w}$$

or

$$\Delta\dot{w} = \frac{\Delta\sigma}{\sqrt{M_{\rm L}\rho}} \tag{4.20}$$

Calculation of Soil Particle Velocity in Various Zones

The velocity of soil particles in various zones shown in Figure 4.6 can now be calculated by using Eq. (4.20).

In zone $0Ct_1$:

$$\sigma_{0Ct_1} = p_1$$

and

$$\dot{w}_{0CT_1} = p_1/\sqrt{M_{\rm L}\rho} \tag{4.21}$$

In zone t_1Ct_2:

$$\sigma_{t_1Ct_2} = p_2$$

From Eq. (4.20),

$$\Delta\dot{w}_{t_1Ct_2} = (p_1 - p_2)/\sqrt{M_U\rho} \qquad (4.22)$$

Note that the constrained secant modulus M_U has been used in Eq. (4.22) instead of M_L. The particle velocity in the zone can now be given by

$$\dot{w}_{t_1Ct_2} = \dot{w}_{0Ct_1} - \Delta\dot{w}_{t_1Ct_2}$$

$$= p_1/\sqrt{M_L\rho} - (p_1 - p_2)/\sqrt{M_U\rho}$$

$$= (\rho M_L M_U)^{-1/2}\left[\sqrt{M_U}\,p_1 - \sqrt{M_L}\,(p_1 - p_2)\right] \qquad (4.23)$$

In zone t_2CD:

Note that, in this zone, the waves in soil due to the unloading pulse at time t_1 are reflected back with a velocity $v'_{c(U)} = \sqrt{M_U/\rho}$. However, the forward wave moves with a velocity $v'_{c(L)} = \sqrt{M_L/\rho}$.

$$\sigma_{t_2CD} = \left[\dot{w}_{t_2CD}\right]\sqrt{\rho M_L} \qquad (4.24)$$

The change in particle velocity,

$$\Delta\dot{w}_{t_2CD} = \dot{w}_{t_1Ct_2} - \dot{w}_{t_2CD}$$

$$= (\sigma_{t_2CD} - \sigma_{t_1Ct_2})/\sqrt{M_U\rho} \qquad (4.25)$$

Combining the above equations,

$$\dot{w}_{t_2CD} = \left(\dot{w}_{t_1Ct_2}\sqrt{M_U\rho} + \sigma_{t_1Ct_2}\right)/\left[\sqrt{\rho}\left(\sqrt{M_L} + \sqrt{M_U}\right)\right] \qquad (4.26)$$

Calculation of the Displacement of the Ground Surface

Referring to Figure 4.6, the following relations hold between time t and ground displacement S:

$$t = 0, \quad S = 0$$
$$t = t_1, \quad S = \dot{w}_{0Ct_1}t_1$$
$$t = t_2, \quad S = \dot{w}_{0Ct_1}t_1 + \dot{w}_{t_1Ct_2}(t_2 - t_1)$$

Calculation of Displacement at Depth H Below the Ground Surface

Referring again to Figure 4.6, at a depth H, the initial overstress front arrives at time $t = t_3$. At that time, then, displacement is zero. At time $t = t_4$,

$$S_{z=H} = \dot{w}_{0Ct_1}(t_4 - t_3) = \left(p_1/\sqrt{M_L\rho}\right)(t_4 - t_3)$$

At time $t = t_5$,

$$S_{z=H} = \dot{w}_{0Ct_1}(t_4 - t_3) + \dot{w}_{t_1 Ct_2}(t_5 - t_4)$$

$$= \left(p_1 / \sqrt{M_L \rho}\right)(t_4 - t_3)$$

$$+ (M_L M_U \rho)^{-1/2} \left[\sqrt{M_U}\, p_1 - \sqrt{M_L}\,(p_1 - p_2) \right](t_5 - t_4)$$

Example 4.3 The overpressure–time relationship at a point is shown in Figure 4.8. Calculate the ground displacement at various times. Given $M_L = 2.33 \times 10^5$ lb/in.^{2} and $M_U = 6.14 \times 10^5$ lb/in.^{2}; unit weight of soil = 118 lb/ft^3.

Solution

$$v'_{c(L)} = \sqrt{\frac{M_L}{\rho}} = \sqrt{\frac{(2.33 \times 10^5)(144)}{(118/32.2)}} = 3025.8 \text{ ft/sec}$$

$$v'_{c(U)} = \sqrt{\frac{M_U}{\rho}} = \sqrt{\frac{(6.14 \times 10^5)(144)}{(118/32.2)}} = 4911.9 \text{ ft/sec}$$

The overpressure–time relationship has been replaced by a step function in Fig. 4.8.

FIGURE 4.8

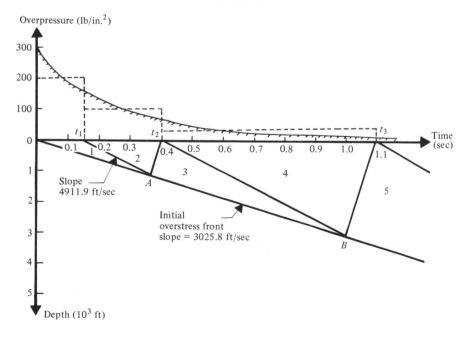

Calculation of Particle Velocity in Various Zones

In Zone 1:

$$\sigma_{z1} = 200 \text{ lb/in.}^2 = 28800 \text{ lb/ft}^2 = p_1$$

$$\dot{w}_{z1} = \sigma_{z1}/\sqrt{M_L \rho} = \frac{28800}{\sqrt{(2.33 \times 10^5)(144)(118/32.2)}} = 2.6 \text{ ft/sec}$$

In Zone 2:

$$\sigma_{z2} = 100 \text{ lb/in.}^2 = 14400 \text{ lb/ft}^2 = p_2$$

From Eq. (4.23),

$$\dot{w}_{z2} = p_1/\sqrt{M_L \rho} - (p_1 - p_2)/\sqrt{M_U \rho}$$

$$= \frac{28,800}{\sqrt{(2.33 \times 10^5)(144)(118/32.2)}} - \frac{14,400}{\sqrt{(6.14 \times 10^5)(144)(118/32.2)}}$$

$$= 2.6 - 0.8 = 1.8 \text{ ft/sec}$$

In Zone 3:
From Eq. (4.26),

$$\dot{w}_{z3} = \left(\dot{w}_{z2}\sqrt{M_U \rho} + \sigma_{z2} \right)/\left[\sqrt{\rho} \left(\sqrt{M_L} + \sqrt{M_U} \right) \right]$$

$$= \frac{1.8\sqrt{(2.33 \times 10^5)(144)\left(\dfrac{118}{32.2}\right)} + 14,400}{\sqrt{\dfrac{118}{32.2}} \left(\sqrt{2.33 \times 10^5 \times 144} \right) + \sqrt{6.14 \times 10^5 \times 144}}$$

$$= 1.18 \text{ ft/sec}$$

$$\sigma_{z3} = \dot{w}_{z3}\sqrt{\rho M_L} = 1.18 \cdot \sqrt{\left(\frac{118}{32.2}\right)(2.33 \times 10^5)(144)}$$

$$= 13,084.4 \text{ lb/ft}^2 = 90.86 \text{ lb/in.}^2$$

In Zone 4:

$$\sigma_{z4} = 30 \text{ lb/in.}^2 = 4320 \text{ lb/ft}^2$$

$$\dot{w}_{z4} = \dot{w}_{z3} - (\sigma_{z3} - \sigma_{z4})/\sqrt{M_U \rho}$$

$$= 1.18 - \frac{(90.86 - 30)144}{\sqrt{(6.14 \times 10^5)(144)(118/32.2)}} = 0.693 \text{ ft/sec}$$

In Zone 5:

$$\dot{w}_{z5} = \frac{\dot{w}_{z4}\sqrt{M_U \rho} + \sigma_{z4}}{\sqrt{\rho}\left(\sqrt{M_L} + \sqrt{M_U}\right)}$$

$$= \frac{0.693 \cdot \sqrt{(2.33 \times 10^5)(144)(118/32.2)} + 4320}{\sqrt{118/32.2}\left(\sqrt{2.33 \times 10^5 \times 144} + \sqrt{6.14 \times 10^5 \times 144}\right)}$$

$$= 0.4127 \text{ ft/sec}$$

$$\sigma_{z5} = \dot{w}_{z5}\sqrt{\rho M_L} = 0.4127 \cdot \sqrt{(118/32.2)(2.33 \times 10^5)(144)}$$

$$= 4576.2 \text{ lb/ft}^2 = 31.78 \text{ lb/in.}^2$$

Calculation of Ground Displacement

At time $t = t_1 = 0.15$ sec,

$$S = \dot{w}_{z1}t_1 = (2.6)(0.15) = 0.39 \text{ ft}$$

At time $t = t_2 = 0.4$ sec,

$$S = 0.39 \text{ ft} + (t_2 - t_1)\dot{w}_{z2}$$
$$= 0.39 + (0.4 - 0.15)1.8 = 0.84 \text{ ft}$$

At time $t = t_3 = 1.1$ sec,

$$S = 0.84 \text{ ft} + (t_3 - t_2)\dot{w}_{z4}$$
$$= 0.84 + (1.1 - 0.4)0.693 = 1.325 \text{ ft}$$

4.4 DETERMINATION OF THE CONSTRAINED MODULUS OF SOIL

In the preceding section it has been shown that, for estimation of the ground displacement due to an overpressure, one needs to have a representative evaluation of the *constrained modulus* of the soil. Although this is an important parameter, there is no single reliable technique by which it can be evaluated; usually, the results developed out of several types of test are combined to obtain representative values. This has been explained by Wilson and Sibley (1962).

There are two extreme limits within which the constrained modulus of soil will generally fall: The lower limit is that derived from triaxial compression test results on *undisturbed specimens*, because the specimens lack lateral restrain; the upper limit of the modulus is that obtained from the seismic explorations in the field. This can be more effectively explained with the aid of the results obtained for a *Playa silt* from Frenchman Flat. Figure 4.9

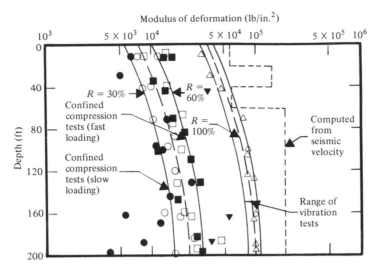

FIGURE 4.9 Modulus of deformation vs depth for Playa Silt. Moduli: ● Initial tangent, triaxial test; ▼ tangent, repetitive loading triaxial test; ○ constrained, static confined compression tests (slow loading); □ constrained, dynamic confined compression tests (medium loading); ■ constrained, dynamic confined compression tests (fast loading); △ constrained, vibration tests; --- constrained, seismic velocity. Note: 1 m = 3.28 ft; 1 lb/in.² = 6.9 kN/m². [Wilson, S. D., and Sibley, E. A. (1962). "Ground Displacement from Air-Blast Loading," *Journal of the Soil Mechanics and Foundations Division, ASCE*, 88 (SM6), Fig. 9, p. 14.]

shows the results of the *initial tangent modulus* obtained from *consolidated-quick* triaxial tests on undisturbed specimens obtained from various depths. These specimens were first consolidated with an all-around pressure equal to the *effective overburden pressure*. Also shown in Figure 4.9 is the *tangent modulus* determined from repetitive loading triaxial tests. For these tests, the specimens were first consolidated under an all-around pressure equal to the effective overburden pressure. *Tangent moduli* were determined at the end of repetitive loading cycles during which small on-and-off load increments were applied to the specimen to eliminate creep effects.

Static and dynamic confined compression tests on undisturbed specimens (i.e., zero lateral expansion) can be conducted. The rate of vertical stress application is slow for *static tests* and faster for *dynamic tests*. The rate of vertical stress application has a large effect on the stress–strain relation of the specimen and hence the *constrained secant modulus*. This is shown in Figure 4.10 for specimens of *weathered Pierre shale*. Note that these specimens were initially consolidated under a vertical pressure $p' = 40$ lb/in.² (276 kN/m²). It can be seen from Figure 4.10 that the constrained modulus increases with the rate of loading. For comparison purposes, the

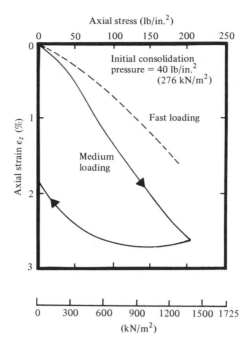

FIGURE 4.10 Stress–strain relationship from dynamic confined compression tests on weathered Pierre shale. [Wilson, S. D., and Sibley, E. A. (1962). "Ground Displacement from Air-Blast Loading," *Journal of the Soil Mechanics and Foundations Division, ASCE*, 88 (SM6), Fig. 5, p. 9.]

constrained modulus of various specimens of *Playa silt* from Frenchman Flat for slow, medium, and fast-loading rates are also shown in Figure 4.9.

The *constrained modulus* may also be determined from *resonant column tests*. It has been shown in Section 2.5 that Poisson's ratio can be determined as

$$\mu = E/2G - 1 \qquad (2.45)$$

The above value of μ can be used to determine the constrained modulus from the equation

$$M = E(1 - \mu)/[(1 - 2\mu)(1 + \mu)]$$

The constrained moduli for the *Playa silt* specimens from Frenchman Flat determined in the above manner are also shown in Figure 4.9.

The values of the Young's modulus determined by field seismic survey for the Playa silt deposit at Frenchman Flat are also shown in Figure 4.9 ($E = \rho v_p^2$). It may be seen that this is the upper limit of *moduli* determined by all techniques.

Strain Recovery

For any of the test procedures described above (i.e., triaxial compression, confined compression, resonant column, or field seismic survey) there is a recovery of strain after the release of applied stress. The *strain recovery ratio* may be defined as (Fig. 4.11)

$$R = \frac{\text{strain recovered after the release of stress}}{\text{maximum strain during the peak stress application}}$$

$$= \epsilon_{z(\text{re})}/\epsilon_z = M_L/M_U \tag{4.27}$$

The general range of the strain recovery ratio for various types of test are shown in Figure 4.9.

Selection of the Value of M_L for Calculation of Soil Deformation

For calculation of the deformation of a soil layer, caution should be used when estimating the value of M_L. The following procedure may generally be used for that purpose (Wilson and Sibley, 1962).

1. Draw the variation of the effective overburden pressure p' (or the preconsolidation pressure p_c), with depth as shown in Figure 4.12a.
2. Superimposed on this, draw the variation of the maximum anticipated overstress σ_{peak} with depth (also see Figure 4.12a). The calculation of maximum anticipated overstress with depth is discussed in Section 4.5.

FIGURE 4.11 Definition of strain recovery ratio.

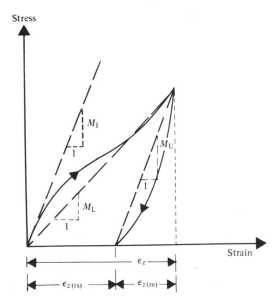

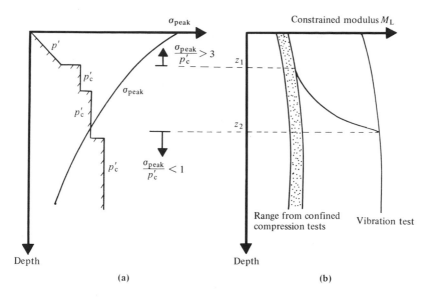

FIGURE 4.12 Selection of the value of constrained modulus. Effective overburden pressure (p'); preconsolidation pressure (p_c').

3. Draw the variation of the constrained modulus M_L with depth as shown in Figure 4.12b. These are values determined from confined compression tests and vibration (resonant column) tests.
4. Calculate the ratio of σ_{peak}/p' or σ_{peak}/p_c as they apply for various soil layers (Figure 4.12a).
5. Determine the depth z_1 from the ground surface up to which σ_{peak}/p' or σ_{peak}/p_c' is greater than or equal to 3 (Figure 4.12a).
6. Determine the depth z_2 beyond which σ_{peak}/p' or σ_{peak}/p_c' is less than one.
7. From the ground surface to a depth z_1, follow the upper limit of M_L as determined from the confined compression tests and shown in Figure 4.12b.
8. From depths larger than z_2, follow the M_L values as determined from vibration tests (Figure 4.12b).
9. The values of M_L between depths z_1 and z_2 will follow a smooth curve joining the points showing $M_{L(z_1)}$ and $M_{L(z_2)}$ as shown in Figur 4.12b.

For soils located below the ground water table, a value of $M_L = 150,000$ lb/in.^{2} (1035 MN/m^2) may be adopted.

4.5 ATTENUATION OF PEAK OVERSTRESS WITH DEPTH

The peak value of overstress in a soil medium due to an overpressure decreases with depth. This is called *attenuation*. There are several reasons for this, such as the plastic yielding of soil or travel of unloading waves through the soil as described in Section 4.3. The peak overstress decreases with depth also due to *spatial attenuation*, which may be explained as follows. For purposes of calculation of stresses and displacements of soil layers, a one-dimensional analysis has been adopted in this chapter; however, in reality, that may not be the case. At any given time, the overpressure acts over a part of the ground surface. This results in a decrease of peak overstress with depth.

The nature of attenuation of the overstress in soil with depth is shown in Figure 4.13. Note that the magnitude of the *peak overstress decreases* with depth. Again, as the depth increases, the time required to attain the peak overstress value also increases.

The peak overstress due to spatial attenuation in the superseismic case may be expressed as (Newmark, 1964),

$$\sigma_{\text{peak}} = \alpha p_0 \qquad (4.28)$$

$$\alpha = \left(1 + z/L_W\right)^{-1} \qquad (4.29)$$

where z is the depth below the ground surface and

$$L_W = 230(100/p_0)^{1/2} W^{1/3} \qquad (4.30)$$

FIGURE 4.13 Nature of attenuation of overstress in soil with depth.

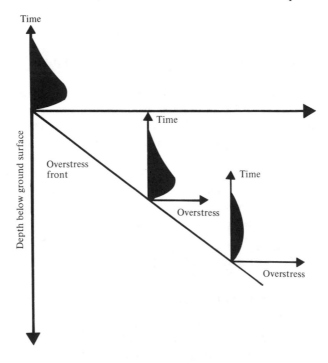

In the above equation, the L_W is in ft, p_0 = maximum overpressure in lb/in.^{2}, and W = yield in megatons. In SI units, L_W can be expressed as

$$L_W = 0.001143(100/p_0)^{1/2}W^{1/3} \qquad (4.31)$$

where L_W is in m, p_0 is in kN/m^2, and W is in Joules.

Example 4.4 Calculate the variation of the peak overstress with depth at a given point $(0 \leqslant z \leqslant 50 \text{ m})$ for an overpressure front with $p_0 = 1380 \text{ kN/m}^2$, given $W = 4.184 \times 10^{15}$ J.

Solution: Determine L_W using Eq. (4.31) as 49.57 m; then prepare a table as shown and determine the value of σ_{peak} at various depths.

z (m)	α [Eq. (4.29)]	σ_{peak} (kN/m^2) [Eq. (4.28)]
0	1	1380.0
5	0.9084	1253.6
10	0.8321	1148.3
15	0.7677	1059.4
20	0.7125	983.3
25	0.6647	917.3
30	0.6225	859.1
40	0.5534	763.7
50	0.4978	687.0

4.6 CALCULATION OF VERTICAL GROUND DISPLACEMENT USING THE SPATIAL ATTENUATION FACTOR (SUPERSEISMIC CASE)

Figure 4.14 shows the variation of the overpressure with time at a given point on the ground surface. Line $0A$ is the initial overstress front showing the initial arrival of the stress waves at various depths below the ground surface. The velocity of propagation of the stress waves can be given by

$$v'_{c(I)} = \sqrt{M_I/\rho} \qquad (4.32)$$

where M_I is the constrained modulus for initial stress conditions (see Figure 4.11 for explanation).

Line BC (Figure 4.14) is the peak overstress front. The slope of the peak overstress front is smaller than that for the initial overstress front because $M_I > M_L$ (Figure 4.11). For practical considerations, the value of $v'_{c(I)}$ can be taken equal to the P-wave velocity as determined from *seismic survey*.

The distribution of overstress with depth at a given time can now be determined considering only the spatial attenuation. The steps involved for

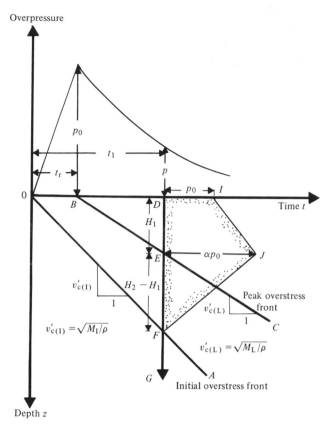

FIGURE 4.14 Overstress due to spatial attenuation.

this purpose are as follows:

1. If the overstress variation at time $t = t_1$ is required, one draws a vertical line DG as shown in Figure 4.14. The line intersects the peak overstress front and initial overstress front at points E and F, respectively. Note

$$\overline{DE} = H_1 = v'_{c(L)}(t_1 - t_r)$$

and

$$\overline{DF} = H_2 = v'_{c(I)}t_1$$

2. At $z = 0$ (i.e., at ground surface), the overstress $\sigma = \overline{DI} = p$ (i.e., overpressure at time $t = t_1$).

3. At $z = H_1$, the overstress

$$\sigma = \sigma_{\text{peak}} = \alpha p_0 = \left(1 + H_1/L_W\right)^{-1} p_0 = \overline{EJ}$$

4. At $z = H_2$, the overstress is equal to zero.
5. *DEFJI* is the distribution of the overstress with depth.

In some cases, if the rise time of the overpressure t_r is equal to zero, then H_1 and H_2 in Figure 4.14 may be assumed to be equal to $v'_{c(L)}t_1$ and $v'_{c(U)}t_1$, respectively.

Once the distribution of overstress with depth at any given time is determined, the displacement of the ground can be easily calculated. The procedure for this is described below with reference to Figure 4.15.

1. At a depth $H_2 \geqslant z \geqslant H_1$, the vertical strain can be given as

$$\epsilon_{z(1)} = \sigma_1/M_L = \sigma_1/\left[\rho\left(v'_{c(L)}\right)\right]^2 \qquad (4.33)$$

2. At a depth $z < H_1$, the vertical strain is

$$\epsilon_{z(2)} = \frac{\sigma_{peak}}{M_L} - \frac{\sigma_{peak} - \sigma_2}{M_U} = \frac{\sigma_2}{M_U} + \frac{\sigma_{peak}}{M_L} - \frac{\sigma_{peak}}{M_U}$$
$$= (1/M_L)\left[(M_L/M_U)\sigma_2 + \sigma_{peak} - (M_L/M_U)\sigma_{peak}\right]$$
$$= (1/M_L)\left[R\sigma_2 + (1-R)\sigma_{peak}\right]$$
$$= \left[1/\rho\left(v'_{c(L)}\right)^2\right]\left[R\sigma_2 + (1-R)\sigma_{peak}\right] \qquad (4.34)$$

where R is the strain recovery ratio [Eq. (4.27)].
3. The vertical ground displacement S can be obtained as

$$S = \sum_{z=0}^{z=H_1} \epsilon_{z(2)}\Delta z + \sum_{z=H_1}^{z=H_2} \epsilon_{z(1)}\Delta z \qquad (4.35a)$$

FIGURE 4.15 Calculation of ground displacement considering only spatial attenuation.

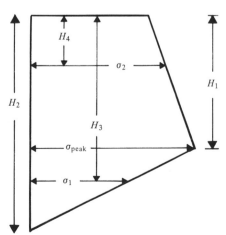

4. If the vertical displacement at a depth H_3 is required ($H_2 \geqslant H_3 \geqslant H_1$), then

$$S_{z=H_3} = \sum \epsilon_{z(1)} \Delta z = \sum_{z=H_3}^{z=H_2} \frac{\sigma}{\rho \left[v'_{c(L)} \right]^2} \Delta z \qquad (4.35b)$$

5. The vertical displacement at a depth H_4 ($H_1 \geqslant H_4 \geqslant 0$) is

$$S_{z=H_4} = \sum_{z=H_1}^{z=H_2} \frac{\sigma}{\rho \left[v'_{c(L)} \right]^2} \Delta z$$

$$+ \sum_{z=H_4}^{z=H_1} \frac{1}{\rho \left[v'_{c(L)} \right]^2} \left[R\sigma + (1 - R)\sigma_{peak} \right] \Delta z \qquad (4.36)$$

Example 4.5 For an overpressure front at a given point, given peak overpressure $p_0 = 300$ lb/in.^{2}, $W = 10$ megatons, unit weight of soil $= 128$ lb/ft^3. Assume strain recovery ratio $= 0.7$ for $0 \leqslant z \leqslant 150$ ft and $R = 1$ for $z > 150$ ft.

a. Determine t_+, t_{50}, and t_0 and draw the variation of overpressure with time.
b. The variation of effective overburden pressure or preconsolidation pressure for the soil profile is shown in Figure 4.16. Calculate the variation of σ_{peak} with depth and plot it on Figure 4.16.
c. Based on the variation of σ_{peak}/p' or σ_{peak}/p'_c, assume a general variation of $v'_{c(I)}$ and $v'_{c(L)}$ with depth.
d. Using the results of Part (c), draw a graph showing (i) the arrival time of initial overstress front vs depth and (ii) the arrival time of peak stress front vs depth.
e. Determine the variation of overstress with depth at time $t = 0.03$ sec.
f. Using the results of Part (e), calculate the ground displacement at time $t = 0.03$ sec.
g. Calculate the displacement of ground at a depth of 40 ft at time $t = 0.03$ sec.

Solution:

a. Calculation of t_+, t_0, and t_{50}. From Eq. (4.1),

$$t_{+(yield = W)} = t_{+(W=1)} W^{1/3}$$

Referring to Figure 4.4 (p. 124), for $p_0 = 300$ lb/in.^{2} and $W = 1$ MT, one obtains t_+ as 0.97 sec. Thus,

$$t_{+(W=10 \text{ megatons})} = 0.97(10)^{1/3} = 2.09 \text{ sec.}$$

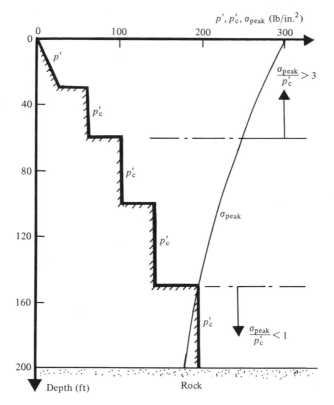

FIGURE 4.16

From Eq. (4.4),

$$t_{50} = 0.2(100/p_0)^{7/8} W^{1/3} = 0.2(100/300)^{7/8}(10)^{1/3} \equiv 0.165 \text{ sec}$$

From Eq. (4.6),

$$t_0 = 0.15(100/p_0) W^{1/3} = 0.15(100/300)(10)^{1/3} = 0.108 \text{ sec}$$

Using the above values of t_+, t_{50}, and t_0, the approximate overpressure–time variation has been constructed (Figure 4.17).

b. *Calculation of the Variation of σ_{peak} with Depth.* From Eq. (4.28),

$$\sigma_{\text{peak}} = \alpha p_0, \qquad \alpha = 1/(1 + z/L_W)$$

$$L_W = 230(100/p_0)^{1/2} W^{1/3} = 230(100/300)^{1/2}(10)^{1/3} = 286 \text{ ft}$$

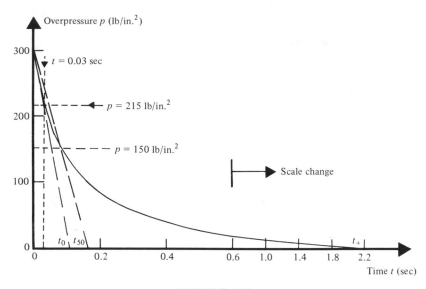

FIGURE 4.17

Now the variation of σ_{peak} with depth can be calculated as shown in tabular form.

z (ft)	α	$\sigma_{peak} = \alpha p_0$ (lb/in.^{2})
0	0	300
10	0.966	290
20	0.935	281
40	0.877	263
60	0.827	248
80	0.78	234
100	0.741	222
120	0.704	211
150	0.656	197
200	0.588	177

The variation of σ_{peak} has been plotted in Figure 4.16.

   c. Assumed Variations of $v'_{c(I)}$ and $v'_{c(L)}$ with depth. These are shown in Figure 4.18a.

   d. Calculation of the Arrival Time of Overstress and Peak Stress Front. The profiles of velocities $v'_{c(I)}$ and $v'_{c(L)}$ are given in Figure 4.18a. Dividing the soil layer into 20-ft-thick layers and using the average velocities ($v'_{c(I)}$ and $v'_{c(L)}$), the arrival time of the initial and peak overstress fronts at various depths have been calculated. These are shown in Figure 4.18b.

147

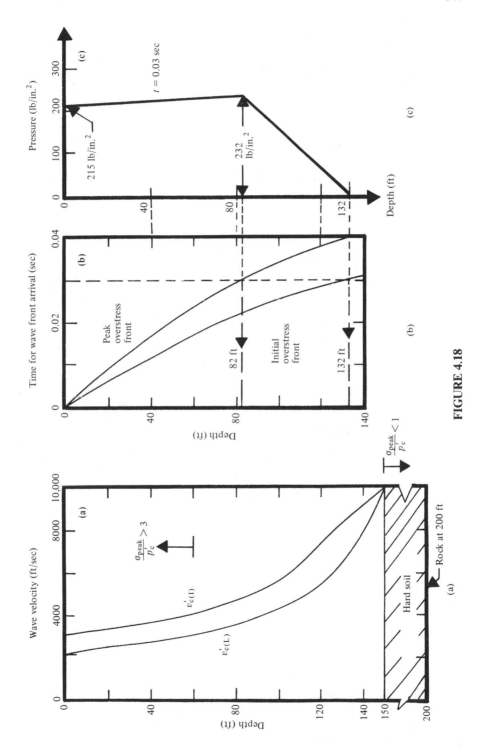

FIGURE 4.18

e. Determination of the Variation of Overstress with Depth at Time t = 0.03 *sec.* From Figure 4.18b, it can be seen that at time $t = 0.03$ sec, the initial overstress front reaches a depth of $H_2 = 132$ ft and the peak overstress front reaches a depth of $H_1 = 82$ ft. Thus

$$\text{at } z = 0, \qquad \sigma = 215 \text{ lb/in.}^2 \quad \text{(from Fig. 4.17)}$$
$$\text{at } z = H_2 = 132 \text{ ft}, \quad \sigma = 0$$
$$\text{at } z = H_1 = 82 \text{ ft}, \quad \sigma = \sigma_{\text{peak}} = 232 \text{ lb/in.}^2$$

The appropriate variation of overstress σ with depth is shown in Figure 4.18c.

f. Calculation of Ground Displacement at Time t = 0.03 sec. See the following table.

z (ft)	Δz (in.)	$v'_{c(L)}$ (ft/sec)[a]	Average σ (lb/in.²)[b]	Average σ (lb/ft²)	ϵ_z	$\epsilon_z \Delta z$ (in.)
0–20	240	2240	217.1	31262.4	0.001621[c]	0.389
20–40	240	2420	221.2	31852.8	0.001402[c]	0.337
40–60	240	2720	225.4	32457.6	0.001120[c]	0.269
60–82	264	3200	229.7	33076.8	0.000817[c]	0.216
82–100	216	3840	190.2	27388.8	0.000467[d]	0.101
100–120	240	4800	102.1	14702.4	0.000161[d]	0.039
120–132	144	6480	27.8	4003.2	0.000024[d]	0.0035

Displacement $S = \Sigma \epsilon_z \, dz = 1.355$ in.

[a]Average value from Figure 4.18a.
[b]Average value from Figure 4.18c.
[c]Eq. (4.34), $R = 0.7$, $\sigma_{\text{peak}} = 232$ lb/in.² $= 33408$ lb/ft².
[d]Eq. (4.33), $\rho = 128/32.2 = 3.975$ lb-sec²/ft⁴.

g. Calculation of Displacement at a Depth of 40 ft.

$$S_{z=40 \text{ ft}} = \sum_{z=132 \text{ ft}}^{z=40 \text{ ft}} \epsilon_z \Delta z$$
$$= 0.269 + 0.216 + 0.101 + 0.039 + 0.0035$$
$$= 0.629 \text{ in.}$$

Note. The ground surface displacement at various times can be calculated in a similar manner. The nature of variation of ground displacement with time is as shown in Figure 4.19.

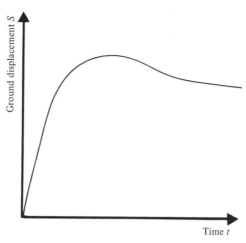

FIGURE 4.19 Nature of variation of ground displacement with time at a given point due to overpressure.

4.7 GROUND DISPLACEMENT DUE TO OVERPRESSURE (SUBSEISMIC CASE)

In Section 4.2, the general conditions under which outrunning occurs were discussed. The approximate values of peak overpressure on various types of ground surfaces below which outrunning occurs have been given by Newmark (1964). These peak overpressures are about 40 lb/in.^{2} for alluvium, 10–100 lb/in.^{2} for dry gravel, 40–500 lb/in.^{2} for wet gravel, and 100–500 lb/in.^{2} for sandy clay. For sandstone, shale, limestone, and granite, the approximate peak overpressures are 500–2000, 650–2500, 1500, and 300 lb/in.^{2}, respectively.

It is generally recommended (U.S. Air Force, 1962) that, for subseismic cases, the peak ground displacement values are not very much different from those calculated for superseismic cases. However, the peak velocity and acceleration values should be multiplied by an amplification factor η. The values of η are given in Table 4.2.

TABLE 4.2 Amplification Factor[a]

Condition	η
Superseismic	
$v_p < V$	1
Outrunning	
Transeismic	
$V < v_p < 1.5V$	v_p/V
$1.5V < v_p < 2V$	1.5
Subseismic	
$2V < v_p$	$1 + V/v_p$

[a]U.S. Air Force (1962), p. 4-40.

PROBLEMS

4.1. Define the following terms:
  a. peak overpressure
  b. rise time
  c. positive phase of overpressure

4.2. For a peak overpressure p_0 of 2415 kN/m^2 at a given point and weapon yield W of 41.84×10^{15} J, calculate t_i, t_{50}, and t_0 in seconds.

4.3. For Problem 4.2, calculate the positive phase t_+ of overpressure. Using the results of Problem 4.2 and t_+, draw an approximate graph for overpressure vs time (in seconds).

4.4. Given a weapon yield of 5 megatons, calculate the peak overpressures at distances of 2000, 3000, 4000, 5000, 7500, and 10,000 ft from ground zero. Plot a graph of p_0 vs distance from ground zero.

4.5. Given weapon yield $W = 5$ MT. Determine:
  a. the peak overpressure at a distance of 3000 ft from ground zero
  b. the duration of positive phase at 3000 ft from ground zero
  c. the overpressure (at a distance of 3000 ft from ground zero) after 1.2 sec of the arrival of the overpressure front [use Eq. (4.7)].

4.6. Solve Problem 4.5c by using Eq. (4.8).

4.7. A step function for overpressure–time relationship and soil properties are given below.

Time (sec)	Overpressure (kN/m^2)
0–0.1	1200
0.1–0.4	600
0.4–1.5	300

$$\text{Unit weight} = 17.1 \text{ kN/m}^3$$
$$M_L = 12.5 \times 10^5 \text{ kN/m}^2$$
$$M_U = 25 \times 10^5 \text{ kN/m}^2$$

Calculate the ground displacement at time $t = 0.1$, 0.4, and 1.5 sec. Use the procedure described in Section 4.3.

4.8. Repeat Problem 4.7 for the following:

Time (sec)	Overpressure (lb/in.^{2})
0–0.1	100
0.1–0.4	75
0.4–1.5	20

$$\text{Unit weight of soil} = 120 \text{ lb/ft}^3$$

$$M_L = 0.80 \times 10^5 \text{ lb/in.}^2$$

$$M_U = 1.5 \times 10^5 \text{ lb/in.}^2$$

4.9. Define and explain strain recovery ratio. What is the strain recovery ratio for the soil described in Problems 4.7 and 4.8?

4.10. Calculate and plot the variation of the peak overstress at a given point (for depths varying from zero to 30 m) for an overpressure front with $p_0 = 2000$ kN/m² [use Eq. (4.28)].

4.11. Refer to Figure 4.15. At a given time, the variation of the overstress with depth in a given soil is as follows:

$$\text{Stress at ground surface} = 180 \text{ lb/in.}^2$$

$$\text{At } H_1 = 50 \text{ ft,} \quad \sigma_{peak} = 220 \text{ lb/in.}^2$$

$$\text{At } H_2 = 120 \text{ ft,} \quad \sigma - 0$$

for this soil, assume the unit weight is 112 lb/ft³, $v'_{c(L)} = 2000$ ft/sec, and the strain recovery ratio is 0.6. Calculate the ground displacement at this time.

4.12. Repeat Problem 4.11 for the following:

$$\text{Stress at ground surface} = 1400 \text{ kN/m}^2$$

$$\text{At } H_1 = 20 \text{ m,} \quad \sigma_{peak} = 1800 \text{ kN/m}^2$$

$$\text{At } H_2 = 60 \text{ m,} \quad \sigma = 0$$

For this soil, assume the unit weight is 18 kN/m³, $v'_{c(L)} = 600 + 10z$ m/sec, and the strain recovery ratio is 0.7.

REFERENCES

Anderson, F. E. (1960). "Blast Phenomena from a Nuclear Burst," *Proceedings, ASCE* 125, 667–672.

Borde, H. L. (1964). "A Review of Nuclear Explosion Phenomena Pertinent to Protective Construction," *R-425-PR*, The Rand Corporation, Santa Monica, California.

Hendron, A. J., and Auld, H. E. (1967). "The Effect of the Soil Properties on the Attenuation of Air-Blast Induced Ground Motions," *Proceedings*, International Symposium on Wave Propagation and Dynamic Properties of Earth Material, University of New Mexico Press, pp. 29–47.

Heierli, W. (1962). "Inelastic Wave Propagation in Soil Columns," *Journal of the Soil Mechanics and Foundations Division, ASCE* 88 (SM6), 33–63.

Newmark, N. M. (1964). "The Basis of Current Criteria for the Design of Underground Protective Construction," Opening Address—Session I, *Proceedings*, Symposium on Soil-Structure Interaction, University of Arizona, Tucson, pp. 1–24.

U.S. Air Force Special Weapons Center (1962). "Principles in Practices for Design of Hardened Structures," Air Force Design Manual, *Technical Documentary Report No. AFSMCTDR-62-138*, Kirtland Air Force Base, New Mexico.

Wilson, S. D., and Sibley, E. A. (1962). "Ground Displacement From Air-Blast Loading," *Journal of the Soil Mechanics and Foundations Division*, *ASCE* 88 (SM6), 1–31.

5

FOUNDATION VIBRATION: THEORIES OF ELASTIC HALF SPACE

In this chapter, the fundamentals of the vibration of foundations supported on an elastic medium are developed. The elastic medium which supports the foundation is considered to be a homogeneous, isotropic, and semi-infinite body. In general, the behavior of soils departs considerably from that of an elastic material; only at low strain levels may it be considered as a reasonable approximation to an elastic material. Hence, the theories developed here should be held as applicable only to those cases where foundations undergo low amplitudes of vibration.

5.1 VERTICAL VIBRATION OF CIRCULAR FOUNDATIONS RESTING ON ELASTIC HALF SPACE

In 1904, Lamb studied the problem of vibration of a single oscillating force acting at a point on the surface of an elastic half space. This study included cases in which the oscillating force acts in the vertical direction and those in which it acts horizontally, as shown in Figure 5.1a, b. This is generally referred to as the "dynamic Boussinesq problem."

In 1936, Reissner analyzed the problem of vibration of a *uniformly loaded flexible circular area* resting on an elastic half space. This was done by integration of Lamb's solution for a point load. Based on Reissner's work, the vertical displacement at the *center* of the flexible loaded area (Figure 5.2a) can be given by

$$z = \left(Q_0 e^{i\omega t} / G r_0 \right) \left(f_1 + i f_2 \right) \tag{5.1}$$

where

Q_0 = amplitude of the total load acting on the foundation

z = periodic displacement at the center of the loaded area

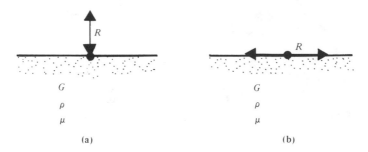

FIGURE 5.1 Oscillating force on the surface of an elastic half space.

ω = circular frequency of the load application

r_0 = radius of the loaded area

G = shear modulus of the soil

$Q = Q_0 e^{i\omega t}$ = total load applied with an amplitude of Q_0

f_1, f_2 = displacement functions

   Consider a flexible circular foundation of weight W (mass $= m = W/g$) which is resting on an elastic half space and is subjected to a vibration by a force of magnitude $Q_0 e^{i(\omega t + \alpha)}$ as shown in Figure 5.2b. (Note: α is the phase difference between the exciting force and the displacement of the foundation.)

FIGURE 5.2 (a) Vibration of a uniformly loaded circular flexible area; **(b)** flexible circular foundation subjected to forced vibration.

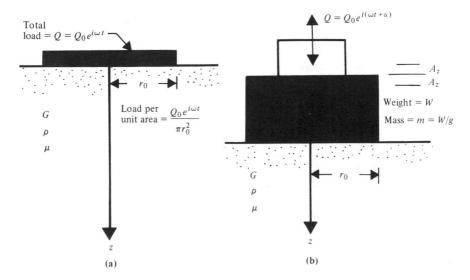

Using the displacement relation given in Eq. (5.1), and solving the equation of equilibrium of forces, Reissner obtained the following relations:

$$A_z = (Q_0/Gr_0)Z \qquad (5.2)$$

where A_z is the amplitude of the foundation motion and Z the dimensionless amplitude.

$$Z = \sqrt{\frac{f_1^2 + f_2^2}{\left(1 - ba_0^2 f_1\right)^2 + \left(ba_0^2 f_2\right)^2}} \qquad (5.3)$$

where dimensionless mass ratio

$$b = \frac{m}{\rho r_0^3} = \frac{W}{g} \cdot \frac{1}{(\gamma/g)r_0^3} = \frac{W}{\gamma r_0^3} \qquad (5.4)$$

ρ is the density of the elastic material, γ is the unit weight of the elastic material (here, soil), dimensionless frequency

$$a_0 = \omega r_0 \sqrt{\rho/G} = \omega r_0/v_s \qquad (5.5)$$

and v_s is the velocity of shear waves in the elastic material on which the foundation is resting

The classical work of Reissner was further extended by Quinlan (1953) and Sung (1953). As mentioned before, Reissner's work only related to the case of *flexible circular foundations* where the soil reaction is uniform over the entire area (Figure 5.3a). Both Quinlan and Sung considered the cases of rigid circular foundations, the contact pressure distribution of which is shown in Figure 5.3b, flexible foundations (Figure 5.3a), and the types of foundation for which the contact pressure distribution is parabolic, as

FIGURE 5.3 Contact pressure distribution under a circular foundation of radius r_0: **(a)** uniform pressure distribution; **(b)** pressure distribution under rigid foundation; **(c)** parabolic pressure distribution.

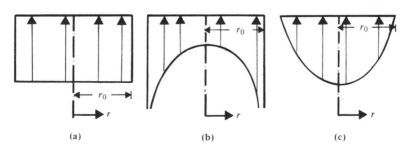

(a)　　　　　　(b)　　　　　　(c)

shown in Figure 5.3c. The distribution of contact pressure for all the three cases may be expressed as follows:

For flexible circular foundations (Figure 5.3a):

$$q = \frac{Q_0 e^{i(\omega t + \alpha)}}{\pi r_0^2} \qquad \text{for} \quad r \leqslant r_0 \tag{5.6}$$

For rigid circular foundations (Figure 5.3b)

$$q = \frac{Q_0 e^{i(\omega t + \alpha)}}{2\pi r_0 \sqrt{r_0^2 - r^2}} \qquad \text{for} \quad r \leqslant r_0 \tag{5.7}$$

For foundations with parabolic contact pressure distribution (Figure 5.3c):

$$q = \frac{2(r_0^2 - r^2) Q_0 e^{i(\omega t + \alpha)}}{\pi r_0^4} \qquad \text{for} \quad r \leqslant r_0 \tag{5.8}$$

where q is the contact pressure at a distance r measured from the center of the foundation.

Quinlan only derived the equations for the rigid circular foundation; however, Sung presented the solutions for all the three cases described above. Since most foundations are rigid (Figure 5.3b), that case is considered in more detail here.

For a *rigid foundation* subjected to a *constant force* excitation $Q_0 e^{i(\omega t + \alpha)}$, the amplitude of motion can also be given by the same relation given by Eqs. (5.2)–(5.5). Note, however, that the displacement functions f_1, f_2 change depending on the type of foundation. The values of f_1 and $-f_2$ for a *flexible foundation* are given in Table 5.1 as a power series of a_0 (0–1.5). Table 5.2 gives the values of f_1 and $-f_2$ for a *rigid foundation* as a power

TABLE 5.1 Values of f_1 and $-f_2$, Flexible Foundation[a]

Poisson's ratio μ	Values of f_1
0	$0.318310 - 0.092841\, a_0^2 + 0.007405\, a_0^4$
0.25	$0.238733 - 0.059683\, a_0^2 + 0.004163\, a_0^4$
0.5	$0.159155 - 0.039789\, a_0^2 + 0.002432\, a_0^4$
	Values of $-f_2$
0	$0.214474\, a_0 - 0.029561\, a_0^3 + 0.001528\, a_0^5$
0.25	$0.148594\, a_0 - 0.017757\, a_0^3 + 0.000808\, a_0^5$
0.5	$0.104547\, a_0 - 0.011038\, a_0^3 + 0.000444\, a_0^5$

[a]From *Foundation Analysis and Design*, Second Edition, by J. E. Bowles. Copyright © 1977 McGraw-Hill. Used with the permission of McGraw-Hill Book Company.

TABLE 5.2 Values of f_1 and $-f_2$, Rigid Foundation[a]

Poisson's ratio μ	Values of f_1
0	$0.250000 - 0.109375\, a_0^2 + 0.010905\, a_0^4$
0.25	$0.187500 - 0.070313\, a_0^2 + 0.006131\, a_0^4$
0.5	$0.125000 - 0.046875\, a_0^2 + 0.003581\, a_0^4$
	Values of $-f_2$
0	$0.21447\, a_0 - 0.039416\, a_0^3 + 0.002444\, a_0^5$
0.25	$0.148594\, a_0 - 0.023677\, a_0^3 + 0.001294\, a_0^5$
0.5	$0.104547\, a_0 - 0.014717\, a_0^3 + 0.00717\, a_0^5$

[a]From *Foundation Analysis and Design*, Second Edition, by J. E. Bowles. Copyright © 1977 McGraw-Hill. Used with the permission of McGraw-Hill Book Company.

series of a_0. Figure 5.4 gives a plot of Z vs a_0 for various values of b for a rigid base oscillator (μ = Poisson's ratio = 0.25).

Vibration of Circular Foundations by Rotating Mass Oscillator

Foundations, on some occasions, may be subjected to a *frequency-dependent* excitation in contrast to the *constant-force* type of excitation discussed above in this section. Figure 5.5 shows a rigid foundation excited by two rotating masses. The amplitude of the external oscillating force can be given as

$$Q_0 = 2m_e e \omega^2 = m_1 e \omega^2 \tag{5.9}$$

where m_1 is the total of rotating masses and ω is the circular frequency of the rotating masses.

For this condition, the amplitude of vibration A_z may be given by the relation

$$A_z = \frac{m_1 e \omega^2}{G r_0} \sqrt{\frac{f_1^2 + f_2^2}{\left(1 - b a_0^2 f_1\right)^2 + \left(b a_0^2 f_2\right)^2}} \tag{5.10}$$

From Eq. (5.5)

$$a_0 = \omega r_0 \sqrt{\rho/G}$$

or

$$\omega^2 = a_0^2 G/\left(\rho r_0^2\right) \tag{5.11}$$

Substituting Eq. (5.11) into (5.10), we obtain

$$A_z = \frac{m_1 e a_0^2}{\rho r_0^3} \sqrt{\frac{f_1^2 + f_2^2}{\left(1 - b a_0^2 f_1\right)^2 + \left(b a_0^2 f_2\right)^2}} = \frac{m_1 e}{\rho r_0^3} Z' \tag{5.12}$$

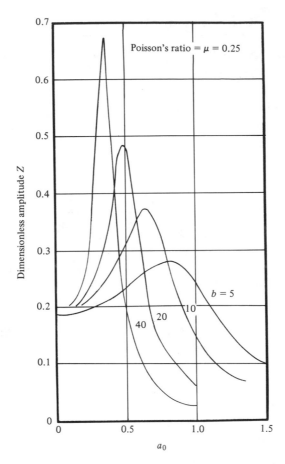

FIGURE 5.4 Plot of Z vs a_0 for rigid circular foundation. [Richart, F. E., Jr. (1962). "Foundation Vibration," *Transactions, ASCE*, 127, Part I, Fig. 7, p. 873.]

where

$$Z' = a_0^2 \sqrt{\frac{f_1^2 + f_2^2}{\left(1 - ba_0^2 f_1\right)^2 + \left(ba_0^2 f_2\right)^2}} \tag{5.13}$$

Figure 5.5 gives a plot of Z' vs a_0 for $b = 5$, 10, 20, and 40 ($\mu = 0.25$). Note that the curves shown in Figures 5.4 and 5.5 are similar to the frequency–amplitude curves shown in Figures 1.9 and 1.10 inasmuch as a_0 is the dimensionless frequency. Figure 5.6a demonstrates the nature of variation of Z' with a_0 for various types of contact pressure distribution; i.e., uniform, rigid, and parabolic (for $b = 5$ and $\mu = 0.25$). The effect of Poisson's ratio on the variation of Z' with a_0 can be seen in Figure 5.6b

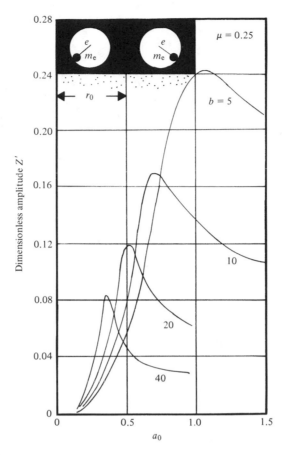

FIGURE 5.5 Variation of Z' with a_0 for rigid circular foundation. [Richart, F. E., Jr. (1962). "Foundation Vibration," *Transactions, ASCE*, 127, Part I, Fig. 7, p. 873.]

(rigid foundation, $b = 5$). Note that, with the increase of μ, the peak value of Z' decreases. Also, the value of the nondimensional frequency a_0 at which maximum Z' occurs increases with the increase of Poisson's ratio.

Resonance Condition for Vibrations of Rigid Circular Foundations

From the amplitude–frequency curves in Figures 5.4 and 5.5, one can pick up the value of a_0 for maximum amplitude (i.e., resonance condition) and the corresponding mass ratio b as shown in Figure 5.7a. These values can be plotted in a graphical form as shown in Figure 5.7b (i.e., b vs a_0 at resonance). Figure 5.8 shows similar plots (for $\mu = 0$, 0.25, and 0.5) for constant force and rotating mass oscillators, respectively.

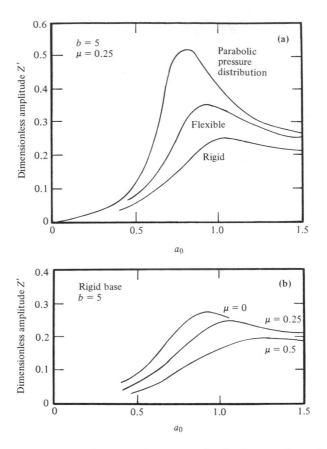

FIGURE 5.6 (a) Effect of the contact pressure distribution on the variation of Z' with a_0; (b) effect of Poisson's ratio on the variation of Z' with a_0. [Richart, F. E., Jr., and Whitman, R. V. (1967). "Comparison of Footing Vibration Tests with Theory," *Journal of the Soil Mechanics and Foundations Division, ASCE*, 83 (SM6), Fig. 4, p. 149.]

Again, referring to Figure 5.7a, one can pick up the value of Z at resonance and the corresponding b and plot them in a graphical form as shown in Figure 5.7c. These types of plot for constant force and rotating mass oscillators are shown in Figure 5.9.

Figures 5.8 and 5.9 are in convenient forms for use in the design of foundations subjected to vertical vibration (see Example 5.1, p. 164).

Design of Rigid Rectangular Foundations

The design curves developed for vertical oscillation in this section are for rigid foundations circular in plan. If a foundation is rectangular in plan with

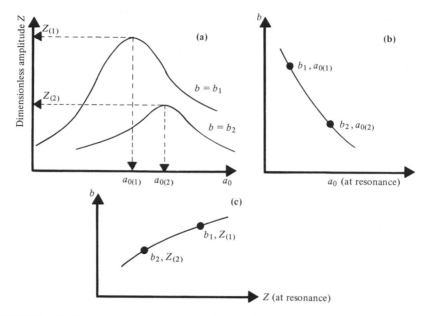

FIGURE 5.7 Procedure for preparing the graphs for b vs a_0 and b vs Z for resonance condition, vertical vibration (constant force oscillation).

FIGURE 5.8 Plot of mass ratio b vs a_0 for resonance condition, vertical oscillation: --- rotating mass; ——— constant force. [Richart, F. E., Jr. (1962). "Foundation Vibration," *Transactions*, *ASCE*, 127, Part I, Fig. 8, p. 875.]

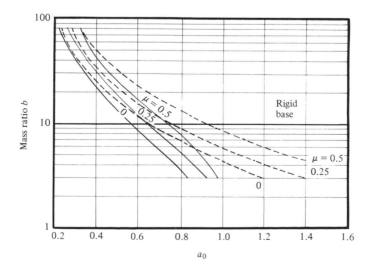

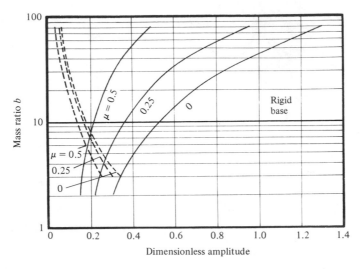

FIGURE 5.9 Plot of mass ratio vs nondimensional amplitude at resonance, vertical oscillation: – – – rotating mass, ——— constant force. [Richart, F. E., Jr. (1962). "Foundation Vibration," *Transactions*, *ASCE*, 127, Part I, Fig. 8, p. 875.]

length L and width B, one must obtain an equivalent radius to use the design curves. This can be done by equating the area of the given foundation to an area of an equivalent circle. Thus

$$\pi r_0^2 = BL$$

or

$$r_0 = \sqrt{BL/\pi} \qquad\qquad (5.14)$$

where r_0 is the radius of the equivalent circle.

5.2 ALLOWABLE VERTICAL VIBRATION AMPLITUDES

It is obviously impossible to eliminate vibration near a foundation. However, an attempt can be made to reduce the vibration problem as much as possible. Richart (1962) has compiled a guideline for allowable vertical vibration amplitude for a particular frequency of vibration. This is given in Fig. 5.10. The data presented refer to maximum allowable amplitudes of vibration. This can be converted to maximum allowable accelerations as

$$\text{maximum acceleration} = (\text{maximum displacement})\,\omega^2 \qquad (5.15)$$

For example, in Fig. 5.10, the limiting amplitude of displacement at an operating frequency of 2000 cycles/min (cpm) is about 0.005 in. (0.127 mm). So, the maximum operating acceleration for a frequency of 2000 cpm

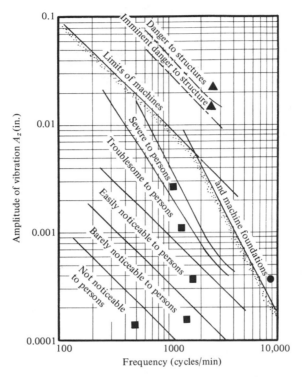

FIGURE 5.10 Allowable vertical vibration amplitudes: ■ Reiher and Meister (1931), steady-state vibration; ● Rausch (1943), steady-state vibration; ▲ Crandell (1949), blasting. Note: 1 in. = 25.4 mm. [Richart, F. E., Jr. (1962). "Foundation Vibration," *Transactions, ASCE*, 127, Part I, Fig. 2, p. 866.]

is

$$(0.005 \text{ in.})(2\pi \cdot 2000/60)^2 = 219.3 \text{ in./sec}^2 \ (5570 \text{ mm/sec}^2)$$

5.3 SOME COMMENTS ON DESIGN OF MACHINE FOUNDATIONS TO AVOID RESONANCE

In the design of machine foundations, the following general rules may be kept in mind for avoiding possible resonance conditions:

1. The resonant frequency of the foundation–soil system should be less than half the operating frequency for high-speed machines (i.e., operating frequency ⩾ 1000 cpm). For this case, during starting or stopping, the machine will briefly vibrate at resonant frequency.
2. For low-speed machineries (⩾ 350–400 cpm), the resonant frequency of the foundation–soil system should be at least twice the operating frequency.

3. In all types of foundation, the increase of weight decreases the resonant frequency.
4. An increase of r_0 increases the resonant frequency of the foundation.
5. An increase of shear modulus of soil (for example, by grouting) increases the resonant frequency of the foundation.

Example 5.1 A foundation subjected to a *constant force type vertical oscillation*. Given the total weight of the machinery and foundation block $W = 150,000$ lbs; $\gamma = 115$ lb/ft^3; $\mu = 0.4$; $G = 3000$ lb/in.^{2}; the amplitude of the oscillating force $Q_0 = 1500$ lbs; the operating frequency $f = 180$ cpm; and the foundation is 20 ft long and 6 ft wide:

a. Determine the resonant frequency. Check if $f_{res}/f_{oper} > 2$.
b. Determine the amplitude of vibration at resonance.

Solution

a. Resonant Frequency. This is a rectangular foundation, so the equivalent radius [Eq. (5.14)]

$$r_0 = \sqrt{BL/\pi} = \sqrt{(6)(20)/\pi} = 6.18 \text{ ft}$$

The mass ratio

$$b = \frac{m}{\rho r_0^3} = \frac{W}{\gamma r_0^3} = \frac{150,000}{115(6.18)^3} = 5.53$$

From Figure 5.8, corresponding to $b = 5.53$ and $\mu = 0.4$, $a_0 = 0.8$.

$$f_{res} = \frac{a_0}{2\pi r_0} \sqrt{\frac{G}{\rho}} = \frac{0.8}{2\pi(6.18)} \sqrt{\frac{(3000)(144)}{(115/32.2)}}$$

$$= 7.17 \text{ cycles/sec} = 430 \text{ cpm}$$

Hence,

$$f_{res}/f_{oper} = 430/180 = 2.39 > 2$$

b. Amplitude of Vibration. From Eq. (5.2),

$$A_z = (Q_0/Gr_0)Z$$

For amplitude of vibration at resonance, refer to Figure 5.9. With $b = 5.53$, $Z = 0.2$; so

$$A_{z(res)} = \frac{1500}{(3000 \times 144)6.18}(0.2) = 0.00011 \text{ ft}$$

$$= 0.00135 \text{ in.}$$

Example 5.2 Figure 5.11a shows a single cylinder reciprocating engine. The data for the engine are as follows: operating speed $= 1500$ cpm; connecting

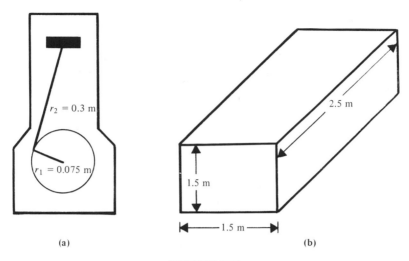

$r_2 = 0.3$ m

$r_1 = 0.075$ m

(a)

2.5 m

1.5 m

1.5 m

(b)

FIGURE 5.11

rod $(r_2) = 0.3$ m; crank $(r_1) = 75$ mm; total reciprocating weight $= 54$ N; total engine weight $= 14$ kN. Figure 5.11b shows the dimensions of the concrete foundation for the engine. The properties of the soil are as follows: $\gamma = 18.5$ kN/m³; $G = 18,000$ kN/m²; and $\mu = 0.5$. Calculate:

   a. primary and secondary unbalanced forces at operating frequency (refer to Appendix A)
   b. the resonance frequency
   c. the vibration amplitude at resonance
   d. the vibration amplitude at operating frequency

Solution

   a. Unbalanced Forces at Operating Frequency. The equations for obtaining the maximum *primary* and *secondary* unbalanced forces for a single cylinder reciprocating engine are given in Appendix A. From Eqs. (A-9) and (A-10),

$$\text{primary} = m_{\text{rec}} r_1 \omega^2 = \frac{54}{1000(9.81)} \left(\frac{75}{1000} \right) \left(\frac{2\pi 1500}{60} \right)^2$$

$$= 10.19 \text{ kN}$$

$$\text{secondary} = \frac{m_{\text{rec}} r_1^2}{r_2^2} \omega^2$$

$$r_1/r_2 = 0.075/0.3 = 0.25$$

so

$$\text{secondary} = (\text{primary}) r_1/r_2 = (10.19)0.25 = 2.55 \text{ kN}$$

b. Resonant Frequency. From Eq. (5.14)

$$r_0 = \sqrt{BL/\pi} = \sqrt{1.5 \times 2.5/\pi} = 1.093 \text{ m}$$

Mass ratio $b = m/\rho r_0^3$; total weight W = weight of foundation + engine; and assume the unit weight of concrete is 23.58 kN/m³. Thus,

$$W = (1.5 \times 2.5 \times 1.5)23.58 + 14 = 146.64 \text{ kN}$$

$$b = \frac{m}{\rho r_0^3} = \frac{W}{\gamma r_0^3} = \frac{146.64}{18.5(1.093)^3} = 6.07$$

From Figure 5.8, for $b = 6.07$, $a_0 = 1.2$; but

$$a_0 = \omega r_0 \sqrt{\rho/G}, \qquad \omega = (a_0/r_0)\sqrt{G/\rho}$$

$$f_{res} = \frac{\omega}{2\pi} = \frac{a_0}{r_0(2\pi)} \sqrt{\frac{G}{\rho}} = \frac{1.2}{(1.093)(2\pi)} \sqrt{\frac{18,000(9.81)}{18.5}}$$

$$= 17.07 \text{ cps} = 1024 \text{ cpm}$$

c. Amplitude at Resonance. From Eq. (5.12),

$$A_z = \left(m_1 e/\rho r_0^3\right) Z'$$

At 1500 cpm, the total unbalanced force = primary force + secondary force = 10.19 + 2.55 = 12.74 kN.

$$Q_{0(1024 \text{ cpm})} = Q_{0(1500 \text{ cpm})} (1024/1500)^2 = 12.74(1024/1500)^2 = 5.94 \text{ kN}$$

$$Q_{0(1024 \text{ cpm})} = m_1 e \omega^2 = 5.94 \text{ kN}$$

Therefore,

$$m_1 e = 5.94/\omega^2, \qquad \omega = 2\pi(1024)/60 = 107.23 \text{ rad/sec}$$

$$m_1 e = 5.94/(107.23)^2$$

From Figure 5.9, for $b = 6.07$, $Z' = 0.18$. Hence

$$A_z = \frac{m_1 e}{(\gamma/g) r_0^3} Z' = \frac{5.94/(107.23)^2}{(18.5/9.81)(1.093)^3} \cdot 0.18 = 0.0000378 \text{ m}$$

$$= 0.0378 \text{ mm}$$

d. Amplitude at Operating Frequency. For operating frequency,

$$a_0 = \omega r_0 \sqrt{\frac{\rho}{G}} = \frac{(2\pi)(1500)}{60}(1.093) \sqrt{\frac{18.5}{(18,000)(9.81)}} = 1.76$$

Referring to Figure 5.5, (for $\mu = 0.25$), by extrapolation, for $b = 6.07$ and $a_0 = 1.76$, $Z' \approx 0.16$. Thus,

$$A_z = \left[m_1 e / (\gamma/g) r_0^3 \right] Z'$$

$$m_1 e = \frac{Q_{0(1500 \text{ cpm})}}{\omega^2} = \frac{12.74}{\left[(2\pi) 1500/60 \right]^2} = 0.000516$$

$$A_z = \frac{0.000516}{(18.5/9.81)(1.093)^3} (0.16) = 0.000033 \text{m} = 0.033 \text{ mm}$$

5.4 ROCKING OSCILLATION OF FOUNDATION

Theoretical solutions for foundations subjected to rocking oscillation have been presented by Arnold et al. (1955) and Bycroft (1956). For *rigid circular foundations* (Figure 5.12), the contact pressure can be described by the

FIGURE 5.12 Rocking oscillation of rigid circular foundation.

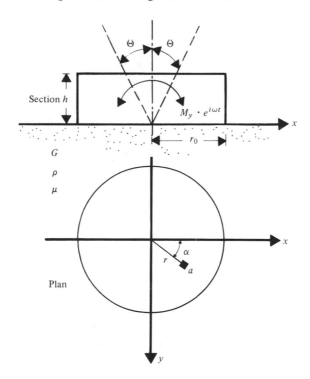

equation

$$q = \frac{3M_y r \cos \alpha}{2\pi r_0^3 \sqrt{r_0^2 - r^2}} e^{i\omega t} \tag{5.16}$$

where M_y is the external moment imposed by the foundation (the moment about the y axis) and q is the contact pressure at any point defined by point a on the plan.

For a static external moment of magnitude M_y, the angular rotation of the foundation can be expressed as

$$\Theta_{\text{stat}} = \tfrac{3}{8}(1 - \mu) M_y / Gr_0^3 \tag{5.17}$$

For the dynamic moment M_y, the amplitude of the angular rotation can be expressed as

$$\Theta = \left(M_y / Gr_0^3 \right) \Theta' \tag{5.18}$$

For $\mu = 0$, the variation of the value of Θ' with the dimensionless frequency a_0 is shown in Figure 5.13a (for $b_i = 2$, 5, 10, and 20). The term b_i is the *inertia ratio* given by the relation

$$b_i = I_0 / \rho r_0^5 \tag{5.19}$$

where I_0 is the mass moment of inertia of the oscillator about the y axis through its base and ρ is the soil density.

The mass moment of inertia can be expressed as

$$I_0 = (W_0 / g)(\tfrac{1}{4} r_0^2 + \tfrac{1}{3} h^2) \tag{5.20}$$

where W_0 is the foundation weight, g is the acceleration due to gravity, and h is the height of the foundation.

It may also be seen from Figure 5.13a that the envelope curve at the top is tangent at the peak points to all the *amplitude vs frequency* curves for different values of b_i. This tangent (envelope) curve can be used to define the relation between a_0 at maximum amplitude (resonant condition) and the values of the inertia ratio b_i (Figure 5.13b).

In the case of rectangular foundations, the design curves (Figure 5.13) can be used by determining an equivalent radius r_0 [as in Eq. (5.14) for vertical vibration] given by the equation

$$r_0 = \sqrt[4]{\tfrac{1}{3} BL^3 / \pi} \tag{5.21}$$

The definitions of B and L are shown in Figure 5.14.

Example 5.3 A horizontal piston type compressor is shown in Figure 5.15. The operating frequency is 600 cpm. The amplitude of the horizontal unbalanced force of the compressor is 30 kN, and it creates a rocking motion of the foundation about point 0 (see Figure 5.15b). The mass

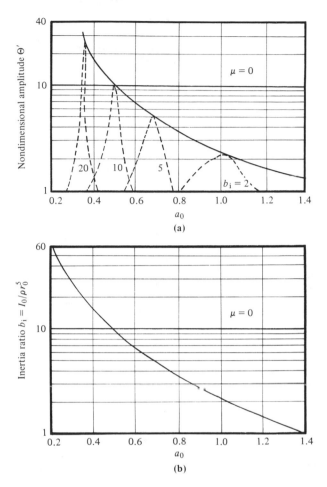

FIGURE 5.13 Rocking oscillation for rigid circular foundation: **(a)** plot of Θ' vs a_0; **(b)** plot of inertia ratio vs a_0. [Richart, F. E., Jr. (1962). "Foundation Vibration," *Transactions, ASCE*, 127, Part I, Fig. 9, p. 876.]

moment of inertia of the compressor assembly about the axis $b'0b'$ (see Figure 5.15c) is 16×10^5 kg-m². Determine

a. the resonant frequency
b. the amplitude of rocking oscillation at resonance

Solution

Moment of Inertia of the Foundation Block and the Compressor Assembly About $b'0b'$.

$$I_0 = (W_{block}/3g)\left[\left(\tfrac{1}{2}L\right)^2 + h^2\right] + 16 \times 10^5 \text{ kg-m}^2$$

170

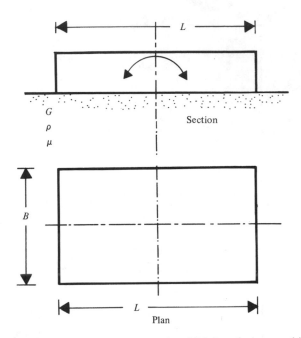

FIGURE 5.14 Equivalent radius of rectangular rigid foundation: rocking motion.

FIGURE 5.15

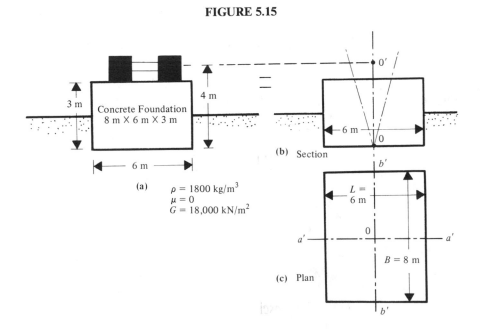

(a) $\rho = 1800 \text{ kg/m}^3$
$\mu = 0$
$G = 18,000 \text{ kN/m}^2$

(b) Section

(c) Plan

Assume unit weight of concrete $= 23.58 \text{ kN/m}^3$:

$$W_{\text{block}} = (8 \times 6 \times 3)(23.58) = 3395.52 \text{ kN}$$

$$= 3395.52 \times 10^3 \text{ N}$$

$$I_0 = \frac{3395.52 \times 10^3}{3(9.81)} [3^2 + 3^2] + 16 \times 10^5 = 36.768 \times 10^5 \text{ kg-m}^2$$

Calculation of Equivalent Radius of the Foundation. From Eq. (5.21), the equivalent radius

$$r_0 = \sqrt[4]{\tfrac{1}{3} BL^3/\pi} = \sqrt[4]{\tfrac{1}{3} (8 \times 6^3)/\pi} = 3.67 \text{ m}$$

Calculation of Inertia Ratio.

$$b_i = \frac{I_0}{\rho r_0^5} = \frac{36.768 \times 10^5}{1800(3.67)^5} = 3.07$$

a. Calculation of Resonant Frequency. From Figure 5.13b, the value of a_0 corresponding to $b_i = 3.07$ is 0.86.

$$a_0 = \omega r_0 \sqrt{\rho/G}$$

$$f_{\text{res}} = \frac{1}{2\pi} \frac{a_0}{r_0} \sqrt{\frac{G}{\rho}} = \frac{1}{2\pi} \frac{0.86}{3.67} \sqrt{\frac{18{,}000(9.81)}{17.66}}$$

$$= 3.73 \text{ cps} = 224 \text{ cpm}$$

b. Calculation of Amplitude of Oscillation at Resonance. From Eq. (5.18),

$$\Theta = \left(M_y/Gr_0^3 \right) \Theta'$$

From Figure 5.13a, for $a_0 = 0.086$, $\Theta' = 3.2$.

$$M_{y(\text{oper freq})} = (\text{unbalanced force}) \times 4$$

$$= 30 \times 4 = 120 \text{ kN-m}$$

Therefore,

$$M_{y(\text{res})} = 120(\text{res freq/oper freq})^2$$

$$= 120(224/600)^2 = 16.72 \text{ kN-m}$$

Hence,

$$\Theta = \left[16.72/18{,}000(3.67)^3 \right] 3.2 = 0.00006 \text{ rad}$$

At a distance 4 m above the point 0, the horizontal amplitude of vibration is equal to

$$(0.00006)(4) = 0.00024 \text{ m} = 0.24 \text{ mm}$$

5.5 SLIDING OSCILLATION OF FOUNDATION

Arnold et al. (1955) have found theoretical solutions for sliding oscillation of rigid circular foundations (Figure 5.16) acted on by a force $Q = Q_0 e^{i\omega t}$. For this type of oscillation, the amplitude of vibration of the foundation can be expressed as follows:

Case I. For $Q_0 = $ Constant

$$A_x = (Q_0/Gr_0) X \qquad (5.22)$$

where A_x is the amplitude of horizontal oscillation and X is the nondimensional amplitude factor.

The variation of the amplitude vs frequency (for $b = 2, 4, 10, 20, 40,$ and 80) for sliding oscillation is shown in Figure 5.17a. Note that these are for the case of $\mu = 0$. The envelope drawn to these curves is used to define the relation between frequency at maximum amplitude (resonant condition) and the mass ratio b. This curve has been used to obtain the plot of $b \ (= m/\rho r_0^3)$ vs a_0 for resonant amplitude as shown in Figure 5.17b.

Case II. For $Q_0 = m_1 e \omega^2$ (In this case, horizontal force is created by eccentric mass oscillators each of mass m_e)

$$A_x = (m_1 e/\rho r_0^3) X' \qquad (5.23)$$

Figure 5.17a shows the envelope curve of X' vs a_0 for resonant condition (similar to the *envelope curve* of X vs a_0 for constant force

FIGURE 5.16 Sliding oscillation of rigid circular foundation.

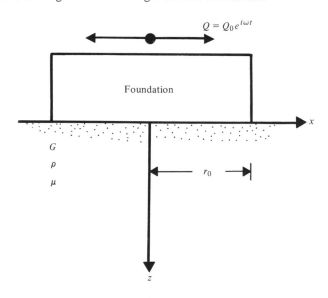

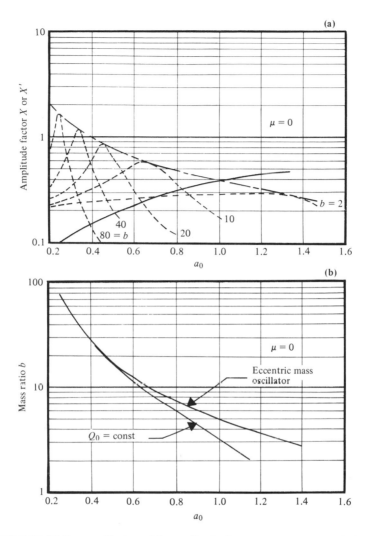

FIGURE 5.17 Sliding oscillation: **(a)** amplitude factor vs a_0; ――― variation of X with a_0 ($Q_0 = $ constant), ――――― variation of X with a_0 at resonance ($Q_0 = $ constant), ――――― variation of X' with a_0 at resonance (eccentric mass oscillator). **(b)** mass ratio b vs a_0 at resonance. [Richart, F. E., Jr. (1962). "Foundation Vibration," *Transactions, ASCE*, 127, Part I, Fig. 10, p. 878.]

shown in Figure 5.17a). The variation of the mass ratio b vs a_0 for resonant condition is drawn in Figure 5.17b.

5.6 TORSIONAL OSCILLATION OF FOUNDATION

Figure 5.18a shows a circular foundation of radius r_0 subjected to a torque $T = T_0 e^{i\omega t}$ about an axis $z-z$. Reissner (1937) solved the oscillation problem of this type considering a linear distribution of shear stress $\tau_{z\theta}$ (shear stress

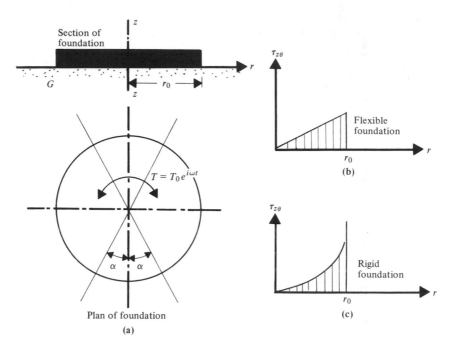

FIGURE 5.18 Torsional oscillation of foundation.

zero at center and maximum at the periphery of the foundation) as shown in Figure 5.18b. This represents the case of a *flexible foundation*. In 1944 Reissner and Sagoli solved the same problem for the case of a *rigid foundation* considering a *linear variation of displacement from the center to the periphery* of the foundation. For this case, the shear stress can be given by (Figure 5.18c)

$$\tau_{z\theta} = (3/4\pi)\left(Tr/r_0^3 \sqrt{r_0^2 - r^2}\right) \qquad \text{for} \quad 0 < r < r_0 \qquad (5.24)$$

For a static torque T the angle of rotation α can be expressed as

$$\alpha = (3/16Gr_0^3)T_{\text{stat}} \qquad (5.25)$$

For a dynamic torque $T = T_0 e^{i\omega t}$ on a rigid foundation, the amplitude of the angle of rotation can be given by

$$\alpha = (T_0/Gr_0^3)\alpha' \qquad (5.26)$$

where α' is the nondimensional amplitude factor.

The variation of amplitude vs frequency (α' vs a_0) curves for rigid foundations with the mass ratio $b_t = 2$, 5, and 10 are shown in Figure 5.19a. The envelope curve drawn to these curves defines the relation between frequency at resonant condition and the mass ratio b_t. This curve has been used to obtain the plot of b_t vs a_0 at resonance as shown in Figure 5.19b.

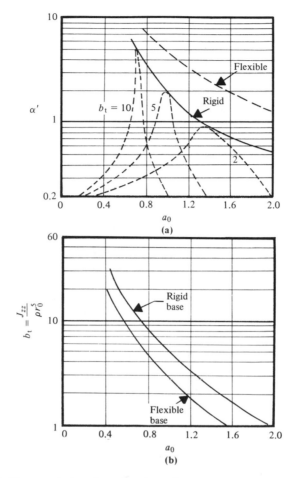

FIGURE 5.19 Characteristics of torsional oscillation of foundation. [Richart, F. E., Jr. (1962). "Foundation Vibration," *Transactions*, *ASCE*, 127, Part I, Fig. 22, p. 923.]

The mass ratio for torsion b_t is defined as

$$b_t = J_{zz}/\rho r_0^5 \tag{5.27}$$

where J_{zz} is the mass moment of inertia of the foundation about the axis $z-z$ (Figure 5.18a).

Figure 5.19 also shows the curve for α' vs a_0 at resonance for a *flexible* foundation. Figure 5.19b also shows the plot of b_t vs a_0 at resonance for a *flexible* circular foundation.

For a rectangular foundation with dimensions $B \times L$, the equivalent radius may be given by

$$r_0 = \sqrt[4]{\tfrac{1}{6}BL(B^2 + L^2)/\pi} \tag{5.28}$$

The torsional vibration of foundations are uncoupled motion and hence can be treated independently of any vertical motion. Also Poisson's ratio does not influence the torsional vibration of foundations.

Example 5.4 A radar antenna foundation is shown in Figure 5.20. For torsional vibration of the foundation, given:

$$T_0 = 18 \times 10^4 \text{ ft-lb} \qquad \text{due to inertia}$$

$$T_0 = 6 \times 10^4 \text{ ft-lb} \qquad \text{due to wind}$$

mass moment of inertia of the tower about the axis $z - z = 10 \times 10^6$ ft-lb-sec^2, and the unit weight of concrete used in the foundation is 150 lb/ft^3. Calculate:

a. the resonant frequency for torsional mode of vibration
b. angular deflection at resonance

FIGURE 5.20

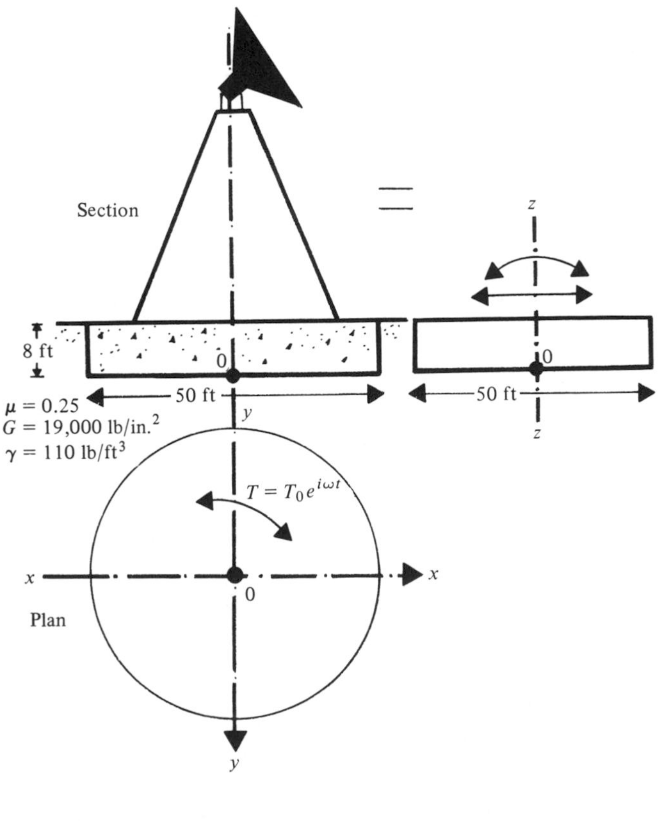

Solution

a. Resonant Frequency.

$$J_{zz} = J_{zz(\text{tower})} + J_{zz(\text{foundat})}$$

$$= 10 \times 10^6 + \tfrac{1}{2}\left[\pi r_0^2 h (150/32.2)\right] r_0^2$$

$$= 10 \times 10^6 + \tfrac{1}{2}\left[(\pi)(25)^2(8)(150/32.2)\right](25)^2$$

$$= 10 \times 10^6 + 22.87 \times 10^6 = 32.87 \times 10^6 \text{ ft-lb-sec}^2$$

From Eq. (5.27)

$$b_t = \frac{J_{zz}}{\rho r_0^5} = \frac{32.87 \times 10^6}{(110/32.2)(25)^5} = 0.985$$

For finding the resonance frequency, refer to Figure 5.19b. For $b_t = 0.895$, $a_0 \approx 2.1$ (by extrapolation) for rigid foundation. Hence,

$$\text{resonance frequency} = \frac{a_0}{2\pi r_0}\sqrt{\frac{G}{\rho}} = \frac{2.1}{2\pi(25)}\sqrt{\frac{19,000 \times 144}{(110/32.2)}}$$

$$= 11.96 \text{ cps}$$

b. Angular Deflection at Resonance Frequency. From Figure 5.19a, for $a_0 = 2.1$, $\alpha' \approx 0.5$. Hence,

$$\alpha_{\text{inert}} = \left[\frac{T_{0(\text{inert})}}{Gr_0^3}\right]\alpha' = \left[\frac{18 \times 10^4}{(19,000)(144)(25)^3}\right]0.5$$

$$= 0.21 \times 10^{-5} \text{ rad}$$

If the torque due to wind (T_0) is to be treated as a static torque, then from Eq. (5.25)

$$\alpha_{\text{stat}} = \frac{3}{16Gr_0^3}T_{0(\text{stat})} = \left[\frac{3}{16(19,000)(144)(25)^3}\right](6 \times 10^4)$$

$$= 0.0263 \times 10^{-5} \text{ rad}$$

At resonance, the total angular deflection is

$$\alpha = \alpha_{\text{inert}} + \alpha_{\text{stat}} = (0.21 + 0.0263) \times 10^{-5}$$

$$= 0.2363 \times 10^{-5} \text{ rad} = 0.135 \times 10^{-3} \text{ deg}$$

5.7 COMPARISON OF FOOTING VIBRATION TESTS WITH THEORY

Richart and Whitman (1967) conducted a comprehensive study regarding the applicability of the theoretical findings presented in this chapter to actual field problems. Ninety-four large-scale field test results for large footings 5–16 ft (1.52–4.88 m) in diameter subjected to *vertical vibration* have been reported by Fry (1963). Of these 94 test results, 55 were conducted at the U.S. Army Waterways Experiment Station, Vicksburg, Mississippi. The remaining 39 were conducted at Eglin Field, Florida. The classification of the soils for the Vicksburg site and the Eglin site were CL and SP, respectively (Unified soil classification system). For these tests, the vertical dynamic force on footings was generated by rotating mass oscillators. Figure 5.21 shows a comparison of the theoretical amplitudes of vibration A_z with the experimental results obtained for two bases at the Vicksburg site. The respective nondimensional mass ratios b of these two bases were 5.2 and 3.8. For the base with $b = 5.2$, the experimental results fall between the theoretical curves with $\mu = 0.5$ and $\mu = 0.25$. However, for the base with $b = 3.8$, the experimental curve is nearly identical to the theoretical curve with $\mu = 0.5$. Figure 5.22 shows a comparison of the theory and experimental values (reported by Fry) in a nondimensional plot of $A_z m / m_1 e$ at resonance vs b (m_1 = total mass of the rotating oscillators). Similarly, a comparison of these test results with theory in a nondimensional plot of a_0 at resonance vs b is shown in Figure 5.23.

FIGURE 5.21 Vertical oscillation: comparison of test results with theory. [Richart, F. E., Jr., and Whitman, R. V. (1967). "Comparison of Footing Vibration Tests with Theory," *Journal of the Soil Mechanics and Foundations Division*, *ASCE*, 93 (SM6), Fig. 10, p. 156.]

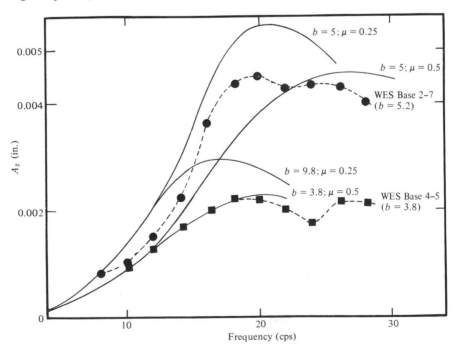

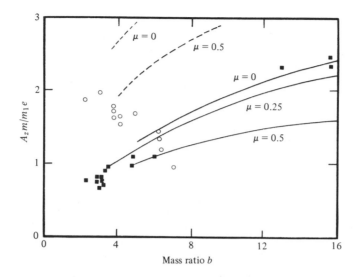

FIGURE 5.22 Motion at resonance for vertical excitation: comparison between theory and experiment. ■ Vicksburg, ○ Eglin; ——— rigid base theory, ––– parabolic theory. [Richart, F. E., Jr., and Whitman, R. V. (1967). "Comparison of Footing Vibration Tests with Theory," *Journal of the Soil Mechanics and Foundations Division, ASCE*, 93 (SM6), Fig. 11, p. 157.]

FIGURE 5.23 Plot of a_0 at resonance vs b: comparison of theory with field test results. ● Vicksburg, ○ Eglin, ——— rigid base theory, ––– parabolic theory. [Richart, F. E., Jr., and Whitman, R. V. (1967). "Comparison of Footing Vibration Tests with Theory," *Journal of the Soil Mechanics and Foundations Division, ASCE*, 93 (SM6), Fig. 12, p. 157.]

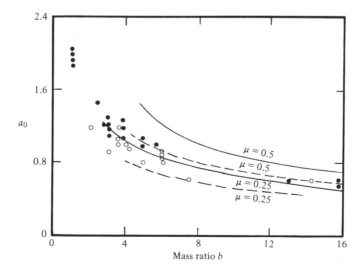

From these two plots it may be seen that the results of the Vicksburg site follow the general trends indicated by the theoretical curve obtained from the elastic half space theory for a rigid base. A considerable scatter, however, exists for the test conducted at Eglin Field. This may be due to the clean fine sand found at that site for which the shear modulus changes with depth (i.e., confining pressure). The fundamental assumption of the theoretical derivation of a homogeneous, elastic, isotropic body is very much different than the actual field conditions. Figure 5.24 shows a summary of all vertical oscillation tests which is a plot of

$$A_{z(\text{computed})}/A_{z(\text{measured})} \text{ vs } A_z\omega^2/g$$

(i.e., nondimensional acceleration, $g =$ acceleration due to gravity). Note that when the nondimensional acceleration reaches 1, probably the footing leaves the ground on the upswing and acts as a hammer. In any case, in actual design problems, a machine foundation is not subjected to an acceleration greater than $0.3g$. However, for dynamic problems of this nature, the general agreement between theory and experiment is fairly good.

Several large-scale field tests have been conducted by the U.S. Army Waterways Experiment Station (Fry, 1963) in which footings were subjected to torsional oscillation. Mechanical vibrators were set to produce pure torque on a horizontal plane. Figure 5.25 shows a plot of the dimensionless amplitude $\alpha J_{zz}/(m_1 e \frac{1}{2} x)$ vs b_t for some of these tests which correspond to the lowest settings of the eccentric masses on the oscillator α is the

FIGURE 5.24 Summary of vertical vibration tests: ● Vicksburg, ○ Eglin. [Richart, F. E., Jr., and Whitman, R. V. (1967). "Comparison of Footing Vibration Tests with Theory," *Journal of the Soil Mechanics and Foundations Division*, *ASCE*, 93 (SM6), Fig. 13, p. 158.]

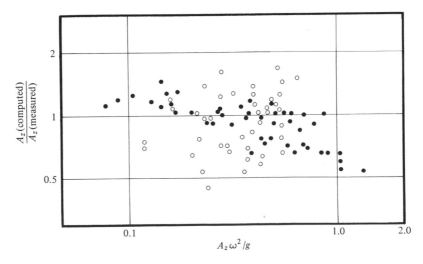

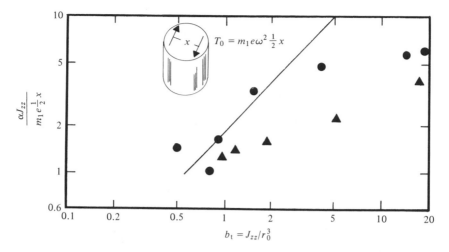

FIGURE 5.25 Comparison of amplitude of vibration for torsional oscillation: ●, Vicksburg, ▲ Eglin. [Richart, F. E., Jr., and Whitman, R. V. (1967). "Comparison of Footing Vibration Tests with Theory," *Journal of the Soil Mechanics and Foundations Division, ASCE*, 93 (SM6), Fig. 15, p. 160.]

amplitude of torsional motion and m_1 is the sum of the oscillating masses (for x see the insert in Figure 5.25). The theoretical curve is also plotted in this figure for comparison purposes. It can be seen that, for low amplitudes of vibration, the agreement between theory and field test results is good. The limiting torsional motion in most practical cases is about 0.1 mil (1×10^{-4} in.). The half space theory therefore generally serves well in most practical design considerations.

Comparisons between theory and experimental results for footing vibration tests in rocking and sliding modes have also been presented by Richart and Whitman (1967). The agreement seemed fairly good.

PROBLEMS

5.1. A concrete foundation is 8 ft in diameter. The foundation is supporting a machine. The weight of the machine and the foundation is 60,000 lb. The machine imparts an oscillating force $Q = Q_0 \sin \omega t$. Given $Q_0 = 6,000$ lb (not frequency dependent); operating frequency of 150 cpm; soil supporting the foundation with

$$\text{unit weight} = 120 \text{ lb/ft}^3$$
$$\text{shear modulus} = 6500 \text{ lb/in.}^2$$
$$\text{Poisson's ratio} = 0.3$$

Determine
a. The resonant frequency
b. The approximate amplitude of vibration at the operating frequency

5.2. Redo Problem 5.1 assuming the foundation to be 8 ft × 6 ft in plan. Assume that the total weight of the foundation and the machine is the same as in Problem 5.1.

5.3. A concrete foundation supporting a machine is 3.5 m × 2.5 m in plan and is subjected to a sinusoidal oscillating force (vertical) having an amplitude of 10 kN (not frequency dependent). The operating frequency is 2000 cpm. The weight of the machine and foundation is 400 kN. Given the following soil properties:

$$\text{unit weight} = 18 \text{ kN/m}^3$$
$$\text{shear modulus} = 38{,}000 \text{ kN/m}^2$$
$$\text{Poisson's ratio} = 0.25$$

Determine
a. The resonant frequency of the foundation
b. The amplitude of vibration at the operating frequency

5.4. Consider the case of a single cylinder reciprocating engine (Figure 5.11a; see p. 165). For the engine:

$$\text{Operating speed} = 1000 \text{ cpm}$$
$$\text{crank } (r_1) = 90 \text{ mm}$$
$$\text{connecting rod } (r_2) = 350 \text{ mm}$$
$$\text{weight of the engine} = 20 \text{ kN}$$
$$\text{reciprocating weight} = 65 \text{ N}$$

The engine is supported by a concrete foundation block ($L \times B \times H$) of 3 m × 2 m × 1.5 m. The properties of the soil supporting the foundation are as follows:

$$\text{unit weight} = 19 \text{ kN/m}^3$$
$$G = 24{,}000 \text{ kN/m}^2$$
$$\mu = 0.25$$

Calculate
a. The resonant frequency
b. The amplitude of vibration at resonance

5.5. Refer to Problem 5.4. What is the amplitude of vibration at operating frequency?

5.6. The concrete foundation of a machine has the following dimensions (refer to Figure 5.14, p. 170):

$$L = 3 \text{ m}, \qquad B = 4 \text{ m}, \qquad H = 1.5 \text{ m}$$

The foundation is subjected to a sinusoidal horizontal force from the machine having an amplitude of 10 kN at a height of 2 m measured from the base of the foundation. The soil supporting the foundation is sandy clay.

Given

$$G = 30,000 \text{ kN/m}^2, \qquad \mu = 0, \qquad \rho = 1700 \text{ kg/m}^3$$

Determine:

a. The resonant frequency for the rocking mode of vibration of the foundation

b. The amplitude of rocking vibration at resonance.

Note: (1) The amplitude of horizontal force is not frequency dependent. (2) Neglect the moment of inertial of the machine.

5.7. Solve Problem 5.6 assuming that the horizontal force is frequency dependent. The amplitude of the force at an operating speed of 800 cpm is 20 kN.

5.8. Refer to Problem 5.6. Determine
a. The resonant frequency for sliding mode of vibration
b. Amplitude for the sliding mode of vibration at resonance
Assume the weight of the machinery on the foundation to be 100 kN.

5.9. Repeat Problem 5.8 assuming that the horizontal force is frequency dependent. The amplitude of the horizontal force at an operating frequency of 800 cpm is 40 kN. The weight of the machinery on the foundation is 100 kN.

5.10. Refer to Figure 5.18 (p. 174). A concrete foundation supporting a machine has the following dimensions:

$$L = 5 \text{ m}, \qquad B - 1 \text{ m}, \qquad H - 2 \text{ m}$$

The machine imparts a torque T on the foundation such that

$$T = T_0 e^{i\omega t}$$

Given $T_0 = 3000$ N-m; the mass moment of inertia of the machine about the $z - z$ axis is 75×10^3 kg-m^2. The soil has the following properties:

$$\mu = 0.25, \qquad \text{unit weight} = 18 \text{ kN/m}^3, \qquad G = 28,000 \text{ kN/m}^2$$

Determine
a. The resonant frequency for the torsional mode of vibration
b. Angular deflection at resonance

REFERENCES

Arnold, R. N., Bycroft, G. N., and Wartburton, G. B. (1955), "Forced Vibrations of a Body on an Infinite Elastic Solid," *Journal of Applied Mechanics, Trans. ASME* 77, 391–401.

Bowles, J. E. (1977), *Foundation Analysis and Design*, McGraw-Hill, New York.

Bycroft, G. N. (1956), "Forced Vibrations of a Rigid Circular Plate on a Semi-infinite Elastic Space and on an Elastic Stratum," *Philosophical Transactions of the Royal Society*, London (Ser. A) 248, 327–368.

Crandell, F. J. (1949), "Ground Vibration Due to Blasting and Its Effects on Structures," *Journal of the Boston Society of Civil Engineers* (April).

Fry, Z. B. (1963), "Report 1: Development and Evaluation of Soil Bearing Capacity, Foundation of Structures, Field Vibratory Tests Data," *Technical Report No. 3-632*, U.S. Army Engineers Waterways Experiment Station, Vicksburg, Mississippi.

Lamb, H. (1904), "On the Propagation of Tremors Over the Surface of an Elastic Solid," *Philosophical Transactions of the Royal Society*, London (Ser. A) 203,1–42.

Quinlan, P. M. (1953), "The Elastic Theory of Soil Dynamics," Symposium on Dynamic Testing of Soils, ASTM, *STP* 156, pp. 3–34.

Rausch, E. (1943), "Maschinenfundamente und Andere Dynamische Bauaufgaben," Vartrieb VDI, Verlag G.J.B.H. (Berlin).

Reiher, H., and Meister, F. J. (1931), "Die Empfindlichkeit der Menschen gegen Ershutterungen," *Forsch. Gebiete Ingenieurwesen* 2 (11), 381–386.

Reissner, E. (1937), "Freie und erzwungene Torsionschwingungen des elastischen Halbraumes," *Ingenieur-Archiv.* 8 (4), 229–245.

Reissner, E. (1936), "Stationare, axialsymmetrische durch eine Schuttelnde Masseerregte Schwingungen eines homogenen elastischen halbraumes," *Ingenieur-Archiv.* 7 (Part 6), 381–396.

Reissner, E., and Sagochi, H. F. (1944), "Forced Torsional Oscillations of an Elastic Half Space," *Journal of Applied Physics* 15, 652–662.

Richart, F. E., Jr. (1962), "Foundation Vibrations," *Trans. ASCE* 127 (Part 1), 863–898.

Richart, F. E., Jr., and Whitman, R. V. (1967), "Comparison of Footing Vibration Tests with Theory," *Journal of the Soil Mechanics and Foundations Division, ASCE* 83 (SM6), 143–167.

Sung, T. Y. (1953), "Vibration in Semi-infinite Solids Due to Periodic Surface Loadings," Symposium on Dynamic Testing of Soils, ASTM, *STP* 156, pp. 35–54.

6

ANALYSIS OF
FOUNDATION VIBRATION:
LUMPED PARAMETER SYSTEM

In Chapter 5, the fundamentals of the elastic half-space theory for the analysis of the vibration of foundations were presented. The elastic half-space problems are rather difficult to follow. A more versatile technique would be to use a mass–spring–dashpot system (i.e., a lumped parameter system) for analysis of foundation vibrations. The fundamentals of the mathematical treatment of mass–spring–dashpot problems were presented in Chapter 1. This approach would be easier than the more fundamentally correct half-space theory provided proper choices of mass, spring constant, and dashpot coefficient (viscous damping) were made. In this chapter, the procedures for proper evaluation of the above parameters for various modes of foundation vibration are given.

6.1 LUMPED PARAMETER SYSTEM FOR VERTICAL MOTION

6.1.1 General Relationships

Lysmer and Richart (1966) have developed the fundamentals of the lumped parameter approach for steady-state motion. This approach is used in this section to develop the general relationships. A linear dynamic system S that is excited by a periodic vertical force P is shown in Figure 6.1a.

The periodic vertical force can be given by the relation

$$P = P_0 e^{i\omega t} \tag{6.1}$$

where P_0 is the amplitude of the vertical force and ω is the circular frequency of vibration. The displacement z can therefore be written as

$$z = (P_0/k) F e^{i\omega t} \tag{6.2}$$

where F is a dimensionless function and k is a spring constant and can be equal to the static spring constant of system S.

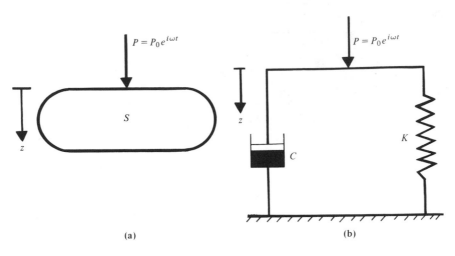

FIGURE 6.1 Typical linear dynamic system.

The dimensionless function can be expressed as

$$F = F_1 + iF_2 \tag{6.3}$$

where F_1 and F_2 are functions of the frequency ω. [Note the similarity between Eq. (6.2) and Eq. (5.1).]

Figure 6.1b shows a spring–dashpot analog for the massless system S. The equation of motion for the massless system shown in Figure 6.1b can be expressed in the form

$$C\dot{z} + Kz = P_0 e^{i\omega t} \tag{6.4}$$

From Eq. (6.2)

$$\dot{z} = (P_0/k)F(i\omega)e^{i\omega t} \tag{6.5}$$

Combining Eqs. (6.2), (6.4), and (6.5),

$$C(P_0/k)Fi\omega e^{i\omega t} + K(P_0/k)Fe^{i\omega t} = P_0 e^{i\omega t} \tag{6.6}$$

Substituting Eq. (6.3) into Eq. (6.6), and separating the real and imaginary parts,

$$-\omega F_2 C + F_1 K = k \tag{6.7}$$

and

$$\omega F_1 C + F_2 K = 0 \tag{6.8}$$

Equations (6.7) and (6.8) can be solved to evaluate the expressions for K and C as

$$K = \left[F_1/(F_1^2 + F_2^2) \right] k \tag{6.9}$$

$$C = \left[-(F_2/\omega)/(F_1^2 + F_2^2) \right] k \tag{6.10}$$

Also

$$F_1 = Kk/(K^2 + \omega^2 C^2) \qquad (6.11)$$

and

$$F_2 = -\omega kC/(K^2 + \omega^2 C^2) \qquad (6.12)$$

If a rigid mass m is added to the system S as shown in Figure 6.2a, the equivalent mass–spring–dashpot analog for the system is as shown in Figure 6.2b. For this system

$$z = (Q_0/k)\overline{F}e^{i\omega t} \qquad (6.13)$$

where $\overline{F}$ is a function of frequency.

The equation of motion for the mass m can be given by

$$m\ddot{z} = Q_0 e^{i\omega t} - P_0 e^{i\omega t} \qquad (6.14)$$

From Eq. (6.2)

$$m\ddot{z} = -m(\omega^2/k)FP_0 e^{i\omega t} \qquad (6.15)$$

Again, from Eq. (6.13)

$$m\ddot{z} = -m(\omega^2/k)\overline{F}Q_0 e^{i\omega t} \qquad (6.16)$$

Combining Eqs. (6.14) and (6.15),

$$-(m\omega^2/k)FP_0 e^{i\omega t} = (Q_0 - P_0)e^{i\omega t}$$

or

$$P_0 = Q_0/[1 - (m\omega^2/k)F] \qquad (6.17)$$

FIGURE 6.2 $S + m$ system.

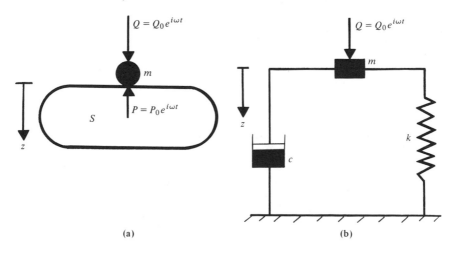

(a)   (b)

Combining Eqs. (6.15)–(6.17),

$$-m\frac{\omega^2}{k}\bar{F}Q_0 e^{i\omega t} = -m\frac{\omega^2}{k}F\left(\frac{Q_0}{1-(m\omega^2/k)F}\right)e^{i\omega t}$$

or

$$\bar{F} = F/\left[1-(m\omega^2/k)F\right] \qquad (6.18)$$

The displacement of the system shown in Figure 6.2a can now be given [from Eq. (6.13)] as

$$z = \frac{Q_0}{k}\bar{F}e^{i\omega t} = \frac{Q_0}{k}\frac{F}{1-(m\omega^2/k)F}e^{i\omega t} \qquad (6.19)$$

If the system $S + m$ (Figure 6.2a) is excited by a real force, i.e., the real part of $Q = Q_0 e^{i\omega t}$, then

$$Q = Q_0\cos\omega t \qquad (6.20)$$

The real part of the displacement [Eq. (6.19)] can then be obtained by the process of separation as

$$z = (Q_0/k)M\cos(\omega t + \alpha) \qquad (6.21)$$

where

$$M = |\bar{F}| = \sqrt{\frac{F_1^2 + F_2^2}{\left[1-(m\omega^2/k)F_1\right]^2 + \left[(m\omega^2/k)F_2\right]^2}} \qquad (6.22)$$

and

$$\alpha = \tan^{-1}\{F_2/\left[F_1 - (m\omega^2/k)(F_1^2 + F_2^2)\right]\} \qquad (6.23)$$

6.1.2 Rigid Circular Foundations

Based on the concept presented in Section 6.1.1, the relations for the vibration of a rigid circular foundation can now be developed. Figure 6.3a shows a massless rigid circular foundation resting on the surface of an elastic half space and being subjected to a force $P = P_0 e^{i\omega t}$. The periodic displacement of this foundation can be given by the equation

$$z = (P_0/k)Fe^{i\omega t} \qquad (6.24)$$

where $k = 4Gr_0/(1-\mu)$ is the static string constant, μ is Poisson's ratio, r_0 is the radius of the foundation, and $F = F_1 + iF_2$ is a dimensionless function. Thus

$$z = \frac{P_0}{4Gr_0/(1-\mu)}(F_1 + iF_2)e^{i\omega t}$$

$$= (P_0/Gr_0)\left[\tfrac{1}{4}(1-\mu)F_1 + \tfrac{1}{4}i(1-\mu)F_2\right]e^{i\omega t} \qquad (6.25)$$

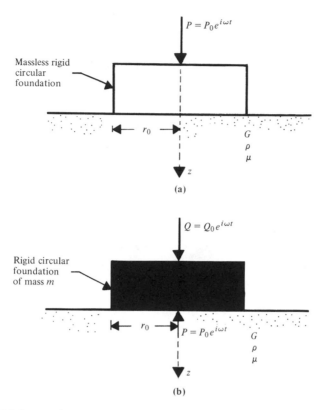

FIGURE 6.3 Lumped parameter system: rigid circular foundation vibration in vertical direction.

Comparing the preceding equation with Eq. (5.1), the dimensionless terms f_1 and f_2 in Eq. (5.1) are

$$f_1 = \tfrac{1}{4}(1 - \mu) F_1 \qquad (6.26)$$

and

$$f_2 = \tfrac{1}{4}(1 - \mu) F_2 \qquad (6.27)$$

The reason for using $F = F_1 + iF_2$ in Eq. (6.25) instead of the function $f = f_1 + if_2$ is that f_1 and f_2, which are functions of the dimensionless frequency a_0 [Eq. (5.5); $a_0 = \omega r_0 \sqrt{\rho/G}$], are highly dependent on the value of the Poisson's ratio μ. However, the values of $F_1 = 4f_1/(1 - \mu)$ and $F_2 = 4f_2/(1 - \mu)$ are practically single valued functions of the dimensionless frequency a_0. This is shown in Figure 6.4. The range of variation of F_1 and F_2 with μ falls within a narrow band.

As in the case of the system shown in Figure 6.1a, one can write the equation of motion of the massless rigid foundation as

$$C\dot{z} + Kz = P_0 e^{i\omega t}$$

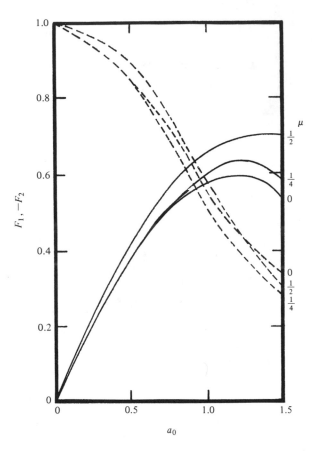

FIGURE 6.4 Plot of F_1 (---) and $-F_2$ (——) for a rigid circular foundation subjected to vertical vibration. [Lysmer, J., and Richart, F. E., Jr. (1966). "Dynamic Response of Footings to Vertical Loading," *Journal of the Soil Mechanics and Foundations Division*, *ASCE*, 92 (SM1), Fig. 8, p. 75.]

The values of C and K in the above equation can also be given by Eqs. (6.9) and (6.10) as

$$K = \left[F_1 / \left(F_1^2 + F_2^2 \right) \right] k = k_1 k \qquad (6.28)$$

where $k_1 = F_1 / (F_1^2 + F_2^2)$ and

$$C = \left[-(F_2 / \omega) / \left(F_1^2 + F_2^2 \right) \right] k \qquad (6.10)$$

However, dimensionless frequency

$$a_0 = \omega r_0 \sqrt{\rho / G} \qquad (5.5)$$

or

$$\omega = (a_0 / r_0) \sqrt{G / \rho} \qquad (6.29)$$

Substitution of Eq. (6.29) into Eq. (6.10) yields

$$C = \frac{-(F_2/a_0)}{F_1^2 + F_2^2} kr_0 \sqrt{\frac{\rho}{G}} = c_1 kr_0 \sqrt{\frac{\rho}{G}} \qquad (6.30)$$

where

$$c_1 = -(F_2/a_0)/(F_1^2 + F_2^2) \qquad (6.31)$$

If the rigid circular foundation under consideration has a mass m as shown in Figure 6.3b, the displacement can be given by the expression

$$z = (Q_0/k)\bar{F}e^{i\omega t} \qquad (6.13)$$

The equation of motion for the foundation of mass m can be given by an equation similar to Eq. (6.14). Thus

$$m\ddot{z} = Q_0 e^{i\omega t} - P_0 e^{i\omega t} \qquad (6.32)$$

However,

$$P_0 e^{i\omega t} = C\dot{z} + Kz = c_1 kr_0 \sqrt{\rho/G}\,\dot{z} + k_1 kz \qquad (6.33)$$

Thus

$$m\ddot{z} + c_1 kr_0 \sqrt{\rho/G}\,\dot{z} + k_1 kz = Q_0 e^{i\omega t} \qquad (6.34)$$

Substituting Eq. (6.13) into Eq. (6.34), the value of $\bar{F}$ can be obtained as

$$\bar{F} = \left[(k_1 - Ba_0^2) + ic_1 a_0\right]^{-1} \qquad (6.35)$$

where dimensionless mass ratio $B = \frac{1}{4}(1-\mu)m/\rho r_0^3$. It may be of some interest to compare Eq. (6.35) with Eq. (5.4):

$$B = \frac{1}{4}(1-\mu)b \qquad (6.36)$$

where $b = m/\rho r_0^3$ is the dimensionless mass ratio defined in Eq. (5.4).

As was shown in Eq. (6.21), the real part of the displacement of the foundation can be expressed as

$$z = (Q_0/k)M\cos(\omega t + \alpha)$$

where

$$M = |\bar{F}| = \sqrt{\left[(k_1 - Ba_0^2)^2 + (c_1 a_0)^2\right]^{-1}} \qquad (6.37a)$$

and

$$\alpha = \tan^{-1}\left[-c_1 a_0/(k_1 - Ba_0^2)\right] \qquad (6.37b)$$

Using the preceding two equations, Lysmer and Richart (1966) have shown that, for all practical purposes, satisfactory results can be obtained if the values of c_1 and k_1 are taken as 0.85 and 1, respectively. Thus, referring to Eq. (6.34),

$$m\ddot{z} + c_1 k r_0 \sqrt{\rho/G}\, \dot{z} \qquad\qquad + k_1 k z \qquad\qquad = Q_0 e^{i\omega t}$$

$$m\ddot{z} + (0.85)\left(\frac{4Gr_0}{1-\mu}\right) r_0 \sqrt{\frac{\rho}{G}}\, \dot{z} + (1)\left(\frac{4G}{1-\mu}\right) z = Q_0 e^{i\omega t}$$

or

$$m\ddot{z} + \frac{3.4}{1-\mu} r_0^2 \sqrt{G\rho}\, \dot{z} \qquad\qquad + \frac{4G}{1-\mu} z \qquad\qquad = Q_0 \cos \omega t \qquad (6.38)$$

If Eq. (6.38) is compared with the mass–spring–dashpot system (Figure 1.2b) equation given by Eq. (1.72), one can see that they are of the same form with

$$c = [3.4/(1-\mu)]\, r_0^2 \sqrt{\rho G} \qquad\qquad (6.39)$$

and

$$k = 4Gr_0/(1-\mu) = \text{static spring constant} \qquad (6.40)$$

Calculation Procedure for Foundation Design Using Eq. (6.38)

Once the equation of motion of a foundation is expressed in the form of Eq. (6.38), it is easy to obtain the resonant frequency and amplitude of vibration based on the mathematical derivations presented in Chapter 1 (Sections 1.2.4 and 1.2.5). The general procedure is outlined below.

Resonant Frequency

Calculation of natural frequency:

From Eq. (1.6)

$$f_n = \frac{1}{2\pi}\sqrt{\frac{k}{m}} = \frac{1}{2\pi}\sqrt{\left(\frac{4Gr_0}{1-\mu}\right)\frac{1}{m}} \qquad (6.41)$$

Calculation of the damping ratio D:

From Eq. (1.47)

$$c_c = 2\sqrt{km} = 2\sqrt{\left(\frac{4Gr_0}{1-\mu}\right)m}$$

$$= 8\sqrt{\frac{Gr_0}{1-\mu}\frac{B\rho r_0^3}{1-\mu}} = \frac{8r_0^2}{1-\mu}\sqrt{GB\rho} \qquad (6.42)$$

From Eq. (1.47b),

$$D = \frac{c}{c_c} = \frac{\text{Eq. (6.39)}}{\text{Eq. (6.42)}} = \frac{[3.4/(1-\mu)]\, r_0^2 \sqrt{G\rho}}{[8r_0^2/(1-\mu)]\sqrt{GB\rho}} = \frac{0.425}{\sqrt{B}} \qquad (6.43)$$

Calculation of the resonance frequency (i.e., frequency at maximum displacement):

From Eq. (1.86), for *constant force oscillation,*

$$f_m = f_n\sqrt{1-2D^2} = \frac{1}{2\pi}\sqrt{\left(\frac{4Gr_0}{1-\mu}\right)\frac{1}{m}} \cdot \sqrt{1-2\left(\frac{0.425}{\sqrt{B}}\right)^2} \qquad (6.44)$$

For rotating mass type excitation,

$$f_m = \frac{f_n}{\sqrt{1-2D^2}} = \frac{\frac{1}{2\pi}\sqrt{[4Gr_0/(1-\mu)](1/m)}}{\sqrt{1-2(0.425/\sqrt{B})^2}} \qquad (6.45)$$

Amplitude of Vibration at Resonance

The amplitude of vibration A_z at resonance for *constant force type oscillation* can be determined from Eq. (1.87) as

$$A_{z(res)} = (Q_0/k)(2D\sqrt{1-D^2})^{-1} \qquad (6.46)$$

where

$$k = 4Gr_0/(1-\mu), \qquad D = 0.425/\sqrt{B}$$

The amplitude of vibration for rotating mass type vertical excitation can be given as [See Eq. (1.99)]

$$A_{z(res)} = (U/m)(2D\sqrt{1-D^2})^{-1} \qquad (6.47)$$

where $U = m_1 e$ (and m_1 is the total rotating mass causing excitation).

Amplitude of Vibration at Frequencies Other than Resonance

Constant force type oscillator:

The plot given in Figure 1.9 has been reproduced again in Figure 6.5, which is a plot of $A_z/(Q_0/k)$ vs f/f_n (or ω/ω_n). So, with known values of D and f/f_n, one can determine the value of $A_z/(Q_0/k)$; and from that, A_z can be determined.

Rotating mass type vertical excitation:

The plot given in Figure 1.10b has been reproduced in Figure 6.6. This is a plot of $A_z/(U/m_1)$ vs f/f_n (or ω/ω_n). The value of A_z can be determined from this figure.

Example 6.1 Redo Example 5.2b–d by the method presented in this chapter. Assume that the Poisson's ratio for the soil supporting the foundation is 0.25.

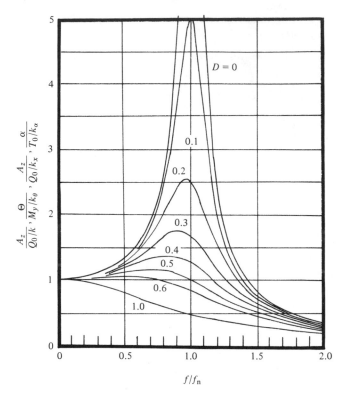

FIGURE 6.5 Plot of $A_z/(Q_0/k)$, $\Theta /(M_y/k_\theta)$, $A_x/(Q_0/k_x)$, $\alpha/(T_0/k_\alpha)$ against f/f_n for constant force oscillator.

Solution: The value of r_0 has been calculated in Example 5.2 to be 1.093 m.

  b. Resonant Frequency. Weight of foundation and engine

$W = 146.64$ kN

$B = \frac{1}{4}(1-\mu)m/\rho r_0^3 = \frac{1}{4}(1-0.25)\left[146.64/18.5(1.093)^3\right] = 1.14$

$D = 0.425/\sqrt{B} = 0.425/\sqrt{1.14} = 0.398$

$f_n = \dfrac{1}{2\pi}\sqrt{\left(\dfrac{4Gr_0}{1-\mu}\right)\dfrac{1}{m}} = \dfrac{1}{2\pi}\sqrt{\dfrac{(4)(18,000)(1.093)}{(1-0.25)}\left(\dfrac{9.81}{146.64}\right)}$

$= 13.33$ cps

$f_m = f_n/\sqrt{1-2D^2} = 13.33/\sqrt{1-2(0.398)^2} = 16.16$ cps $= 968$ cpm

  Note. One could also have found the resonant frequency f_m from Figure 6.6 by looking to the peak point of the curve with $D = 0.398$. The abscissa corresponding to the peak point gives $f/f_n = f_m/f_n = 1.21$, so $f_m = 1.21(13.33) = 16.13$ cps.

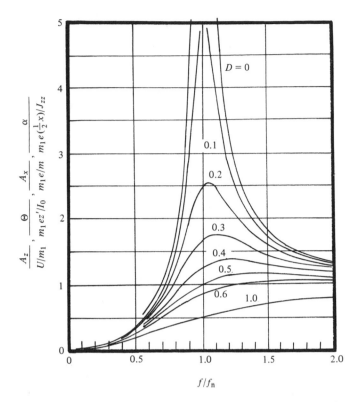

FIGURE 6.6 Plot of $A_z/(U/m_1)$, $\Theta/(m_1 ez'/I_0)$, $A_x/(m_1 e/m)$, $\alpha/m_1 e(\frac{1}{2}x)/J_{zz}$ against f/f_n for rotating mass type excitation.

c. Amplitude of Vibration at Resonance. From Eq. (6.47),

$$A_{z(\text{res})} = (U/m)\left(2D\sqrt{1-D^2}\right)^{-1}$$

The total unbalanced force at 1500 cpm has been calculated to be 12.74 kN (see Problem 5.2). Hence, Q_0 at 968 cpm is

$$Q_{0(968 \text{ cpm})} = Q_{0(1500 \text{ cpm})}(968/1500)^2$$

$$= 12.74(968/1500)^2 = 5.31 \text{ kN}$$

$$Q_0 = m_1 e\omega^2 = U\omega^2$$

$$5.31 = U(2\pi \cdot 968/60)^2$$

$$U = 5.31(60/2\pi \cdot 968)^2$$

$$A_{z(\text{res})} = \left[(5.31)\left(\frac{60}{2\pi \cdot 968}\right)^2\left(\frac{9.81}{146.64}\right)\right]\left(\frac{1}{2(0.398)\sqrt{1-(0.398)^2}}\right)$$

$$= 0.0000473 \text{ m} = 0.0473 \text{ mm}$$

Note. This could have also been determined from Figure 6.6 as follows:

$$f_m / f_n = 1.21$$

For $D = 0.398$ and $f_m / f_n = 1.21$,

$$A_z / (U/m_1) = 1.37$$

$$A_z = 1.37 \left(\frac{U}{m_1} \right) = 1.37 \left[5.31 \left(\frac{60}{2\pi \cdot 968} \right)^2 \right] \left(\frac{9.81}{146.64} \right) = 0.0474 \text{ mm}$$

d. Amplitude of Vibration at Operating Frequency

$$f_{(oper)} = 1500 \text{ cpm}$$

$$f_n = 13.33 \times 60 = 800 \text{ cpm}$$

$$f_{(oper)} / f_n = 1500/800 = 1.875$$

From Figure 6.6, for $f_{(oper)} / f_n = 1.875$, $D = 0.398$, and

$$A_z / (U/m_1) \approx 1.18$$

Thus

$$A_z = 1.18(U/m_1)$$

$$U = \frac{Q_{0(oper)}}{\omega^2} = \frac{12.74}{(2\pi \cdot 1500/60)^2} = 0.000516$$

$$A_z = 1.18(0.000516)(9.81/146.64) = 0.0000407 \text{ m} = 0.0407 \text{ mm}$$

6.2 ROCKING OSCILLATION OF RIGID CIRCULAR FOUNDATIONS

Hall (1967) investigated the problem of developing the lumped parameters for rocking oscillation of rigid circular foundations in the same manner as Lysmer and Richart's work (1966) did for vertical oscillation. According to Hall, the equation of motion for a rocking foundation can be given by (Figure 6.7)

$$I_0 \ddot{\theta} + c_\theta \dot{\theta} + k_\theta \theta = M_y e^{i\omega t} \tag{6.48}$$

where θ is the rotation of the vertical axis of the foundation at any time t, I_0 is the mass moment of inertia about the axis y (through its base) given by Eq. (5.20),

spring constant $\quad k_\theta = \tfrac{8}{3} G r_0^3 / (1 - \mu) \tag{6.49}$

dashpot coefficient $\quad c_\theta = 0.8 r_0^4 \sqrt{G} / [(1 - \mu)(1 + B_i)] \tag{6.50}$

where

inertia ratio $\quad B_i = \tfrac{3}{8}(1 - \mu) I_0 / \rho r_0^5 \tag{6.51}$

Compare the inertia ratio defined by Eq. (6.51) with that defined by Eq.

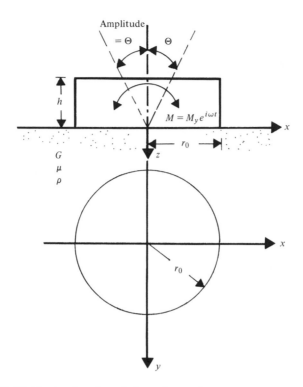

FIGURE 6.7 Rigid circular foundation: lumped parameter system for rocking oscillation.

(5.19). Note

$$B_i = \tfrac{3}{8}(1-\mu)b_i \qquad (6.52)$$

where b_i is the inertia ratio defined by Eq. (5.19).

Calculation Procedure for Foundation Design Using Eq. (6.48)

Resonant Frequency

Calculate the natural frequency:

$$f_n = (1/2\pi)\sqrt{k_\theta/I_0} \qquad (6.53)$$

Calculate the critical damping ratio D:

$$c_{\theta c} = 2\sqrt{k_\theta I_0}$$

$$D = c_\theta/c_{\theta c} = 0.15/\sqrt{B_i}\,(1+B_i) \qquad (6.54)$$

Calculate the resonant frequency:

$$f_m = f_n \sqrt{1 - 2D^2} \qquad \text{for constant force excitation}$$

$$f_m = f_n / \sqrt{1 - 2D^2} \quad \text{for rotating mass type excitation}$$

Amplitude of Vibration at Resonance

$$\Theta_{res} = \left(M_y / k_\theta \right) \left(2D\sqrt{1 - D^2} \right)^{-1} \quad \text{for constant force oscillation}$$

$$(6.55)$$

Refer to Figure 6.5 to obtain Θ_{res}.

$$\Theta_{res} = \left(m_1 e z' / I_0 \right) \left(2D\sqrt{1 - D^2} \right)^{-1} \quad \text{for rotating mass type excitation case}$$

$$\text{(see Figure 6.8)} \qquad (6.56)$$

where m_1 is the total rotating mass causing excitation and e is the eccentricity of each mass. Refer to Figure 6.6 to obtain Θ_{res}.

Amplitude of Vibration at Frequencies Other than Resonance

For constant force type oscillator:

Calculate f / f_n and refer to Figure 6.5 to obtain $\Theta / (M_y / k_\theta)$.

For rotating mass type excitation:

Calculate f / f_n and refer to Figure 6.6 to obtain $\Theta / (m_1 e z' / I_0)$.

Example 6.2 Redo Example 5.3. Assume the Poisson's ratio of the soil to be 0.35.

FIGURE 6.8

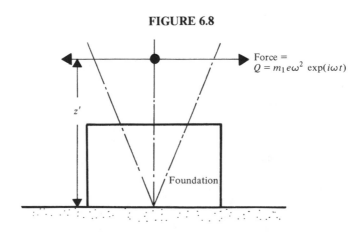

Solution

a. Determination of Resonant Frequency. From Eq. (6.49)

$$k_\theta = \frac{\frac{8}{3}Gr_0^3}{(1-\mu)} = \frac{8(18,000)(3.67)^3}{3(1-0.35)} = 3650279 \text{ kN-m/rad}$$

$$f_n = \frac{1}{2\pi}\sqrt{\frac{k_\theta}{I_0}} = \frac{1}{2\pi}\sqrt{\frac{3650279 \times 10^3 \text{ N-m/rad}}{36.768 \times 10^5}} = 5.01 \text{ cps} = 300 \text{ cpm}$$

$$B_i = \frac{3}{8}(1-\mu)\frac{I_0}{\rho r_0^5} = \frac{3(1-0.35)}{8}\frac{36.768 \times 10^5}{1800(3.67)^5} = 0.748$$

$$D = \frac{0.15}{\sqrt{B_i}(1+B_i)} = \frac{0.15}{\sqrt{0.748}(1+0.748)} = 0.099$$

$$f_m = \frac{f_n}{\sqrt{1-2D^2}} = \frac{300}{\sqrt{1-2D^2}} = \frac{300}{\sqrt{1-2(0.099)^2}} = 303 \text{ cpm}$$

b. Calculation of Amplitude of Oscillation at Resonance

$$f/f_n = f_m/f_n = 303/300 = 1.01$$

$$M_{y(\text{oper freq})} = 120 \text{ kN-m (see Problem 5.3)}$$

$$M_{y(\text{res})} - 120(f_m/f_{\text{oper}})^2 - 120(303/600)^2 = 30.6 \text{ kN-m}$$

$$(m_1 e\omega^2)z' = M_y$$

$$\omega_{\text{res}} = 2\pi \cdot 303/60 = 31.73 \text{ rad/sec}$$

$$m_1 ez' = M_y/\omega^2 = 30.6 \times 10^3 \text{ N-m}/(31.73)^2 = 0.0304 \times 10^3$$

From Figure 6.6, for $f/f_n = 1.01$, $\Theta/(m_1 ez'/I_0) = 5$. Thus

$$\Theta = 5(m_1 ez'/I_0) = 5(0.0304 \times 10^3)/(36.768 \times 10^5) = 0.000041 \text{ rad}$$

6.3 SLIDING OSCILLATION OF RIGID CIRCULAR FOUNDATIONS

Hall (1967) has provided the lumped parameters for the case of sliding oscillation of a rigid circular foundation (Figure 6.9). This equation of motion can be given in the form

$$m\ddot{x} + c_x \dot{x} + k_x x = Q_0 e^{i\omega t} \tag{6.57}$$

where m is the mass of the foundation, static spring constant for sliding

$$k_x = 32(1-\mu)Gr_0/(7-8\mu) \tag{6.58}$$

and dashpot coefficient for sliding

$$c_x = [18.4(1-\mu)/(7-8\mu)]r_0^2\sqrt{\rho G} \tag{6.59}$$

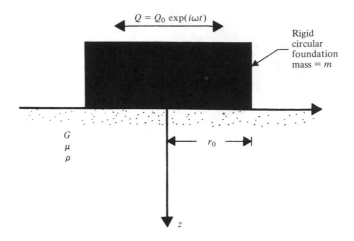

FIGURE 6.9 Sliding oscillation of rigid circular foundation.

Based on Eqs. (6.57)–(6.59), the natural frequency of the foundation for sliding can be calculated as

$$f_n = \frac{1}{2\pi} \sqrt{\frac{k_x}{m}} = \sqrt{\frac{32(1-\mu)Gr_0}{(7-8\mu)m}} \tag{6.60}$$

The critical damping and damping ratio in sliding can be evaluated as

$$c_{xc} = 2\sqrt{k_x m}$$
$$= 2 \cdot \sqrt{32(1-\mu)Gr_0 m /(7-8\mu)} \tag{6.61}$$

$$D = c_x/c_{xc}$$
$$= \frac{[18.4(1-\mu)/(7-8\mu)]r_0^2\sqrt{\rho G}}{2\sqrt{32(1-\mu)Gr_0 m/(7-8\mu)}} = 0.288/\sqrt{B_x} \tag{6.62}$$

where dimensionless mass ratio

$$B_x = [(7-8\mu)/32(1-\mu)]m/\rho r_0^3 \tag{6.63}$$

Calculation Procedure for Foundation Design Using Eq. (6.57)

Resonant Frequency

1. Calculate natural frequency f_n using Eq. (6.60).
2. Calculate critical damping ratio D by using Eq. (6.62). [Note: B_x can be obtained from Eq. (6.63).]
3. For constant force excitation (i.e., $Q_0 = $ const), calculate

$$f_m = f_n\sqrt{1-2D^2}$$

4. For rotating mass type excitation, calculate

$$f_m = f_n / \sqrt{1 - 2D^2}$$

Amplitude of Vibration at Resonance

1. For constant force excitation, amplitude of vibration at resonance

$$A_{x(\text{res})} = (Q_0/k_x)(2D\sqrt{1-D^2})^{-1} \qquad (6.64)$$

2. For rotating mass type excitation,

$$A_{x(\text{res})} = (m_1 e/m)(2D\sqrt{1-D^2})^{-1} \qquad (6.65)$$

where m_1 is the total rotating mass causing excitation and e is the eccentricity of each rotating mass.

Amplitude of Vibration at Frequency Other than Resonance

For constant force excitation, calculate f/f_n and refer to Figure 6.5 to obtain $A_x/(Q_0/k_x)$. For rotating mass type excitation, calculate f/f_n and refer to Figure 6.6 to obtain $A_x/(m_1 e/m)$.

6.4 LUMPED PARAMETER SYSTEM FOR TORSIONAL OSCILLATION OF RIGID CIRCULAR FOUNDATIONS

Similar to the cases of vertical, rocking, and sliding modes of oscillation, the equation of motion for the torsional oscillation of a rigid circular foundation (Figure 6.10) can be written as

$$J_{zz}\ddot{\alpha} + c_\alpha \dot{\alpha} + k_\alpha \alpha = T_0 e^{i\omega t} \qquad (6.66)$$

where J_{zz} is the mass moment of inertia of the foundation about the axis $z-z$, c_α is the dashpot coefficient for torsional vibration, the static spring constant for torsional vibration

$$k_\alpha = \tfrac{16}{3} G r_0^3 \qquad (6.67)$$

and α is the rotation of the foundation at any time due to the application of a torque $T = T_0 e^{i\omega t}$.

The damping ratio D for this mode of vibration has been determined as (Richart et al., 1970)

$$D = 0.5/(1 + 2B_t) \qquad (6.68)$$

where the dimensional mass ratio for torsion

$$B_t = J_{zz}/\rho r_0^5 \qquad (6.69)$$

By comparing Eqs. (6.69) and (5.27), it can be seen that $B_t = b_t$. This is due to the fact that Poisson's ratio does not influence the torsional vibration of foundations.

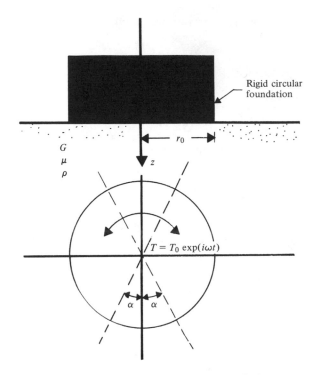

FIGURE 6.10 Torsional oscillation of rigid circular foundation.

Calculation Procedure for Foundation Design Using Eq. (6.66)

Resonant Frequency

1. Calculate the natural frequency of the foundation as

$$f_n = (1/2\pi)\sqrt{k_\alpha/J_{zz}} \qquad (6.70)$$

2. Calculate B_t by using Eq. (6.69) and then D by using Eq. (6.68).
3. For constant force excitation (i.e., $T_0 = \text{const}$),

$$f_m = f_n\sqrt{1 - 2D^2}$$

For rotating mass type excitation

$$f_m = f_n/\sqrt{1 - 2D^2}$$

Amplitude of Vibration at Resonance

For constant force excitation, the amplitude of vibration at resonance is

$$\alpha_{res} = (T_0/k_\alpha)\left(2D\sqrt{1 - D^2}\right)^{-1} \qquad (6.71)$$

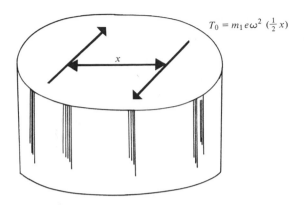

$$T_0 = m_1 e \omega^2 \left(\tfrac{1}{2} x \right)$$

FIGURE 6.11

For rotating mass type excitation

$$\alpha_{\mathrm{res}} = \left[m_1 e \left(\tfrac{1}{2} x \right) / J_{zz} \right] \left(2 D \sqrt{1 - D^2} \right)^{-1} \tag{6.72}$$

For the definition of x in Eq. (6.72), see Figure 6.11.

Amplitude of Vibration at Frequency Other than Resonance

For constant force excitation, calculate f/f_n and then refer to Figure 6.5 to obtain $\alpha/(T_0/k_\alpha)$. For rotating mass type excitation, calculate f/f_n and then refer to Figure 6.6 to obtain $\alpha/[m_1 e (\tfrac{1}{2} x)/J_{zz}]$.

6.5 COMMENTS ON THE LUMPED PARAMETERS USED FOR SOLVING FOUNDATION VIBRATION PROBLEMS

The equations for the lumped parameter systems for various modes of vibration of rigid circular foundations developed in the preceding sections may be summarized as follows:

$$m\ddot{z} + c\dot{z} + kz = Q_0 e^{i\omega t} \quad \text{for vertical oscillation} \tag{6.38}$$

$$I_0 \ddot{\theta} + c_\theta \dot{\theta} + k_\theta \theta = M_y e^{i\omega t} \quad \text{for rocking oscillation} \tag{6.48}$$

$$m\ddot{x} + c_x \dot{x} + k_x x = Q_0 e^{i\omega t} \quad \text{for sliding oscillation} \tag{6.57}$$

$$J_{zz} \ddot{\alpha} + c_\alpha \dot{\alpha} + k_\alpha \alpha = T_0 e^{i\omega t} \quad \text{for torsional oscillation} \tag{6.66}$$

The mathematical approach for solution of the above equations is similar for determination of the natural frequency, resonant frequency, critical damping, damping ratio, and the amplitudes of vibration at various frequencies. However, the agreement of these solutions with the field conditions depends on proper choice of the lumped parameters (m, I_0, J_{zz}, c, c_θ, c_x, c_α, k, k_θ, k_x, k_α). In this next subsection, we take a critical look at these parameters.

Choice of Mass and Mass Moment of Inertia

The mass terms m used in Eqs. (6.38) and (6.57) are actually the sum of

1. Mass of the structural foundation block m_f and
2. Mass of all the machineries mounted on the block m_m

However, during the vibration of foundations, there is a mass of soil under the foundation which vibrates along with the foundation. Thus, it would be reasonable to consider the term m in Eqs. (6.38) and (6.57) to be the sum of

$$m = m_f + m_m + m_s \qquad (6.73)$$

where m_s is the effective mass of soil vibrating with the foundation.

In a similar manner, the mass moment of inertia terms I_0 and J_{zz} included in Eqs. (6.48) and (6.66), includes the contributions of the mass of the foundation and that of the machine mounted on the block. It appears reasonable to add also the contribution of the *effective mass* of the vibrating soil m_s, i.e., the *effective soil mass moment of inertia*. Thus

$$I_0 = I_{0(\text{foundat})} + I_{0(\text{mach})} + I_{0(\text{eff soil mass})} \qquad (6.74)$$

and

$$J_{zz} = J_{zz(\text{foundat})} + J_{zz(\text{mach})} + J_{zz(\text{eff soil mass})} \qquad (6.75)$$

Theoretically, calculated values of m_s, $I_{0(\text{eff soil mass})}$, and $J_{zz(\text{eff soil mass})}$ are given by Hsieh (1962). They are as follows:

Poisson's ratio μ	Vertical oscillation m_s
0.0	$0.5\rho r_0^3$
0.25	$1.0\rho r_0^3$
0.5	$2.0\rho r_0^3$

Poisson's ratio μ	Horizontal oscillation m_s
0.0	$0.2\rho r_0^3$
0.25	$0.2\rho r_0^3$
0.5	$0.1\rho r_0^3$

Rocking oscillation: Poisson's ratio $\mu = 0$; $I_{0(\text{eff soil mass})} = 0.4\rho r_0^5$

Poisson's ratio μ	Torsional oscillation $J_{zz(\text{eff soil mass})}$
0.00	$0.3\rho r_0^5$
0.25	$0.3\rho r_0^5$
0.5	$0.3\rho r_0^5$

In most cases, engineers neglect the contribution of the effective soil mass. In general, this will lead to answers which are *within 30% accuracy*.

Choice of Spring Constants

In Eqs. (6.38), (6.48), (6.57), and (6.66), the spring constants defined were for the cases of *rigid circular foundations*. In example problems where rigid rectangular foundations were encountered, the equivalent radii r_0 were first determined. These values of r_0 were then used to determine the value of the spring constants. However, more exact solutions for spring constants for rectangular foundations derived from the theory of elasticity can be used. These are given in Table 6.1 along with those for circular foundations.

Another fact that also needs to be kept in mind is that the foundation blocks are never placed at the surface. If the bottom of the foundation block is placed at a depth z measured from the ground surface, the spring constants will be higher than that calculated by theory. This fact is demonstrated in Figure 6.13 which is for the case of vertical motion of rigid circular foundations (Kaldjian, 1969). In Figure 6.13, curve 1 is for the case where the sides of the foundation adhere to the vertical surface; curve 2 is for the case where they do not.

Besides the above technique for obtaining the spring constant (which is based on the theory of elasticity), Whitman and Richart (1967) have suggested three other alternatives:

1. Use of small-scale plate bearing tests (to obtain k)
2. Use of small-scale vibrating foundation (to obtain k)

TABLE 6.1 Values of Spring Constants for Rigid Foundations[a]

Motion	Spring constant	Reference
	Circular foundations	
Vertical	$k = 4Gr_0/(1-\mu)$	Timoshenko and Goodier (1951)
Horizontal (sliding)	$k_x = 32(1-\mu)Gr_0^3/(7-8\mu)$	Bycroft (1956)
Rocking	$k_\theta = \frac{8}{3}Gr_0^3/(1-\mu)$	Borowicka (1943)
Torsion	$k_\alpha = \frac{16}{3}Gr_0^3$	Reissner and Sagoci (1944)
	Rectangular foundations	
Vertical	$k = [G/(1-\mu)]F_z\sqrt{BL}$	Barkan (1962)
Horizontal (sliding)	$k_x = 2(1+\mu)GF_x\sqrt{BL}$	Barkan (1962)
Rocking[b]	$k_\theta = [G/(1-\mu)]F_\theta BL^2$	Gorbunov-Possadov and Serebrajanyi (1961)

[a]Whitman, R. V., and Richart, F. E., Jr. (1967). "Design Procedure for Dynamically Loaded Foundations," *Journal of the Soil Mechanics and Foundations Division, ASCE*, 93 (SM6), Table 4, p. 182.
[b]For definition of B and L, refer to Figure 5.14. Refer to Figure 6.12 for values of F_z, F_x, and F_θ.

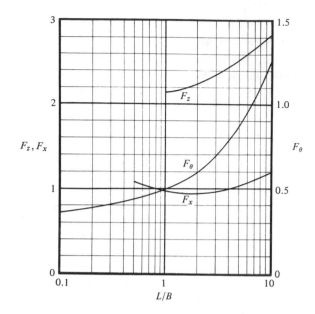

FIGURE 6.12 Plot of F_z, F_x, and F_θ vs L/B. [Whitman, R. V., and Richart, F. E., Jr. (1967). "Design Procedure for Dynamically Loaded Foundations," *Journal of the Soil Mechanics and Foundations Division*, *ASCE*, 93 (SM6), Fig. 4, p. 183.]

FIGURE 6.13 Variation of spring constant with depth of embedment of foundation. [Kaldjian, M. H. (1969). "Discussion on the Design Procedures for Dynamically Loaded Foundation," *Journal of the Soil Mechanics and Foundations Division*, *ASCE*, 95 (SM1), Fig. 9, p. 365.]

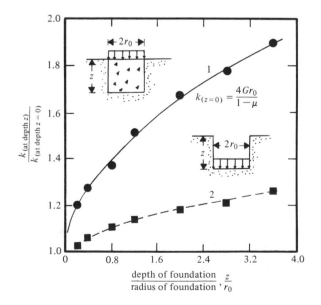

3. Use of the concept of *elastic subgrade modulus* combined with empirical tables or charts (to obtain $k, k_\theta, k_x, k_\alpha$)

The above three alternative procedures are briefly described below.

Small-Scale Plate Bearing Test

This is similar to the plate bearing tests conducted in the field for evaluation of allowable bearing capacity of soils for foundation design purposes. The key here is to use a *load per unit area* (live load + dead load) similar to that expected on the prototype foundation. Also, the load should be cycled several times before measuring the load–deformation relations. The cycling of load should be done as fast as possible to avoid consolidation and secondary compression effects, but should be slow enough to take the readings; an average cycle should take about 10–20 min. The spring constant for the test plate may be given as

$$k_{\text{plate}} = qA/z \qquad (6.76)$$

where q is the load per unit area applied on the plate, A is the area of the plate, and z is the vertical deformation due to the applied load.

The spring constant for vertical loading obtained from plate load tests given by Eq. (6.76) has to be extrapolated to obtain the spring constant for the prototype foundation. This is done as follows (Terzaghi, 1955):

Cohesive soils:

$$k_{\text{foundat}} = k_{\text{plate}} \cdot \left(\frac{\text{foundation width}}{\text{plate width}} \right) \qquad (6.77)$$

Cohesionless soils:

$$k_{\text{foundat}} = k_{\text{plate}} \cdot \left(\frac{\text{foundation width} + \text{plate width}}{2 \times \text{plate width}} \right)^2 \qquad (6.78)$$

Use of Small-Scale Vibrating Footing Test

This method is similar to plate bearing tests. It consists of setting up a small vibrator on a 12–30-in. (0.305–0.762-m) diameter plate operating at various frequencies until resonance frequency is found. The equivalent spring constant can then be back-calculated as

$$f_n = (1/2\pi)\sqrt{k/m}$$

or

$$k = (2\pi f_n)^2 m \qquad (6.79)$$

There are, however, some difficulties in this procedure that should be kept in mind:

1. With small plate vibration, the *effective soil mass* vibrating in phase with the plate becomes important and should be taken into account.

2. In order to obtain the amplitude of vibration large enough for measurement, fairly high accelerations are necessary. However, it should be kept below 0.5 g; otherwise, the soil will be highly stressed and nonlinear effects will be significant.
3. For small plate vibration, radiation damping will probably be large. This may result in imprecise determination of the resonant frequency.
4. The value of k determined from these tests has to be extrapolated to the prototype foundation condition in the manner described previously for plate bearing tests.

Use of the Concept of Elastic Subgrade Modulus, Combined with Empirical Tables or Charts

Barkan (1962) [as quoted by Whitman and Richart (1967)] has suggested that, using the concept of subgrade modulus, the spring constant equations for various modes of vibration can be written as follows:

$$k = S_z A \qquad \text{for vertical oscillation} \tag{6.80}$$

$$k_x = S_x A \qquad \text{for horizontal oscillation} \tag{6.81}$$

$$k_\theta = S_\theta I_0 \qquad \text{for rocking oscillation} \tag{6.82}$$

$$k_\alpha = S_\alpha J_{zz} \qquad \text{for torsional oscillation} \tag{6.83}$$

where A is the area of the foundation and $S_z, S_x, S_\theta, S_\alpha$ are subgrade moduli for vertical, sliding, rocking, and torsional oscillations, respectively.

The approximate relation of S_x, S_θ, and S_α in terms of S_z can be given as

$$S_x = \tfrac{1}{2} S_z \tag{6.84}$$

$$S_\theta = 2 S_z \tag{6.85}$$

$$S_\alpha = 1.5 S_z \tag{6.86}$$

The recommended values of S_z are given in Table 6.2.

It needs to be pointed out that these coefficients $(S_z, S_x, S_\theta, S_\alpha)$ are actually dependent on the soil type and the geometry of the foundation. However, here they are assumed to be the function solely the soil type. Hence, this method may be used for preliminary design purposes only.

Choice of Poisson's Ratio

Whitman and Richart (1967) recommended the following values for Poisson's ratio:

Sand (dry, moist, partially saturated) $\mu = 0.35$–0.4

Clay (saturated) $\mu = 0.5$

A good value for most partially saturated soils is about 0.4.

TABLE 6.2 Recommended Design Values[a] of S_z

Soil group	Allowable static bearing stress[b]		S_z	
	ton/ft^2	kN/m^2	ton/ft^3	kN/m^3
Weak soil (clay and silty clays with sand in a plastic state; clayey and silty sands; also soils of Categories II and III with laminae of organic silts and of peat)	1.5	150	95	29,900
Soils of medium strength (clays and silty clays with sand, close to the plastic limit; sand)	1.5–3.5	150–340	95–155	29,900–48,800
Strong soils (clays and silty clays with sand of hard consistency; gravels and gravelly sands; loess and loessial soils)	3.5–5	340–480	155–310	48,800–97,500
Rocks	5	480	310	97,500

[a]Barkan (1962); from Whitman, R. V., and Richart, F. E., Jr. (1967). "Design Procedure for Dynamically Loaded Foundations," *Journal of the Soil Mechanics and Foundations Division, ASCE*, 93 (SM6), Table 6, p. 188.
[b]Metric figures are rounded off.

Choice of Damping Ratio

In Chapter 3, two types of damping in soils, i.e. *geometric (radiation) damping* and *internal damping*, were discussed. The *radiation damping* depends on parameters such as Poisson's ratio, mass of the foundation, equivalent radius, and the density of the soil. The relations for the damping ratio, which are given in Eqs. (6.43), (6.54), (6.62), and (6.68), are for *radiation damping only*.

Table 6.3 summarizes some available information regarding internal damping of soils at the level of stress changes occurring under a machine foundation. From the values presented in Table 6.3, it can be seen that the damping ratio due to internal damping D_i varies over a wide range (i.e., 0.01–0.1). Thus, an average value of $D_i = 0.05$ would be a good estimate. The damping ratio can then be approximated as

$$D = D_{rad} + 0.05$$

For vertical and sliding motions, the contribution of internal damping can be somewhat neglected. However, for torsional and rocking modes of oscillation, the contribution of the internal damping may be too large to be ignored.

TABLE 6.3 Internal Damping in Soils[a]

Type of soil	Equivalent internal damping ratio D_i	Reference
Dry sand and gravel	0.03–0.07	Weissmann and Hart (1961)
Dry and saturated sand	0.01–0.03	Hall and Richart (1963)
Dry sand	0.03	Whitman (1963)
Dry and saturated sands and gravels	0.05–0.06	Barkan (1962)
Clay	0.02–0.05	Barkan (1962)
Silty sand	0.03–0.1	Stevens (1966)
Dry sand	0.01–0.03	Hardin (1965)

[a]Whitman, R. V., and Richart, F. E., Jr. (1967). "Design Procedure for Dynamically Loaded Foundations," *Journal of the Soil Mechanics and Foundations Division*, *ASCE*, 93 (SM6), Table 2, p. 178.

6.6 COUPLED MOTION FOR ROCKING AND SLIDING TYPE OF OSCILLATION

In several cases of machine foundation, the rocking and sliding oscillations are coupled. This is because of the fact that the center of gravity of the footing and oscillators are not coincident with the center of sliding resistance. This can be seen from Figure 6.14a. This is a case of vibration of foundation with two degrees of freedom and similar to the types described in Section 1.3. The derivation for the coupled motion for rocking and sliding given below is based on the analysis given by Richart and Whitman (1967). It can be seen that the nature of foundation motion shown in Figure 6.14a is equal to the sum of the sliding motion shown in Figure 6.14b and the rocking motion shown in Figure 6.14c. Note that

$$x_b = x_g - h\theta \tag{6.87}$$

For the sliding motion

$$m\ddot{x}_g = P \tag{6.88}$$

where the horizontal resistance to sliding

$$P = -c_x(dx_b/dt) - k_x x_b \tag{6.89}$$

Substituting Eq. (6.87) into (6.89), we get

$$P = -c_x d(x_g - h\theta)/dt - k_x(x_g - h\theta)$$
$$= -c_x\dot{x}_g + c_x h\dot{\theta} - k_x x_g + k_x h\theta \tag{6.90}$$

Now, combining Eqs. (6.88) and (6.90)

$$m\ddot{x}_g + c_x\dot{x}_g + k_x x_g - c_x h\dot{\theta} - k_x h\theta = 0 \tag{6.91}$$

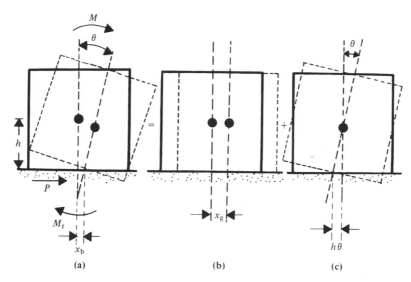

FIGURE 6.14 Coupled rocking and sliding oscillation. [Richart, F. E., Jr., and Whitman, R. V. (1967). "Comparison of Footing Vibration Tests with Theory," *Journal of the Soil Mechanics and Foundations Division, ASCE*, 93 (SM6), Fig. 5, p. 150.]

For rocking motion about the center of gravity

$$I_g \ddot{\theta} - M + M_r - hP \tag{6.92}$$

where I_g is the mass moment of inertia about the horizontal axis passing through the center of gravity (at right angles to the cross section shown) and M_r is the soil resistance to rotational motion. However,

$$M_r = -c_\theta \dot{\theta} - k_\theta \theta \tag{6.93}$$

Substitution of Eqs. (6.90) and (6.93) into Eq. (6.92) gives

$$I_g \ddot{\theta} = M - \left(c_\theta \dot{\theta} + k_\theta \theta \right) - h \left(-c_x \dot{x}_g + c_x h \dot{\theta} - k_x x_g + k_x h \theta \right)$$

or

$$I_g \ddot{\theta} + \left(c_\theta + c_x h^2 \right) \dot{\theta} + \left(k_\theta + k_x h^2 \right) \theta - h \left(c_x \dot{x}_g + k_x x_g \right) = M \tag{6.94}$$

Let

$$x_g = A_{x_1} \sin \omega t + A_{x_2} \cos \omega t \tag{6.95}$$

$$\theta = \Theta_1 \sin \omega t + \Theta_2 \cos \omega t \tag{6.96}$$

$$M = M_y \sin \omega t \tag{6.97}$$

Equations (6.95)–(6.97) can now be substituted into Eqs. (6.91) and (6.94). This establishes four equations and four unknowns. The solution of four simultaneous equations at each value of the frequency provides evaluation of the response.

6.7 POSSIBLE USES OF PILES

In Section 5.3 it was mentioned that, for low-speed machineries subjected to vertical oscillation, the natural frequency of the foundation–soil system should be at least twice the operating frequency. In the design of these types of foundation, if changes in size and mass of the foundation do not lead to a satisfactory design, or there exists a layer of hard soil or rock within a reasonable depth, a pile foundation may be tried. A pile foundation extending to rock will tend to stiffen and thus increase the fundamental natural frequency. This may also reduce the maximum vertical displacement.

The benefit derived from the use of piles depends on several factors such as the type of piles to be used, the length of piles, and the portion of load carried by each pile. Figure 6.15 (see insert) shows a pile driven up to a rock layer. The length of the pile is equal to L. The load on the pile coming from the foundation is W_0. This problem can be treated as a vertical rod *fixed at the base* (i.e., at the rock layer) and *free on top*. For the determination of the natural frequency of vertical vibration of the pile, three possible cases may arise:

1. If W_0 is very small (≈ 0), the natural frequency of pile vibration can be given by

$$f_{n(P)} = (1/4L)\sqrt{E_P/\rho_P} \qquad (2.42)$$

FIGURE 6.15 Plot of Eq. (6.99).

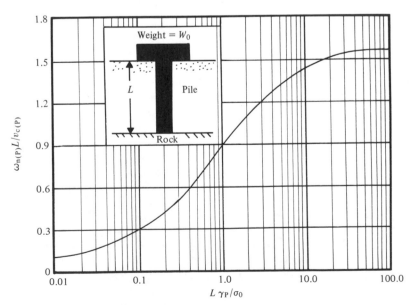

where E_P and ρ_P are the Young's modulus and density of the pile material.

2. If W_0 is of the same order of magnitude as the weight of the pile, the natural frequency of vibration can be given by Eq. (2.56). (Note the similarity of end conditions between Figure 2.10a and Figure 6.15.) Thus

$$\frac{AL\gamma_P}{W_0} = \frac{\omega_{n(P)}L}{v_{c(P)}}\tan\left(\frac{\omega_{n(P)}L}{v_{c(P)}}\right) \tag{6.98}$$

or

$$\frac{L\gamma_P}{\sigma_0} = \frac{\omega_{n(P)}L}{v_{c(P)}}\tan\left(\frac{\omega_{n(P)}L}{v_{c(P)}}\right) \tag{6.99}$$

where A is the area of cross section of the pile, γ_P is the unit weight of the pile material, $\omega_{n(P)}$ is the natural circular frequency of the pile–mass system, $v_{c(P)}$ is the longitudinal wave propagation velocity in the pile, and $\sigma_0 = W_0/A$.

Figure 6.15 is a plot of $\omega_{n(P)}L/v_{c(P)}$ against $L\gamma_P/\sigma_0$ and can be used to determine $\omega_{n(P)}$ and then $f_{n(P)}$ ($f_{n(P)} = \tfrac{1}{2}\omega_{n(P)}/\pi$).

3. If W_0 is large and the weight of the pile is negligible in comparison, then the right-hand side of Eq. (6.98) is small; hence, we can write

$$AL\gamma_P/W_0 = \left(\omega_{n(P)}L/v_{c(P)}\right)^2 \tag{6.100}$$

$$\omega_{n(P)} = \sqrt{AL\gamma_P/W_0}\cdot v_{c(P)}/L = \sqrt{AL\gamma_P/W_0}\cdot\sqrt{E_P/\rho_P}\cdot(1/L)$$

$$= \sqrt{AE_Pg/LW_0}$$

or

$$f_{n(P)} = \frac{1}{2\pi}\sqrt{E_Pg/\sigma_0L} \tag{6.101}$$

Richart (1962) prepared a graph for $f_{n(P)}$ with various values of pile length and σ_0. He considered the cases of steel, concrete, and wooden piles (Figure 6.16). Pile-material properties that have been used in preparing Figure 6.16 are given in Table 6.4.

TABLE 6.4 Properties of Pile Materials Shown in Figure 6.16[a]

Material	E_P		γ_P	
	lb/in.^{2}	kN/m^2	lb/ft^3	kN/m^3
Steel	29.4×10^6	202.9×10^6	480	75.46
Concrete	3.0×10^6	20.7×10^6	150	23.58
Wood	1.2×10^6	8.2×10^6	40	6.29

[a] Conversion factors: 1 lb/in.^{2} = 6.9 kN/m^2; 1 lb/ft^3 = 10.1572 kN/m^3

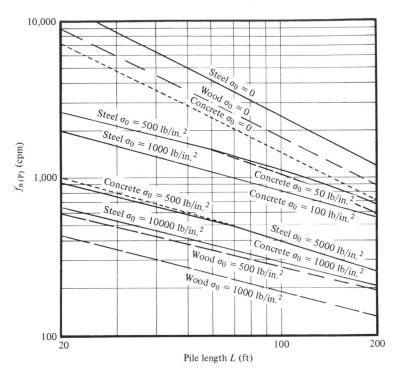

FIGURE 6.16 Resonant frequency of vertical oscillation for a point bearing pile. [Richart, F. E., Jr. (1962). "Foundation Vibrations," *Transactions*, *ASCE*, 127, Part I, Fig. 12, p. 886.]

6.8 VIBRATION SCREENING

In Section 5.2, allowable vertical vibration amplitudes for machine foundations were considered. It is sometimes possible that, for some rugged vibratory equipment, the intensity of vibration may not be objectionable for the equipment itself. However, the vibration may not be within a tolerable limit for a sensitive equipment nearby. Under these circumstances, it is desirable to control the vibrating energy reaching the sensitive zone. This is referred to as *vibration screening*. It needs to be kept in mind that most of the vibratory energy affecting structures nearby is carried by *Rayleigh (surface) waves* traveling from the source of vibration. Effective screening of vibration may be achieved by proper *interception*, *scattering*, and *diffraction* of surface waves by using barriers such as trenches, sheet-pile walls, and piles.

6.8.1 Active and Passive Isolation: Definitions

While studying the problem of vibration screening, it is convenient to group the screening problems into two major categories as follows:

Active Isolation. This type involves screening at the source of vibration, such as shown in Figure 6.17, in which a circular trench of radius R and depth H surrounds the foundation that is the source of disturbance.

Passive Isolation. This process involves providing a barrier at a point remote from the source of disturbance but near a site where vibration has to be reduced. An example of this is shown in Figure 6.18, in which an open trench of length L and depth H is used near a sensitive instrument foundation to protect it from damage.

6.8.2 Active Isolation by Use of Open Trenches

Woods (1968) has reported the results of a field investigation for active isolation by use of open trenches. The field tests were conducted at a site with a deep stratum of silty sand. The experimental study consisted of applying vertical vibrations by a small vibrator [18-lb (80.1-N) maximum force] resting on a circular pad. Trenches were constructed around the circular pad to screen the surface displacement due to the surface waves.

FIGURE 6.17 Schematic diagram of vibration isolation using a circular trench surrounding the source of vibrations: active isolation. [Woods, R. D. (1968). "Screening of Surface Waves in Soils," *Journal of the Soil Mechanics and Foundations Division, ASCE*, 94 (SM4), Fig. 2, p. 954.]

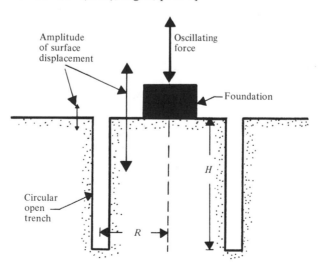

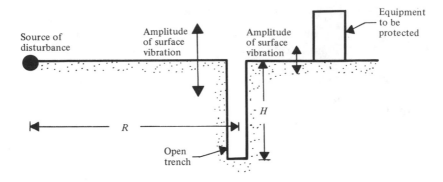

FIGURE 6.18 Passive isolation, using an open trench.

Vertical velocity transducers were used for measurement of surface displacements around the trench over a 25-ft (7.62-m) diameter area. Other conditions remaining the same, measurements for the surface displacements due to the vibration of the circular pad were also taken without the trenches surrounding the pad. Some of the results of this investigation are shown in Figure 6.19 in the form of *amplitude-reduction factor* contour diagrams. The amplitude-reduction factor (ARF) is defined as

$$\text{ARF} = \frac{\text{vertical amplitude of vibration with trench}}{\text{vertical amplitude of vibration without trench}} \qquad (6.102)$$

FIGURE 6.19 Amplitude reduction factor contour diagrams. [Woods, R. D. (1968). "Screening of Surface Waves in Soils," *Journal of the Soil Mechanics and Foundations Division*, *ASCE*, 94 (SM4), Fig. 13, p. 964, and Fig. 14, p. 966.]

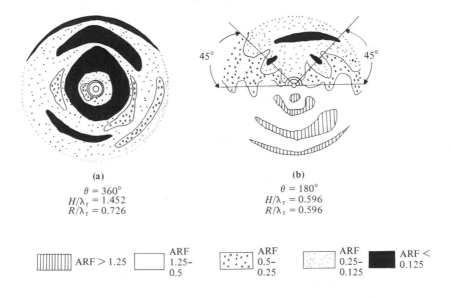

(a)

$\theta = 360°$
$H/\lambda_r = 1.452$
$R/\lambda_r = 0.726$

(b)

$\theta = 180°$
$H/\lambda_r = 0.596$
$R/\lambda_r = 0.596$

ARF > 1.25   ARF 1.25–0.5   ARF 0.5–0.25   ARF 0.25–0.125   ARF < 0.125

Also note that in Figure 6.19, θ is the angular length of the trench (in degrees) and λ_r is the wavelength of Rayleigh waves. The value of λ_r for a given frequency of vibration at a given site can be determined in a manner similar to that described in Section 3.12. The tests of Woods (1968) were conducted for $R/\lambda_r = 0.222-0.910$ and $H/\lambda_r = 0.222-1.82$. For satisfactory isolation, Woods defined that ARF should be less than or equal to 0.25. The conclusions of this study can be summarized as follows:

1. For $\theta = 360°$, a minimum value of $H/\lambda_r = 0.6$ is required to achieve ARFs less than or equal to 0.25.
2. For $360° > \theta > 90°$, the screened zone may be defined to be an area outside the trench bounded on the sides by radial lines from the center of the source through points 45° from the ends of the trench. To obtain ARFs less than or equal to 0.25 in the screened zone, a minimum value of $H/\lambda_r = 0.6$ is required.
3. For $\theta \leqslant 90°$, effective screening of vibration by trenches cannot be obtained.

6.8.3 Passive Isolation by Use of Open Trenches

Woods (1968) has also investigated the case of passive isolation in the field by using open trenches. The plan view of the field site layout used by Woods for screening at a distance is shown in Figure 6.20. The layout consisted of two vibration exciter footings (used one at a time for the tests),

FIGURE 6.20 Plan view of the field site layout for passive isolation by use of an open trench. [Woods, R. D. (1968). "Screening of Surface Waves in Soils," *Journal of the Soil Mechanics and Foundations Division, ASCE*, 94 (SM4), Fig. 17, p. 969.]

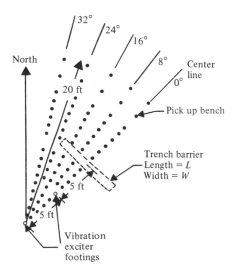

a trench barrier, and 75 pickup benches. For these tests, it was assumed that the zones screened by the trench would be symmetrical about the 0° line. The variables used to study the passive isolations tests are as follows:

1. the distance R from the vibration to the center of the open trench
2. the trench length L
3. the trench width W
4. the trench depth H

In this investigation, the value of R/λ_r was varied from 2.22 to 9.10. For satisfactory isolation, it was defined that ARFs [Eq. (6.102)] should be less than or equal to 0.25 in a *semicircular zone of radius $\frac{1}{2}L$ behind the trench*. Figure 6.21 shows the ARF contour diagram for one of these tests. The conclusions of this study may be summarized as follows:

1. For a satisfactory passive isolation (for $R = 2\lambda_r$ to about $7\lambda_r$), the minimum trench depth H should be about 1.2–$1.5\lambda_r$. This means that, in general, H/λ_r should be about 1.33.
2. The trench width W has practically no influence on the effectiveness of screening.

FIGURE 6.21 Amplitude reduction factor contour diagram for passive isolation. [Woods, R. D. (1968). "Screening of Surface Waves in Soils," *Journal of the Soil Mechanics and Foundations Division*, *ASCE*, 94 (SM4), Fig. 20, p. 973.]

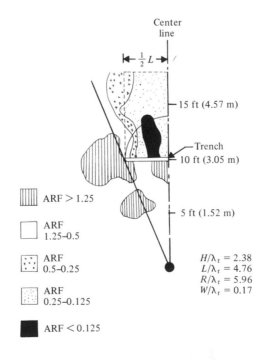

3. To maintain the same degree of isolation, the least area of the trench in the vertical direction i.e., $LH = A_T$, should be as follows:

$$A_T = 2.5\lambda_r^2 \quad \text{at} \quad R = 2\lambda_r$$
$$A_T = 6.0\lambda_r^2 \quad \text{at} \quad R = 7\lambda_r$$

6.8.4 Passive Isolation by Use of Piles

There are several situations in which the Rayleigh waves that emanate from man-made sources may be in the range of 120–150 ft ($\approx$ 40–50 m). For these types of problem, a trench depth of 1.33×120–150 ft (about 50–60 m) is needed for effective passive isolation. Open trenches or bentonite-slurry-filled trenches deep enough to be effective are not practical. At the same time, solidification of the bentonite-slurry also poses a problem. For this reason, possible use of rows of piles as an energy barrier has been studied by Woods et al. (1974) and also by Liao and Sangrey (1978). Woods et al. used the principle of holography and observed vibrations in a model half-space in order to develop the criteria for *void cylindrical obstacles* (Figure 6.22) for passive isolation. The model half-space was prepared in a fine-sand medium in a box. Note that in Figure 6.22 the diameter of the void cylindrical obstacle is D, and the net space for the energy to penetrate between two consecutive void obstacles is equal to S_n. The numerical evaluation of the barrier effectiveness was made by obtaining the average ARFs from several lines beyond the barrier in a sector $\pm 15°$ of both sides

FIGURE 6.22 Void cylindrical obstacles for passive isolation.

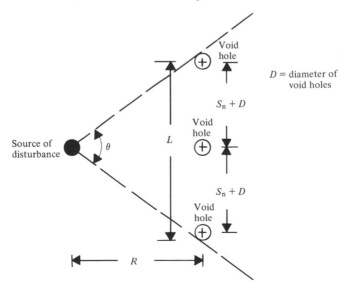

of an axis through the source of disturbance and perpendicular to the barrier. For all tests, H/λ_r and L/λ_r were kept at 1.4 and 2.5, respectively. These values of H/λ_r and L/λ_r are similar to those suggested for open trenches in Section 6.8.3. A nondimensional plot of the *isolation effectiveness* developed from these tests is given in Figure 6.23. The isolation effectiveness is defined as

$$\text{Effectiveness} = 1 - \text{ARF} \qquad (6.103)$$

Based on these test results, Woods et al. (1974) have suggested that a row of void cylindrical holes may act as an isolation barrier if

$$D/\lambda_r \geqslant \tfrac{1}{6} \qquad (6.104)$$

and

$$S_n/\lambda_r < \tfrac{1}{4} \qquad (6.105)$$

Liao and Sangrey (1978) used an acoustic model employing sound waves in a fluid medium to evaluate the possibility of the use of rows of piles as passive isolation barriers. Model piles for the tests were made out of aluminum, steel, styrofoam, and polystyrene plastic. Based on their study,

FIGURE 6.23 Isolation effectiveness as a function of hole diameter and spacing. [Woods, R. D., Barnett, N. E., and Sagessar, R. (1974). "Holography—A New Tool for Soil Dynamics," *Journal of the Geotechnical Engineering Division, ASCE*, 100 (GT11), Fig. 7, p. 1240.]

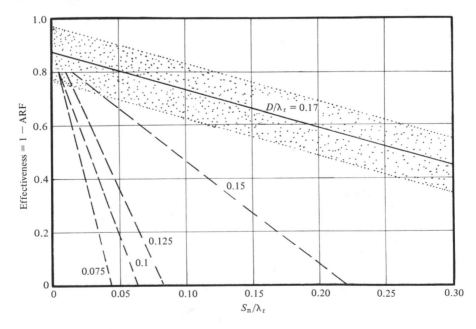

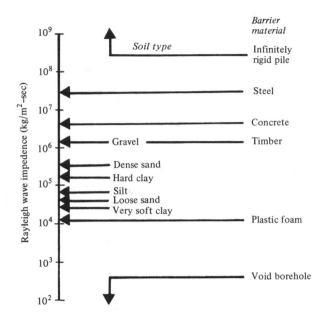

FIGURE 6.24 Estimated values of Rayleigh wave impedance for various soils and pile materials. [Liao, S., and Sangrey, D. A. (1978). "Use of Piles of Isolation Barriers," *Journal of the Geotechnical Engineering Division, ASCE*, 104 (GT9), Fig. 12, p. 1149.]

Liao and Sangrey determined that Eqs. (6.104) and (6.105) as suggested by Woods et al. are generally valid. Also they determined that $S_n = 0.4\lambda_r$ may be the upper limit for a barrier to have some effectiveness. However, the degree of effectiveness of the barrier depends on whether the piles are soft or hard *as compared to the soil* in which they are embedded. The degree of *softness* or *hardness* may be determined by the term *impedance ratio* (IR) defined as

$$IR = \frac{\rho_P v_{r(P)}}{\rho_S v_{r(S)}} \tag{6.106}$$

where ρ_P and ρ_S are the respective densities of the pile material and soil, and $v_{r(P)}$ and $v_{r(S)}$ are the velocities of Rayleigh waves in the two media. The piles are considered *soft* if IR < 1 and *hard* if IR > 1. Soft piles are the more efficient of the two as isolation barriers. Figure 6.24 gives a general range of the Rayleigh wave impedance ($= \rho v_r$) for various soils and pile materials. For a more detailed discussion, the readers are referred to the original paper of Liao and Sangrey.

PROBLEMS

6.1. Redo Problem 5.1 using the method outlined in Section 6.1.2.

6.2. Redo Problem 5.2 using the method outlined in Section 6.1.2.

6.3. Redo Problem 5.3 using the method outlined in Section 6.1.2.

6.4. Redo Problem 5.4 using the method outlined in Section 6.1.2.

6.5. Redo Problem 5.5 using the method outlined in Section 6.1.2.

6.6. Redo Problem 5.6 using the method outlined in Section 6.2.

6.7. Redo Problem 5.7 using the method outlined in Section 6.2.

6.8. Redo Problem 5.8 using the method outlined in Section 6.3.

6.9. Redo Problem 5.9 using the method outlined in Section 6.3.

6.10. Redo Problem 5.10 using the method outlined in Section 6.4.

6.11. Redo Example 5.4 using the method outlined in Section 6.4.

6.12. A machine foundation is supported by four prestressed concrete piles driven to bedrock. The lengths of the piles are 80 ft, and they are 12×12 in. in section. The weight of the machine and the foundation is 300 kips (1 kip = 1000 lb). Given the unit weight of concrete, 160 lb/ft^3; and Young's modulus of concrete used in piles, 3.5×10^6 lb/in.^{2}. Determine the natural frequency of the pile foundation system.

REFERENCES

Barkan, D. D. (1962). *Dynamic Bases and Foundations*, McGraw-Hill, New York.

Borowicka, H. (1943). "Uber ausmittig belastere Starre Platten auf elastischisotropem untergrund," *Ingenieur-Archiv* 1, 1–8.

Bycroft, G. N. (1956). "Forced Vibrations of a Rigid Circular Plate on a Semi-infinite Elastic Space and on an Elastic Stratum," *Philosophical Transactions of the Royal Society*, London (Ser. A) 248, 327–368.

Gorbunov-Possadov, M. I., and Serebrajanyi, R. V. (1961). "Design of Structures upon Elastic Foundations," *Proceedings 5th International Conference on Soil Mechanics and Foundation Engineering*, Vol. 1, pp. 643–648.

Hall, J. R., Jr. (1967). "Coupled Rocking and Sliding Oscillations of Rigid Circular Footings," *Proceedings, International Symposium on Wave Propagation and Dynamic Properties of Earth Materials*, Albuquerque, New Mexico, p. 139.

Hall, J. R., Jr., and Richart, F. E., Jr. (1963). "Dissipation of Elastic Wave Energy in Granular Soils," *Journal of the Soil Mechanics and Foundations Division*, ASCE 89, (SM6), 27–56.

Hardin, B. O. (1965). "The Nature of Damping in Sands," *Journal of the Soil Mechanics and Foundations Division*, ASCE 91 (SM1), 63–97.

Hsieh, T. K. (1962). "Foundation Vibrations," *Proceedings Institute of Civil Engineers*, London, Vol. 22, pp. 211–226.

Kaldjian, M. J. (1969). "Discussion on Design Procedures for Dynamically Loaded Foundations," *Journal of the Soil Mechanics and Foundations Division*, ASCE 95 (SM1), 364–366.

Liao, S., and Sangrey, D. A. (1978). "Use of Piles as Isolation Barriers," *Journal of the Geotechnical Engineering Division*, ASCE 104 (GT9), 1139–1152.

Lysmer, J., and Richart, F. E., Jr. (1966). "Dynamic Response of Footings to Vertical Loading," *Journal of the Soil Mechanics and Foundations Division*, ASCE 92 (SM1), 65–91.

Reissner, E., and Sagoci, H. F. (1944). "Forced Torsional Oscillations of an Elastic Half-Space," *Journal of Applied Physics* 15, 652–662.

Richart, F. E., Jr. (1962). "Foundation Vibrations," *Transactions*, ASCE 127 (Part I), 863–898.

Richart, F. E., Hall, J. R., and Woods, R. D. (1970). *Vibration of Soils and Foundations*, Prentice-Hall, Englewood Cliffs, N.J.

Richart, F. E., Jr., and Whitman, R. V. (1967). "Comparison of Footing Vibration Tests with Theory," *Journal of the Soil Mechanics and Foundations Division*, ASCE 93 (SM6), 143–168.

Stevens, H. W. (1966). "Measurement of Complex Moduli and Damping of Soils Under Dynamic Loads," Tech. Report No. 173, U.S. Army Cold Region Research and Engineering Laboratories, Hanover, N.H.

Terzaghi, K. (1955). "Evaluation of Coefficient of Subgrade Reaction," *Geotechnique* 5, 297–326.

Timoshenko, S. P., and Goodier, J. N. (1951). *Theory of Elasticity*, McGraw-Hill, New York.

Weissmann, G. F., and Hart, R. R. (1961). "The Damping Capacity of Some Granular Soils," ASTM, *STP* 305, 45–54.

Whitman, R. V. (1963). "Stress–Strain–Time Behavior of Soil in One Dimensional Compression," Report R63-25, Department of Civil Engineering, MIT.

Whitman, R. V., and Richart, F. E., Jr. (1967). "Design Procedures for Dynamically Loaded Foundations," *Journal of the Soil Mechanics and Foundations Division*, ASCE 93 (SM6), 169–193.

Woods, R. D. (1968). "Screening of Surface Waves in Soils," *Journal of the Soil Mechanics and Foundations Division*, ASCE 94 (SM4), 951–979.

Woods, R. D., Barnett, N. E., and Sagesser, R. (1974). "Holography—A New Tool for Soil Dynamics," *Journal of the Geotechnical Engineering Division*, ASCE 100 (GT11), 1231–1247.

7

DYNAMIC BEARING CAPACITY OF SHALLOW FOUNDATIONS

The static bearing capacity of shallow foundations has been extensively studied and reported in the literature. However, foundations can be subjected to single pulse dynamic loads which may be in vertical or horizontal directions. The dynamic loads due to nuclear blasts are mainly vertical. Horizontal dynamic loads on foundations are due mostly to earthquakes. These types of loading may induce large permanent deformations in foundations. A fundamental definition of the *dynamic bearing capacity* has not yet been found. However, one must keep in mind that, during the analysis of the time-dependent motion of a foundation subjected to dynamic loading, several factors need to be considered. Most important of these factors are

nature of variation of the magnitude of the loading pulse

duration of the pulse

strain-rate response of the soil during deformation

A rather limited amount of information on the dynamic bearing capacity of foundations is available in literature at this time. Most of the important works on this topic are summarized in this chapter.

7.1 ULTIMATE DYNAMIC BEARING CAPACITY IN SAND

The static ultimate bearing capacity of shallow foundations subjected to vertical loading (Figure 7.1) can be given by the equation

$$q_u = cN_cS_cd_c + qN_qS_qd_q + \tfrac{1}{2}\gamma BN_\gamma S_\gamma d_\gamma \tag{7.1}$$

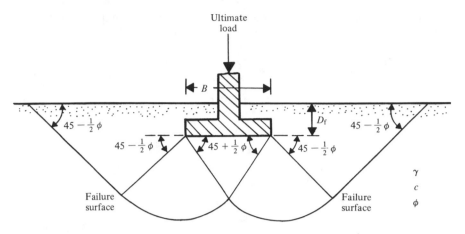

FIGURE 7.1 Static ultimate bearing capacity of continuous shallow foundations.

where

$\quad q_{\mathrm{u}}$ = ultimate load per unit area of the foundation

$\quad \gamma$ = effective unit weight of soil below the foundation

$\quad q = \gamma D_{\mathrm{f}}$

$\quad D_{\mathrm{f}}$ = depth of foundation

$\quad B$ = width of foundation

N_c, N_q, N_γ = bearing capacity factors which are only functions
$\quad\quad\quad$ of the soil friction angle ϕ

S_c, S_q, S_γ = shape factors

d_c, d_q, d_γ = depth factors

In sands, with $c = 0$, Eq. (7.1) becomes

$$q_{\mathrm{u}} = qN_q S_q d_q + \tfrac{1}{2}\gamma BN_\gamma S_\gamma d_\gamma \qquad (7.2)$$

The values of N_q (Reissner, 1924) and N_γ (Caquot and Kerisel, 1953; Vesic, 1973) can be represented by the following equations:

$$N_q = e^{\pi \tan\phi}\tan^2(45 + \phi/2) \qquad (7.3)$$

$$N_\gamma = 2(N_q + 1)\tan\phi \qquad (7.4)$$

where ϕ is the angle of friction of soil. The values of N_q and N_γ for various soil friction angles are given in Table 7.1. The shape and depth factors have been proposed by DeBeer (1970) and Brinch Hanson (1970):

Shape Factors:

$$S_q = 1 + (B/L)\tan\phi \tag{7.5}$$

$$S_\gamma = 1 - 0.4(B/L) \tag{7.6}$$

Depth Factors:

$$\text{For } D_f/B \leqslant 1, \qquad d_q = 1 + 2\tan\phi(1 - \sin\phi)^2(D_f/B) \tag{7.7}$$

$$d_\gamma = 1 \tag{7.8}$$

$$\text{For } D_f/B > 1, \qquad d_q = 1 + 2\tan\phi(1 - \sin\phi)^2\tan^{-1}(D_f/B) \tag{7.9}$$

$$d_\gamma = 1 \tag{7.10}$$

In Eqs. (7.5)–(7.10), B and L are the width and length of rectangular foundations, respectively. For circular foundations, B is the diameter, and $B = L$.

The above equations for static ultimate bearing capacity evaluation are valid for dense sands where the failure surface in the soil extends to the ground surface as shown in Figure 7.1. This is what is referred to as the case of *general shear failure*. For shallow foundations (i.e., $D_f/B \leqslant 1$), if the relative density of granular soils R_D is less than about 70%, *local* or *punching shear failure* may occur. Hence, for static ultimate bearing capacity calculation, if $0 \leqslant R_D \leqslant 0.67$, the values of ϕ in Eqs. (7.3)–(7.10) should be replaced by the modified friction angle

$$\phi' = \tan^{-1}\left[(0.67 + R_D - 0.75R_D^2)\tan\phi\right] \tag{7.11}$$

The facts described above relate to the static bearing capacity of shallow foundations. However, when load is applied rapidly to a foundation to cause failure, the ultimate bearing capacity changes somewhat. This fact has been shown experimentally by Vesic et al. (1965), who conducted several laboratory model tests with a 4-in. (101.6-mm) diameter rigid rough model footing placed on the surface of a dense river sand (i.e., $D_f = 0$), both dry and saturated. The rate of loading to cause failure was varied in a range of 10^{-5} in./sec to over 10 in./sec. Hence, the rate was in the range of static (10^{-5} in./sec) to impact (10 in./sec) loading conditions. All but the four most rapid tests in submerged sand [loading velocity, 0.576–0.790 in./sec (14.63–20.07 mm/sec)] showed peak failure loads as obtained in the case of general shear failure of soil. The four most rapid tests in *submerged* sand gave the load–displacement plots as obtained in the case of *punching shear* failure, where the failure planes do not extend to the ground surface.

For surface footings ($D_f = 0$) in sand, $q = 0$ and $d_\gamma = 1$. So

$$q_u = \tfrac{1}{2}\gamma BN_\gamma S_\gamma \tag{7.12}$$

TABLE 7.1 Values[a] of Bearing Capacity Factors, N_q and N_γ

ϕ	N_q	N_γ	ϕ	N_q	N_γ
0	1.00	0.00	26	11.85	12.54
			27	13.20	14.47
1	1.09	0.07	28	14.72	16.72
2	1.20	0.15	29	16.44	19.34
3	1.31	0.24	30	18.40	22.40
4	1.43	0.34			
5	1.57	0.45	31	20.63	25.99
			32	23.18	30.22
6	1.72	0.57	33	26.09	35.19
7	1.88	0.71	34	29.44	41.06
8	2.06	0.86	35	33.30	48.03
9	2.25	1.03			
10	2.47	1.22	36	37.75	56.31
			37	42.92	66.19
11	2.71	1.44	38	48.93	78.03
12	2.97	1.69	39	55.96	92.25
13	3.26	1.97	40	64.20	109.41
14	3.59	2.29			
15	3.94	2.65	41	73.90	130.22
			42	85.38	155.55
16	4.34	3.06	43	99.02	186.54
17	4.77	3.53	44	115.31	224.64
18	5.26	4.07	45	134.88	271.76
19	5.80	4.68			
20	6.40	5.39	46	158.51	330.35
			47	187.21	403.67
21	7.07	6.20	48	222.31	496.01
22	7.82	7.13	49	265.51	613.16
23	8.66	8.20	50	319.07	762.89
24	9.60	9.44			
25	10.66	10.88			

[a]Vesic, A. S. (1973). "Analysis of Ultimate Loads of Shallow Foundations," *Journal of the Soil Mechanics and Foundations Division, ASCE*, 99 (SM1), Table 1, p. 54.

or

$$q_u/\tfrac{1}{2}\gamma B = N_\gamma S_\gamma \qquad (7.13)$$

The variation of $q_u/\tfrac{1}{2}\gamma B$ with load velocity for the tests of Vesic et al. is shown in Figure 7.2. It may be seen that, for any given series of tests, the value of $q_u/\tfrac{1}{2}\gamma B$ gradually decreases with the loading velocity to a minimum value and then continues to increase. This, in effect, corresponds to a decrease in the angle of friction of soil by about 2° when the loading velocity reached a value of about 2×10^{-3} in./sec (50.8×10^{-3} mm/sec). Such effects of strain-rate in reducing the angle of friction of sand has also been observed by Whitman and Healy (1962).

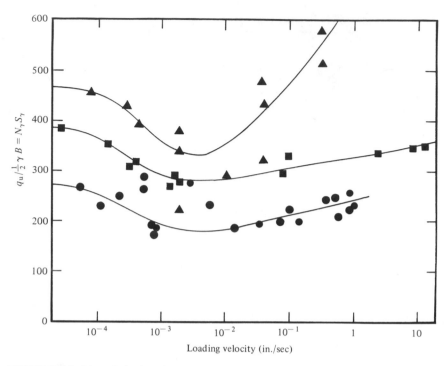

FIGURE 7.2 Plot of the bearing capacity factor vs loading velocity: ■ Series I, dry sand; ● Series II, dry sand; ▲ Series III, submerged sand. [Vesic, A. S., Banks, D. C., and Woodward, J. M. (1965, Fig. 7, p. 212).]

Based on the experimental results available at this time, the following general conclusions regarding the ultimate dynamic bearing capacity of shallow foundations in sand can be drawn:

1. For a foundation resting on sand and subjected to an acceleration level of $a_{max} \leqslant 13$ g, it is possible for general shear type of failure to occur in soil (Heller, 1964).
2. For a foundation on sand subjected to an acceleration level of $a_{max} > 13$ g, the nature of soil failure is by punching (Heller, 1964).
3. The difference in the nature of failure in soil is due to the inertial restrain of the soil involved in failure during the dynamic loading. The restrain has almost a similar effect as the overburden pressure as observed during the dynamic loading which causes the punching shear type failure in soil.
4. The *minimum value* of the *ultimate dynamic bearing capacity* of shallow foundations on dense sands obtained between static to impact loading range can be estimated by using a friction angle ϕ_{dy}, such that (Vesic, 1973)

$$\phi_{dy} = \phi - 2° \tag{7.14}$$

The value of ϕ_{dy} can be substituted in place of ϕ in Eqs. (7.2)–(7.10). However, if the soil strength parameters with proper strain rate are known from laboratory testing, they should be used instead of the approximate equation [Eq. (7.14)].

5. The increase of the ultimate bearing capacity at high loading rates as seen in Figure 7.2 is due to the fact that the soil particles in the failure zone do not always follow the *path of least resistance*. This results in a higher shear strength of soil, which leads to a higher bearing capacity.

6. In the case of foundations resting on loose submerged sands, transient liquefaction effects may exist (Vesic, 1973). This may result in unreliable prediction of ultimate bearing capacity.

7. The rapid increase of the ultimate bearing capacity in dense saturated sand at fast loading rates is due to the development of negative pore water pressure in the soil.

Example 7.1 A square foundation with dimensions $B \times B$ has to be constructed on a dense sand. Its depth is $D_f = 1$ m. The unit weight and the static angle of friction of the soil can be assigned representative values of 18 kN/m³ and 39°, respectively. The foundation may occasionally be subjected to a maximum dynamic load of 1800 kN increasing at a moderate rate. Determine the size of the foundation using a safety factor of 3.

Solution: Given that $\phi = 39°$ in the absence of any other experimental data, for minimum ultimate dynamic bearing capacity

$$\phi_{dy} = \phi - 2° = 39 - 2 = 37°$$

From Eq. (7.2)

$$q_u = qN_qS_qd_q + \tfrac{1}{2}\gamma BN_\gamma S_\gamma d_\gamma$$

$$q = \gamma D_f = (18)(1) = 18 \text{ kN/m}^2$$

For $\phi_{dy} = 37°$, $N_q = 42.92$ and $N_\gamma = 66.19$.

$$S_q = 1 + (B/L)\tan\phi = 1 + \tan 37° = 1.754$$

$$S_\gamma = 1 - 0.4(B/L) = 1 - 0.4 = 0.6$$

$$d_q = 1 + 2\tan\phi(1 - \sin\phi)^2(D_f/B)$$

$$= 1 + 2\tan 37°(1 - \sin 37°)^2(1/B) = 1 + 0.239/B$$

$$d_\gamma = 1$$

Thus

$$q_u \text{ (kN/m}^2) = (18)(42.92)(1.754)(1 + 0.239/B)$$

$$+ \tfrac{1}{2}(18)(B)(66.19)(0.6)(1)$$

$$= 1355 + 323.9/B + 357.4B \qquad \text{(E7.1a)}$$

Given

$$q_u = (1800 \times 3)/B^2 \text{ kN/m}^2 \qquad \text{(E7.1b)}$$

Combining Eqs. (E7.1a) and (E7.1b),

$$5400/B^2 = 1355 + 323.9/B + 357.4B \qquad \text{(E7.1c)}$$

Following is a table to determine the value of B by trial and error. Clearly, $B \approx 1.6$ m.

B (m)	$5400/B^2$ (kN/m^2)	$1355 + 323.9/B + 357.4B$ (kN/m^2)
2	1350	2331.75
1.5	2400	2107
1.6	2109	2133

7.2 ULTIMATE DYNAMIC BEARING CAPACITY IN CLAY

For foundations resting on saturated clays ($\phi = 0$ and $c = c_u$; i.e., undrained condition), Eq. (7.1) transforms to the form

$$q_u = c_u N_c S_c d_c + q N_q S_q d_q \qquad (7.15)$$

(Note: $N_\gamma = 0$ for $\phi = 0$ in Table 7.1.)

$$N_c = 5.14 \qquad (7.16)$$

and

$$N_q = 1 \qquad (7.17)$$

The values for S_c and S_q (DeBeer, 1970) and d_c and d_q (Brinch Hansen, 1970) are as follows:

$$S_c = 1 + (B/L)(N_q/N_c)$$

For $\phi = 0$,

$$S_c = 1 + (B/L)(1/5.14) = 1 + 0.1946(B/L) \qquad (7.18)$$

$$S_q = 1 + \tan\phi$$

$$S_q = 1 \qquad (7.19)$$

$$d_c = 1 + 0.4(D_f/B) \qquad \text{for} \quad D_f/B \leqslant 1 \qquad (7.20)$$

$$d_c = 1 + 0.4\tan^{-1}(D_f/B) \qquad \text{for} \quad D_f/B > 1 \qquad (7.21)$$

$$d_q = 1 \qquad (7.22)$$

Substituting Eqs. (7.16)–(7.22) into Eq. (7.15),

$$q_u = 5.14c_u\left[1 + 0.1946\left(\frac{B}{L}\right)\right]\left[1 + 0.4\left(\frac{D_f}{B}\right)\right] + q \qquad \text{for} \quad \frac{D_f}{B} \leqslant 1 \qquad (7.23)$$

and

$$q_u = 5.14c_u\left[1+0.1946\left(\frac{B}{L}\right)\right]\left[1+0.4\tan^{-1}\left(\frac{D_f}{B}\right)\right]+q \qquad \text{for} \quad \frac{D_f}{B}<1$$

$$(7.24)$$

The ultimate dynamic bearing capacity of foundations resting on saturated clay soils can be estimated by using Eqs. (7.23) and (7.24), provided the strain-rate effect due to dynamic loading is taken into consideration in

FIGURE 7.3 Unconsolidated–undrained triaxial test results on Buckshot clay. Note: 1 lb/ft^2 = 47.88 N/m^2. [Carroll, W. F. (1963, Fig. 2.12, p. 92).]

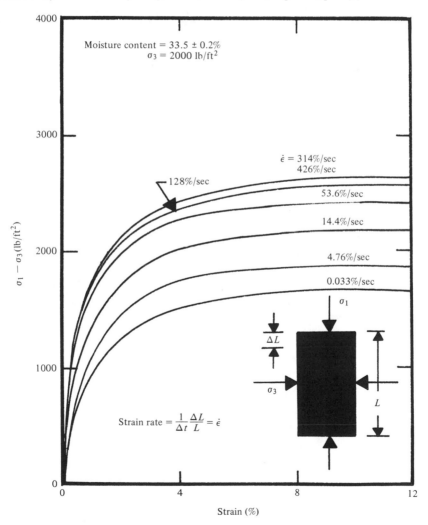

determination of the undrained cohesion. Unlike the case in sand, the undrained cohesion of saturated clays increases with the increase of the strain rate. This fact is shown in Figure 7.3 which shows the results of unconsolidated–undrained triaxial tests on *Buckshot* clay. Note that the undrained cohesion c_u for any given *strain rate* can be obtained as $\frac{1}{2}(\sigma_1 - \sigma_3)_{fail}$. From Eq. (7.3), it can also be seen that the values of undrained cohesion for strain rates between 50% and 425% (which is the dynamic loading range) are not too different and can be approximated to be a single value. Carrol (1963) suggested that $c_{u(dyn)}/c_{u(stat)}$ may be approximated to be about 1.5.

For a given foundation, the strain rate $\dot{\epsilon}$ can be approximated as (Figure 7.4)

$$\dot{\epsilon} = (1/\Delta t)(\tfrac{1}{2}\Delta S/B) \tag{7.25}$$

where B is the width of the foundation.

FIGURE 7.4 Definition of strain rate under a foundation.

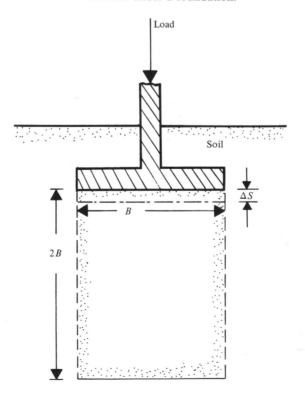

7.3 RIGID PLASTIC ANALYSIS OF FOUNDATIONS

7.3.1 Transient Vertical Load on Foundation (Rotational Mode of Failure)

Triandafilidis (1965) has presented a solution for dynamic response of continuous footing supported by *saturated cohesive soil* ($\phi = 0$ condition) and subjected to a transient load. Now, let the transient stress pulse be expressed in the form

$$q_d = q_0 e^{-\beta t} = \lambda q_u e^{-\beta t} \tag{7.26}$$

where

q_d = stress at time t

q_u = *static* bearing capacity of continuous footing

β = decay function

t = time

$q_0 = \lambda q_u$ = *instantaneous* peak intensity of the stress pulse

λ = overload factor

FIGURE 7.5 Transient vertical load on continuous foundation resting on saturated clay (rotational mode of failure).

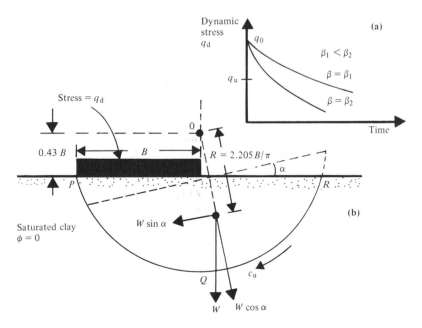

The nature of variation of the exponentially decaying stress pulse represented by Eq. (7.26) with time is shown in Figure 7.5a. The assumptions made by Triandafilidis in the analysis are as follows:

A condition of $\phi = 0$ exists.

The saturated cohesive soil behaves as a rigid plastic material.

The failure surface of soil is cylindrical for evaluation of the bearing capacity under static condition.

Such a failure surface is shown in Figure 7.5b. The center of rotation of the failure surface (point 0) is located at a height of $0.43B$ above the ground surface; the static bearing capacity can be given by the relation

$$q_u = 5.54c_u \qquad (7.27)$$

where c_u is the undrained condition.

With the stress pulse on the foundation, the equation of motion can be given by (Figure 7.5b)

moment of the driving forces per unit length about point 0
       = moment of the restoring forces per unit length about point 0

$$(7.28)$$

Moment of the *driving force* is due to the pulse load and is equal to $q_d(B \times 1) \times (\frac{1}{2}B) = \frac{1}{2}q_d B^2$. The moment of restoring forces is the sum of the following:

Moment of the soil resistance, which is equal to $q_u(B \times 1)(\frac{1}{2}B) = \frac{1}{2}q_u B^2$.

Soil inertia—resisting moment due to the rigid body motion of the *failed soil mass PQR* (Figure 7.5b)

$$M_I = J_0 \ddot{\alpha} \qquad (7.29)$$

where

$\dot{\alpha}$ = angular acceleration

J_0 = polar mass moment of inertia = $WB^2/(1.36g)$ (7.30)

where

W = weight per unit length of the cylindrical soil mass

$$PQR = 0.31\pi\gamma B^2 \qquad (7.31)$$

g = acceleration due to gravity

γ = unit weight of soil

Substituting Eq. (7.30) into Eq. (7.29) yields

$$M_I = \left[WB^2/(1.36g)\right]\ddot{\alpha} \qquad (7.32)$$

Restoring moment due to the displaced soil mass *PQR* which is equal to $W\overline{R}\sin\alpha$ (see Figure 7.5b for definition of $\overline{R}$).

Substitution of the above moments into Eq. (7.28) gives

$$\tfrac{1}{2}q_{\mathrm{d}}B^2 = \tfrac{1}{2}q_{\mathrm{u}}B^2 + \left[WB^2/(1.36g)\right]\ddot{\alpha} + W\overline{R}\sin\alpha \tag{7.33}$$

Assuming $\sin\alpha \approx \alpha$, rearrangement of the preceding equation yields

$$\ddot{\alpha} + (3g/\pi B)\alpha = (0.68g/W)(q_{\mathrm{d}} - q_{\mathrm{u}}) \tag{7.34}$$

Substituting the expression for q_{d} [Eq. (7.26)] into the above equation gives

$$\ddot{\alpha} + (3g/\pi B)\alpha = (0.68g/W)q_{\mathrm{u}}(\lambda e^{-\beta t} - 1) \tag{7.35}$$

It may be noted that Eq. (7.35) is of a form similar to that of Eq. (1.23). The natural period of foundations can be given by the relation

$$T = 2\pi/\sqrt{3g/\pi B} = 2\pi\sqrt{\pi B/3g}$$

Solution of Eq. (7.34) yields the following relation

$$\frac{W}{0.68gq_{\mathrm{u}}}(\alpha) = \frac{T^2}{4\pi^2 + \beta^2 T^2}\left[\left(1 - \lambda + \frac{\beta^2 T^2}{4\pi^2}\right)\cos\left(\frac{2\pi t}{T}\right)\right.$$

$$\left. + \frac{\beta\lambda T}{2\pi}\sin\left(\frac{2\pi t}{T}\right) + \lambda e^{-\beta t} - \frac{\beta^2 T^2}{4\pi^2} - 1\right] \tag{7.36}$$

The above relation can be used to trace the history of motion of the foundation. For determination of the maximum angular deflection α, Eq. (7.36) can be differentiated with respect to time. Thus

$$\frac{W}{0.68gq_{\mathrm{u}}}\dot{\alpha} = \frac{2\pi T}{4\pi^2 + \beta^2 T^2}\left[\left(\lambda - 1 - \frac{\beta^2 T^2}{4\pi^2}\right)\sin\left(\frac{2\pi t}{T}\right)\right.$$

$$\left. + \frac{\beta\lambda T}{2\pi}\cos\left(\frac{2\pi t}{T}\right) - \frac{\beta\lambda T}{2\pi}e^{-\beta t}\right] \tag{7.37}$$

For obtaining the critical time $t = t_{\mathrm{c}}$ which corresponds to $\alpha = \alpha_{\max}$, the right-hand side of Eq. (7.37) is equated to zero. Since $2\pi T/(4\pi^2 + \beta^2 T^2) \neq 0$,

$$\left(\lambda - 1 - \frac{\beta^2 T^2}{4\pi^2}\right)\sin\left(\frac{2\pi t}{T}\right) + \frac{\beta\lambda T}{2\pi}\cos\left(\frac{2\pi t}{T}\right) - \frac{\beta\lambda T}{2\pi}e^{-\beta t} = 0 \tag{7.38}$$

By using small increments of time t in Eq. (7.38), the value of t_{c} can be obtained. This value of $t = t_{\mathrm{c}}$ can then be substituted into Eq. (7.36) with known values of β, λ, and B to obtain $(W/0.68gq_{\mathrm{u}})\alpha_{\max} = K$. Figure 7.6a–c gives the values of $(W/0.68gq_{\mathrm{u}})\alpha_{\max} = K$ (sec^2) for $B = 2$, 5, and 10 ft (0.61, 1.52, and 3.05 m), respectively, with $\lambda = 1$–5 and $\beta = 0$–50 sec^{-1}.

However, it needs to be pointed out that the influence of the *strain rate* on the shear strength of soil and the *dead weight of the foundation* have not been introduced in obtaining the above equations. Another factor of concern is the assumption of the nature of failure surface (general shear failure).

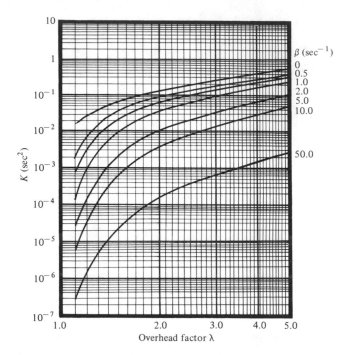

FIGURE 7.6 Variation of the dynamic load factor $K = [W/(0.68g\,q_u)]\alpha_{max}$ with overload factor λ for various values of decay function B: **(a)** $B = 2$ ft (0.61 m); **(b)** $B = 5$ ft (1.52 m); **(c)** $B = 10$ ft (3.05 m). [Triandafilidis, G. E. (1965, Figs. 3, 4, 5, p. 207).]

It has been pointed out by Triandafilidis that the above results should not be applied for rotation $\alpha > 15°–20°$.

Example 7.2 A 5-ft-wide strip foundation is subjected to a transient stress pulse which can be given as $q_d = 10{,}000\,e^{-\beta t}$ lb/ft², $\beta = 10$ sec^{-1}. Determine the maximum angular rotation the footing might undergo. The soil supporting the foundation is saturated clay with undrained cohesion = 900 lb/ft². The saturated unit weight of the soil is 115 lb/ft³.

Solution: Given $B = 5$ ft, $c_u = 900$ lb/ft²; thus the static bearing capacity

$$q_u = c_u N_c = 5.54 c_u = (5.54)(900) = 4986 \text{ lb/ft}^2$$

Overload factor $= \lambda = q_0/q_u = 10{,}000/4986 = 2$

Referring to Figure 7.6b, for $\lambda = 2$, $\beta = 10$ sec^{-1};

$$(W/0.68 g q_u)\,\alpha_{max} \approx 0.003 \text{ sec}^2$$

237

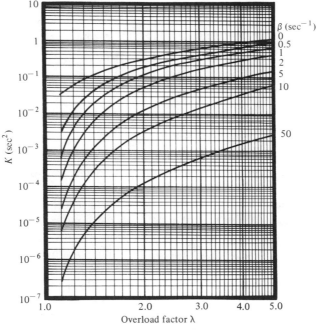

FIGURE 7.6(b)

FIGURE 7.6(c)

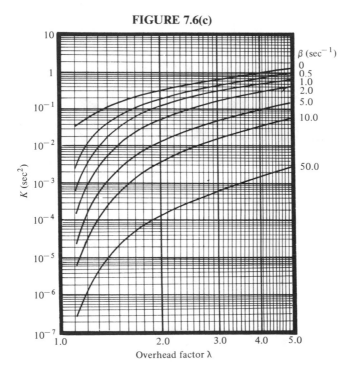

From Eq. (7.31)

$$W = 0.31\pi\gamma B^2 = (0.31)(\pi)(115)(5)^2 = 2800 \text{ lb/ft}$$
$$\alpha_{max} = (0.003)(0.68)gq_u/W = (0.003)(0.68)(32.2)(4986)/2800$$
$$= 0.117 \text{ rad} = 6.7°$$

7.3.2 Transient Horizontal Load on Foundation (Rotational Mode of Failure)

The rigid plastic analysis for the bearing capacity in cohesive soils presented in the preceding section has been extended for determination of the bearing capacity of continuous foundations resting on a $c - \phi$ soil and subjected to a transient horizontal load as shown in Figure 7.7a (Prakash and Chummar, 1967). In Figure 7.7b, Q is the vertical load per unit length of the foundation and $\lambda Q = Q_d$ is the horizontal transient force per unit length (λ is the overload factor). The variation of the dynamic force with time considered here is shown in Figure 7.7a. The nature of the failure surface in the soil due to the loadings is assumed to be a logarithmic spiral with its

FIGURE 7.7 Transient horizontal load on a continuous foundation resting on ground surface (rotational mode of failure).

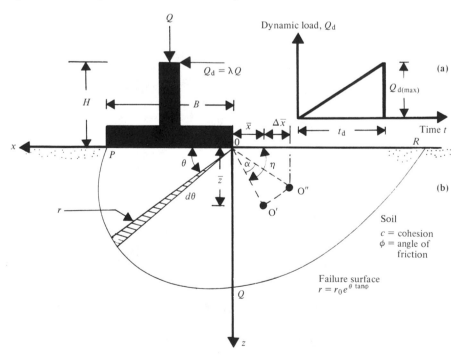

center located at the corner of the foundation base. The static ultimate bearing capacity of such a foundation can be given by

$$q_u = cN_c + \tfrac{1}{2}\gamma BN_\gamma \tag{7.39}$$

where c is cohesion, γ is the unit weight of soil, and B is the foundation width. Using such a simplified assumption, Prakash and Chummar have determined the bearing capacity factors as follows:

$$N_c = (e^{2\pi \tan\phi} - 1)/\tan\phi \tag{7.40}$$

where ϕ is the soil friction angle, and

$$N_\gamma = \left[4\tan\phi(e^{3\pi \tan\phi} + 1)\right]/(9\tan^2\phi + 1) \tag{7.41}$$

Note that these bearing capacity factors are somewhat different than those presented in Table 7.1 due to the difference in the assumption of the failure surface.

For a purely cohesive soil ($\phi = 0$), the log-spiral failure surface becomes a semicircle and

$$N_c = 2\pi \tag{7.42}$$

With a factor of safety of 2, the static vertical force on the foundation per unit length can be given as

$$Q = \tfrac{1}{2}B\left(cN_c + \tfrac{1}{2}\gamma BN_\gamma\right) \tag{7.43}$$

Dynamic Equilibrium

For consideration of the dynamic equilibrium of the foundation with the horizontal transient load, the moment of each of the forces (per unit length) about the center of the log spiral needs to be considered:

1. Moment due to the vertical load Q

$$M_1 = \tfrac{1}{2}QB \tag{7.44}$$

2. Moment due to the horizontal force Q_d

$$M_2 = Q_d H = \frac{Q_{d(max)} Ht}{t_d} = \frac{M_{d(max)} t}{t_d} \tag{7.45}$$

where $M_{d(max)} = Q_{d(max)} H$.

3. Moment due to the cohesive force acting along the failure surface, which can be given as (Figure 7.7)

$$M_3 = \left(\tfrac{1}{2}c/\tan\phi\right)\left(r_1^2 - r_0^2\right) \tag{7.46}$$

In this case, $r_0 = \overline{OP} = B$ and

$$r_1 = r_0 e^{\theta \tan\phi} = \overline{OR} = Be^{\pi \tan\phi}$$

Substitution of these values of r_0 and r_1 into Eq. (7.46) gives

$$M_3 = \psi c B^2 \tag{7.47}$$

where

$$\psi = \tfrac{1}{2}(e^{2\pi \tan \phi} - 1)/\tan \phi \tag{7.48}$$

4. Moment due to the frictional resistance along the failure surface: One of the properties of a log spiral is that the radial line at any point on it makes an angle ϕ with the normal at that point. The resultant of the normal and shear forces developed due to friction at any point of a log spiral makes an angle ϕ with the normal at that point; hence, the direction is the same as the radial line which passes through the origin 0. Thus

$$M_4 = 0 \tag{7.49}$$

5. Moment due to the weight of soil mass in the failure wedge PQR:

$$M_5 = \int_0^\pi \left[\tfrac{1}{2}(r\,d\theta)(r)\right]\left[\tfrac{2}{3}r\cos\theta\right]\gamma$$

Substituting $r = r_0 e^{\theta \tan \phi} = B e^{\theta \tan \phi}$ and integrating with limits of $\theta = 0$ to $\theta = \pi$ yields

$$M_5 = -\epsilon \gamma B^3 \tag{7.50}$$

where

$$\epsilon = \tan\phi(e^{3\pi \tan \phi} + 1)/(9\tan^2\phi + 1) \tag{7.51}$$

The negative sign in Eq. (7.50) indicates that the moment is clockwise. For a purely cohesive soil ($\phi = 0$)

$$M_5 = 0 \tag{7.52}$$

6. Moment of the force due to displacement of the center of gravity of the failure wedge (0′ in Figure 7.7b) from its initial position:

$$M_6 = W\Delta \bar{x} \tag{7.53}$$

where the weight of the failure wedge

$$W = \gamma B^2(e^{2\pi \tan \phi} - 1)/(4\tan \phi) \tag{7.54}$$

$$\Delta \bar{x} = R\cos(\eta - \alpha) - \bar{x} \tag{7.55}$$

and $R = \overline{00'}$ (Figure 7.7b). When α is small, Eq. (7.55) can be written as

$$\Delta \bar{x} = (R\cos\eta)\alpha \tag{7.56}$$

However

$$R = \sqrt{\bar{x}^2 + \bar{z}^2} \tag{7.57}$$

where

$$\bar{x} = \frac{-4B\tan^2\phi\,(e^{3\pi\tan\phi}+1)}{(9\tan^2\phi+1)(e^{2\pi\tan\phi}-1)} \tag{7.58}$$

and

$$\bar{z} = \frac{4B\tan\phi\,(e^{3\pi\tan\phi}+1)}{3\left(\sqrt{9\tan^2\phi+1}\right)(e^{2\pi\tan\phi}-1)} \tag{7.59}$$

Combining Eqs. (7.53)–(7.59),

$$M_6 = \beta B^3 (\sin\eta)\alpha \tag{7.60}$$

where

$$\beta = (e^{3\pi\tan\phi}+1)\Big/\left(3\sqrt{9\tan^2\phi+1}\right) \tag{7.61}$$

when

$$\phi = 0,\ \beta = \tfrac{2}{3}. \tag{7.62}$$

7. Moment due to inertia force of soil wedge:

$$M_7 = (d^2\alpha/dt^2)J \tag{7.63}$$

where J is the mass moment of inertia of the soil wedge about the axis of rotation

$$J = \left[\gamma B^4/(16g\tan\phi)\right](e^{4\pi\tan\phi}-1) \tag{7.64}$$

and g is the acceleration due to gravity. Substitution of Eq. (7.64) into Eq. (7.63) yields

$$M_7 = \frac{\mu\gamma B^4}{g}\frac{d^2\alpha}{dt^2} \tag{7.65}$$

where

$$\mu = (e^{4\pi\tan\phi}-1)/(16\tan\phi) \tag{7.66}$$

For $\phi = 0$,

$$\mu = \tfrac{1}{4}\pi \tag{7.67}$$

Now for the equation of motion,

$$M_1 + M_2 = M_3 + M_4 + M_5 + M_6 + M_7 \tag{7.68}$$

Substitution of the proper terms for the moments in Eq. (7.68) gives

$$(d^2\alpha/dt^2)+k^2\alpha = A\left[(M_{d(\max)}t/t_d)+\tfrac{1}{2}QB - E\right] \tag{7.69}$$

where

$$k = \sqrt{g\beta\sin\eta/\mu B} \tag{7.70}$$

$$A = g/\gamma B^4\mu \tag{7.71}$$

$$E = \psi cB^2 + \epsilon\gamma B^3 \tag{7.72}$$

Solution of the differential equation of motion [Eq. (7.69)] with proper boundary conditions yields the following results:

For $t \leqslant t_d$

$$\alpha = \frac{A}{k^2}\left(E - \frac{1}{2}QB\right)\cos(kt) - \frac{A}{k^3}\frac{M_{d(max)}}{t_d}\sin(kt)$$

$$+ \frac{A}{k^2}\left[\frac{M_{d(max)}t}{t_d} + \frac{1}{2}QB - E\right] \qquad (7.73)$$

For $t > t_d$

$$\alpha = (1/k)\left[Gk\cos(kt_d) - H'\sin(kt_d)\right]\cos(kt)$$

$$+ (1/k)\left[Gk\sin(kt_d) + H'\cos(kt_d)\right]\sin(kt) + (A/k^2)(\tfrac{1}{2}QB - E) \qquad (7.74)$$

where

$$G = \frac{A}{k^2}\left(E - \frac{1}{2}QB\right)\cos(kt_d) - \frac{A}{k^3}\frac{M_{d(max)}}{t_d}\sin(kt_d) + \frac{AM_{d(max)}}{k^2} \qquad (7.75)$$

and

$$H' = -\frac{A}{k}\left(E - \frac{1}{2}QB\right)\sin(kt_d) - \frac{A}{k^2}\frac{M_{d(max)}}{t_d}\cos(kt_d) + \frac{A}{k^2}\frac{M_{d(max)}}{t_d} \qquad (7.76)$$

FIGURE 7.8 Variation of the angle of rotation of continuous foundation due to a horizontal transient load: $B = 2$ m, $t_d = 0.25$ sec, factor of safety $= 2$ [Eq. (7.43)]. [Prakash, S., and Chummar, A. V. (1967). "Response of Footings to Lateral Loads," *Proceedings, International Symposium on Wave Propagation and Dynamic Properties of Earth Materials*, University of New Mexico, Albuquerque, Fig. 5, p. 689.]

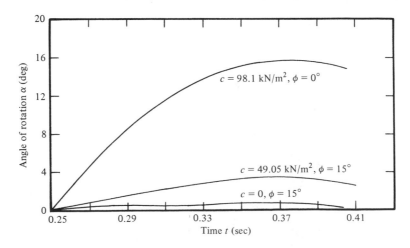

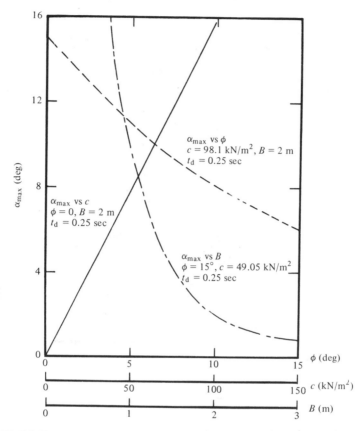

FIGURE 7.9 Dependence of various parameters on the value of the maximum rotation of the foundation. Factor of safety = 2 [Eq. (7.43)]. [Prakash, S., and Chummar, A. V. (1967). "Response of Footings to Lateral Lodas," *Proceedings, International Symposium on Wave Propagation and Dynamic Properties of Earth Materials*, University of New Mexico, Albuquerque, Figs. 6, 7, 8, p. 689.]

Equations (7.73) and (7.74) can be used to determine the angle of rotation α of a foundation with time. This can be done as follows.

1. Determine the soil parameters, c, ϕ, and γ.
2. Determine Q. Equation (7.43) is based on a factor of safety of 2; however, any other factor of safety can be used.
3. Determine B, H, and t_d.
4. Determine ψ, ϵ, μ, and β. They are only functions of the friction angle ϕ.
5. Determine $\sin \eta$.

$$\sin \eta = \bar{z}/\sqrt{\bar{x}^2 + \bar{z}^2}$$

6. Determine the values of k, A, and E.

7. $M_{d(max)} = HQ_{d(max)} = H\lambda Q$
8. Substitute the values of A, k, E, Q, B, and $M_{d(max)}$ in Eq. (7.73). Note that λ is unknown now, so the expression takes a

$$\alpha = f(\lambda, t) \tag{7.77}$$

9. Substitute $t = t_d$ in Eq. (7.77) and increase λ gradually starting at zero. The value at which α becomes positive is the critical value of $\lambda = \lambda_{cr}$. For values $\lambda < \lambda_{cr}$, α is negative. This means that the failure surface in soil has not developed.
10. Using $\lambda = \lambda_{cr}$, calculate $M_{d(max)}$, G, and H.
11. Go to Eq. (7.74). For various values of $t > t_d$, calculate α.

Figure 7.8 shows such a calculation as outlined above for $B = 6.56$ ft (2 m) and $t_d = 0.25$ sec for various values of c and ϕ. Note that the maximum value of $\alpha = \alpha_{max}$ occurs at approximately the same time for various combinations of c and ϕ. Figure 7.9 shows a plot of α_{max} vs ϕ for given values of c, B, and t_d. The value of α_{max} decreases with the increase of friction angle. Figure 7.9 also shows the variation of α_{max} with the cohesion of soil for given values of ϕ, B, and t_d. Also, the value of the maximum angle of rotation is shown to decrease with increase of the width of the foundation (α_{max} vs B in Figure 7.9) for given values of c, ϕ, and t_d.

It needs to be pointed out that λ_{cr} was about 0.38 in all combinations shown in Figures 7.8 and 7.9.

7.3.3 Vertical Transient Load on Strip Foundation (Punching Mode of Failure)

In Sections 7.3.1 and 7.3.2, the analyses for strip foundations presented involved a rotational mode of failure. However, it is possible that a foundation may fail by vertically punching into the soil mass due to the application of a vertical transient load. Wallace (1961) has presented a procedure for the estimation of the vertical displacement of a strip foundation with the assumption that the soil behaves as a *rigid plastic material*. In this analysis, the failure surface in the soil mass is assumed to be of similar type as suggested by Terzaghi (1943) for the evaluation of static bearing capacity of strip foundations. This is shown in Figure 7.10. Note that bd is an arc of a logarithmic spiral with its center at 0 which is defined by the equation $r = r_0 e^{\theta \tan\phi}$. In Figure 7.10, r_0 is the distance $0b$.

The static ultimate bearing capacity for such a failure surface can be expressed as

$$q_u = cN_c + qN_q + \tfrac{1}{2}\gamma BN_\gamma \tag{7.78}$$

where B is the foundation width and $q = \gamma D_f$. Comparing Eq. (7.78) with Eq. (7.1), it is apparent that the depth factors (d_c, d_q, and d_γ) have all been assumed to be equal to one.

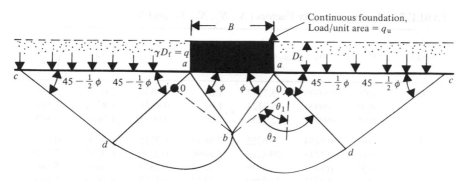

FIGURE 7.10 Failure surface is a soil mass for determination of static bearing capacity of continuous foundation: bd is an arc of a log spiral with its center at 0; dc is a straight line.

If $q = 0$ (i.e., $D_f = 0$) and $\gamma = 0$ (weightless soil),

$$q_u = cN_c \tag{7.79}$$

Again, if $c = 0$ and $\gamma = 0$

$$q_u = qN_q \tag{7.80}$$

Also, if $c = 0$ and $q = 0$ (i.e., $D_f = 0$)

$$q_u = \tfrac{1}{2}BN_\gamma \tag{7.81}$$

Using this method and assuming 0 to be the center of the log spiral, Wallace determined the following relations for the bearing capacity factors N_c, N_q, and N_γ.

$$N_c = \left(\tfrac{1}{2}b^2 + mb^2\sin\delta\right)^{-1}\left[\left(r_2^2 - m^2b^2\right)\left(\tfrac{1}{2}\cos\phi\right) + (2\tan\phi)^{-1}\left(r_2^2 - r_0^2\right)\right.$$

$$\left. - \frac{mb^2}{2\cos\delta}\right] + \tan\phi, \qquad \text{for} \quad \phi \neq 0 \tag{7.82}$$

$$N_c = \left(\tfrac{1}{2}b^2 + mb^2\sin\delta\right)^{-1}\left[\tfrac{1}{2}\left(r_0^2 - m^2b^2\right) + \left(\tfrac{3}{4}\pi + \alpha\right)r_0^2 - \frac{mb^2}{2\cos\delta}\right],$$

$$\text{for} \quad \phi = 0 \tag{7.83}$$

$$N_q = \left(\tfrac{1}{2}b^2 + mb^2\sin\delta\right)^{-1}\left(r_2^2 - m^2b^2\right)\sin^2\delta \tag{7.84}$$

$$N_\gamma = \left(\tfrac{2}{3}b^3 + mb^3\sin\delta\right)^{-1}\left\{\tfrac{1}{3}(r_2 + mb)^2\left(r_2 - \tfrac{1}{2}mb\right)\cos\phi\sin\delta\right.$$

$$+ \left(27\tan^2\phi + 3\right)^{-1}$$

$$\times\left[r_2^3(3\tan\phi\sin\delta - \cos\delta) + r_0^3(\cos\theta_1 + 3\tan\phi\sin\theta_1)\right]$$

$$- \tfrac{1}{6}r_0\cos(\phi - \alpha)\left[\tan\phi - \tan(\phi - \alpha)\right]$$

$$\left.\times\left[r_0^2\cos^2(\phi - \alpha) - m^2b^2\sin^2\delta\right]\right\} - \tfrac{1}{2}\tan\phi \tag{7.85}$$

TABLE 7.2 Bearing Capacity Factors $(N_c, N_q, N_\gamma, N_I, \text{ and } N_R)^a$

ϕ (deg)	m	N_γ	N_c	N_q	N_I	N_R	$\sqrt{\dfrac{N_R}{N_I}}$
0	−0.05	0.0000	5.7277	1.0	0.0633	2.0125	5.6366
	0.00	0.0000	5.7124	1.0	0.0631	1.9723	5.5887
	+0.05	0.0000	5.7258	1.0	0.0633	1.9433	5.5394
5	−0.65	0.1454	79.6255	7.9664	0.3755	8.9076	4.8709
	−0.60	0.1445	29.8163	3.6086	0.2280	6.4362	5.3126
	−0.55	0.1481	18.9958	2.6619	0.1579	5.0332	5.6460
	−0.50	0.1553	14.3469	2.2552	0.1213	4.1699	5.8636
	−0.45	0.1655	11.8179	2.0339	0.1011	3.6088	5.9750
	−0.40	0.1786	10.2699	1.8985	0.0897	3.2299	6.0020
	−0.35	0.1945	9.2580	1.8100	0.0833	2.9674	5.9698
	−0.30	0.2131	8.5723	1.7500	0.0799	2.7828	5.9005
	−0.25	0.2344	8.1007	1.7087	0.0786	2.6523	5.8108
	−0.20	0.2585	7.7778	1.6805	0.0785	2.5604	5.7116
	−0.15	0.2855	7.5629	1.6617	0.0793	2.4969	5.6099
	−0.10	0.3154	7.4291	1.6500	0.0809	2.4547	5.5096
	−0.05	0.3483	7.3580	1.6437	0.0829	2.4288	5.4128
	0.00	0.3843	7.3366	1.6419	0.0853	2.4155	5.3205
	+0.05	0.4233	7.3553	1.6435	0.0881	2.4122	5.2330
10	−0.60	0.5700	53.9491	10.5127	0.1120	5.7922	7.1922
	−0.55	0.5588	28.9945	6.1125	0.0935	4.8411	7.1948
	−0.50	0.5645	20.5266	4.6194	0.0833	4.2238	7.1228
	−0.45	0.5832	16.3539	3.8837	0.0779	3.8095	6.9932
	−0.40	0.6127	13.9337	3.4569	0.0757	3.5264	6.8273
	−0.35	0.6521	12.4031	3.1870	0.0755	3.3323	6.6445
	−0.30	0.7008	11.3881	3.0080	0.0767	3.2008	6.4587
	−0.25	0.7586	10.7004	2.8868	0.0790	3.1147	6.2781
	−0.20	0.8253	10.2345	2.8046	0.0821	3.0625	6.1071
	−0.15	0.9012	9.9267	2.7503	0.0858	3.0360	5.9474
	−0.10	0.9863	9.7361	2.7167	0.0901	3.0294	5.7994
	−0.05	1.0807	9.6352	2.6990	0.0948	3.0386	5.6676
	0.00	1.1848	9.6049	2.6936	0.0999	3.0604	5.5360
	+0.05	1.2986	9.6313	2.6983	0.1053	3.0923	3.4187
15	−0.55	1.5462	46.5473	13.4724	0.0707	5.2677	8.6324
	−0.50	1.5198	30.2759	9.1124	0.0696	4.7177	8.2310
	−0.45	1.5342	23.2038	7.2175	0.0707	4.3564	7.8481
	−0.40	1.5806	19.3483	6.1844	0.0734	4.1189	7.4903
	−0.35	1.6540	16.9964	5.5542	0.0773	3.9669	7.1622
	−0.30	1.7520	15.4722	5.1458	0.0823	3.8766	6.8645
	−0.25	1.8730	14.4550	4.8732	0.0881	3.8322	6.5961
	−0.20	2.0166	13.7730	4.6905	0.0947	3.8232	6.3542
	−0.15	2.1825	13.3257	4.5706	0.1020	3.8418	6.1361
	−0.10	2.3710	13.0501	4.4968	0.1101	3.8825	5.9388
	−0.05	2.5823	12.9048	4.4579	0.1183	3.9413	5.7596
	0.00	2.8168	12.8613	4.4462	0.1282	4.0149	5.5961
	+0.05	3.0750	12.8991	4.4563	0.1383	4.1008	5.4463

^{a}Wallace, W. F. (1961). "Displacement of Long Footings by Dynamic Loads," *Journal of the Soil Mechanics and Foundations Division*, ASCE, 87 (SM5), Table 1, p. 58.

TABLE 7.2 (*continued*)

φ (deg)	m	N_γ	N_c	N_q	N_I	N_R	$\sqrt{\dfrac{N_R}{N_I}}$
20	−0.50	3.6745	46.2884	17.8477	0.0673	5.6658	9.1768
	−0.45	3.6419	33.8986	13.3381	0.0728	5.3067	8.5380
	−0.40	3.6943	27.6099	11.0492	0.0796	5.0886	7.9941
	−0.35	3.8151	23.9213	9.7067	0.0877	4.9684	7.5267
	−0.30	3.9952	21.5875	8.8572	0.0970	4.9199	7.1214
	−0.25	4.2298	20.0542	8.2992	0.1076	4.9258	6.7672
	−0.20	4.5161	19.0369	7.9289	0.1194	4.9746	6.4552
	−0.15	4.8533	18.3742	7.6877	0.1325	5.0582	6.1783
	−0.10	5.2413	17.9678	7.5398	0.1470	5.1704	5.9309
	−0.05	5.6804	17.7542	7.4620	0.1629	5.3068	5.7084
	0.00	6.1717	17.6903	7.4368	0.1802	5.4638	5.5072
	+0.05	6.7161	17.7457	7.4589	0.1989	5.6486	5.3243
25	−0.50	8.5665	73.8778	35.4499	0.0732	7.2346	9.9384
	−0.45	8.3599	51.2706	24.9079	0.0835	6.8363	9.0503
	−0.40	8.3728	40.7056	19.9814	0.0954	6.6214	8.3291
	−0.35	8.5541	34.7663	17.2119	0.1094	6.5339	7.7297
	−0.30	8.8760	31.1015	15.5029	0.1254	6.5404	7.2223
	−0.25	9.3230	28.7315	14.3977	0.1437	6.6199	6.7864
	−0.20	9.8871	27.1750	13.6720	0.1646	6.7584	6.4075
	−0.15	10.5646	26.1681	13.2024	0.1882	6.9462	6.0748
	−0.10	11.3542	25.5533	12.9157	0.2148	7.1761	5.7803
	−0.05	12.2569	25.2309	12.7654	0.2445	7.4429	5.5178
	0.00	13.2745	25.1345	12.7205	0.2775	7.7423	5.2825
	+0.05	14.4095	25.2180	12.7594	0.3139	8.0710	5.0704
30	−0.45	19.3095	80.8644	47.6872	0.1064	9.3123	9.3540
	−0.40	19.1315	62.4470	37.0539	0.1267	9.0899	8.4705
	−0.35	19.3718	52.5548	31.3426	0.1506	9.0494	7.7518
	−0.30	19.940	46.6067	27.9084	0.1787	9.1446	7.1533
	−0.25	20.8187	42.8208	25.7226	0.2116	9.3473	6.6458
	−0.20	21.9566	40.3597	24.3017	0.2500	9.6392	6.2095
	−0.15	23.3512	38.7778	23.3884	0.2944	10.0081	5.8303
	−0.10	24.9984	37.8159	22.8330	0.3456	10.4452	5.4979
	−0.05	26.8993	37.3127	22.5425	0.4041	10.9441	5.2044
	0.00	29.0580	37.1624	22.4558	0.4706	11.4998	4.9436
	+0.05	31.4810	37.2926	22.5309	0.5457	12.1084	4.7107
35	−0.45	46.2942	134.3023	95.0397	0.1527	13.4981	9.4021
	−0.40	45.4427	100.6609	71.4837	0.1887	13.2639	8.3844
	−0.35	45.6687	83.4477	59.4308	0.2323	13.3114	7.5703
	−0.30	46.7356	73.3676	52.3727	0.2849	13.5708	6.9017
	−0.25	48.5145	67.0529	47.9511	0.3481	14.0015	6.3419
	−0.20	50.9356	62.9887	45.1052	0.4237	14.5786	5.8661
	−0.15	53.9640	60.3926	43.2874	0.5133	15.2859	5.4569
	−0.10	57.5868	58.8199	42.1862	0.6191	16.1127	5.1018
	−0.05	61.8051	57.9989	41.6113	0.7428	17.0515	4.7911
	0.00	66.6296	57.7539	41.4398	0.8868	18.0970	4.5175
	+0.05	72.0773	57.9662	41.5884	1.0529	19.2451	4.2753

(*continued*)

TABLE 7.2 Bearing Capacity Factors $(N_c, N_q, N_\gamma, N_I, \text{ and } N_R)^a$ (*continued*)

ϕ (deg)	m	N_γ	N_c	N_q	N_I	N_R	$\sqrt{\dfrac{N_R}{N_I}}$
40	−0.40	115.7097	172.8231	146.0161	0.3229	20.8738	8.0404
	−0.35	115.5504	141.1002	119.3973	0.4107	21.1138	7.1701
	−0.30	117.6386	123.0124	104.2199	0.5195	21.7125	6.4650
	−0.25	121.5875	111.8576	94.8599	0.6536	22.6077	5.8817
	−0.20	127.1879	104.7472	88.8935	0.8175	23.7619	5.3914
	−0.15	134.3346	100.2323	85.1051	1.0168	25.1570	4.9741
	−0.10	142.9868	97.5069	82.8181	1.2572	26.7775	4.6152
	−0.05	153.1451	96.0866	81.6263	1.5450	28.6173	4.3038
	0.00	164.839	95.6630	81.2709	1.8870	30.6724	4.0317
	+0.05	178.1176	96.0303	81.5791	2.2904	32.9409	3.7924
45	−0.40	327.6781	322.2748	323.2752	0.6576	36.2961	7.4295
	−0.35	325.4943	259.1345	260.1349	0.8611	37.0113	6.5559
	−0.30	329.9752	224.0769	225.0772	1.1194	38.3965	5.8568
	−0.25	339.8627	202.7837	203.7840	1.4447	40.3468	5.2846
	−0.20	354.4804	189.3358	190.3361	1.8515	42.8070	4.8083
	−0.15	373.4971	180.8450	181.8452	2.3565	45.7496	4.4062
	−0.10	393.7473	175.7358	176.7361	2.9784	49.1634	4.0628
	−0.05	424.2605	173.0775	174.0778	3.7386	53.0475	3.7669
	0.00	456.1177	172.2851	173.2853	4.6607	57.4067	3.5096
	+0.05	492.4763	172.9729	173.9732	5.7709	62.2499	3.2843

where

$$b = \tfrac{1}{2}B$$
$$\delta = 45 + \tfrac{1}{2}\phi$$
$$\alpha = \measuredangle ab0$$
$$m = (\text{length } 0a)/b$$
$$\theta_1 = \text{internal } \measuredangle \text{ between } r_0 \text{ and the vertical line through } 0$$
$$\theta_2 = \text{internal } \measuredangle \text{ between } r_0 \text{ and line } 0d$$
$$r_0 = \text{length } 0b \text{ (Figure 7.10)}$$
$$r_2 = \text{length } 0d \text{ (Figure 7.10)}$$

The values of N_c, N_q, and N_γ for various values of ϕ and m are given in Table 7.2.

Differential Equation of Motion

In order to obtain the equation of motion for the foundation with a transient load on it, one needs to consider *only one half of the foundation* (Figure 7.11a). The forces per unit length of the foundation involved in here are as follows:

1. Force due to dynamic loading (refer to Figure 7.11b):

$$\text{Dynamic load} = q_{d(\max)}b(1 - t/t_d) \qquad \text{for} \quad 0 \leqslant t \leqslant t_d \quad (7.86)$$

$$\text{Dynamic load} = 0 \qquad\qquad\qquad \text{for} \quad t_d \geqslant 0 \qquad (7.87)$$

2. Force due to soil resistance $= q_u b$.
3. Inertia force (IF): With the transient loading, the soil mass in the failure wedge is displaced. The nature of displacement assumed here is shown with broken lines in Figure 7.11a. Note that the vertical displacement of the foundation is equal to Δ. Using energy considerations, Wallace (1961) has expressed that

$$\text{IF} = N_I \gamma b^2 (d^2 \Delta / dt^2) \qquad (7.88)$$

   The variation of N_I with m and the soil friction angle ϕ is given in Table 7.2.
4. The downward displacement of the foundation results in the movement of soil mass in the failure wedge. This soil mass contributes to a restoring moment about 0 (Figure 7.11a). The restoring force (RF) on

FIGURE 7.11 (a) Nature of displacement of the foundation and the soil failure wedge due to transient loading; (b) variation of the transient load on the foundation with time.

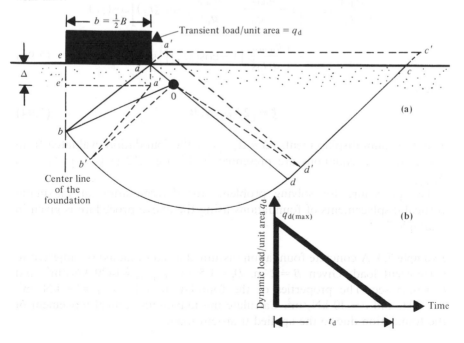

the foundation can be expressed as

$$\mathrm{RF} = N_R b \gamma \Delta \qquad (7.89)$$

The values of N_R are given in Table 7.2.

Combining the above forces, the differential equation of motion can be written as

$$q_u b + N_I \gamma b^2 \left(d^2\Delta/dt^2 \right) + N_R \gamma b \Delta = q_{d(max)} b \left(1 - t/t_d \right) \qquad \text{for} \quad 0 \leqslant t \leqslant t_d \qquad (7.90)$$

and

$$q_u b + N_I \gamma b^2 \left(d^2\Delta/dt^2 \right) + N_R \gamma b \Delta = 0 \qquad \text{for} \quad t \geqslant t_d \qquad (7.91)$$

The solutions of the above differential equations with proper boundary conditions take the form

For $0 \leqslant t \leqslant t_d$:

$$\left(\frac{N_R \gamma}{q_u} \right) \Delta = \left[\frac{q_{d(max)}}{q_u} - 1 \right] \left[1 - \cos(\zeta t) \right] + \frac{q_{d(max)}}{q_u t_d \zeta} \left[\sin(\zeta t) - \zeta t \right] \quad (7.92)$$

For $t \geqslant t_d$:

$$\left(\frac{N_R \gamma}{q_u} \right) \Delta = \left[1 - \frac{q_{d(max)}}{q_u} + \frac{q_{d(max)}}{q_u t_d \zeta} \sin(\zeta t_d) \right] \cos(\zeta t)$$

$$+ \left\{ \frac{q_{d(max)}}{q_u t_d \zeta} \left[1 - \cos(\zeta t_d) \right] \right\} \sin(\zeta t) - 1 \qquad (7.93)$$

where

$$\zeta = \sqrt{2 N_R / N_I B} \qquad (7.94)$$

The maximum displacements ($\Delta = \Delta_{max}$) of the foundation evaluated from the above two equations are presented in Figure 7.12 [$t_d \zeta$ vs ($\gamma N_R / q_u$) Δ_{max}].

The procedure for solving problems for determination of maximum vertical displacements of foundations using the above procedure is given in Example 7.3.

Example 7.3 A concrete foundation (assumed as continuous) is subjected to a transient load. Given $B = 2$ m, $D_f = 1.5$ m, $q_{d(max)} = 1420$ kN/m^2, and $t_d = 0.25$ sec. The properties of the foundation soil are $\gamma = 19$ kN/m^3, $\phi = 20°$, and $c = 40$ kN/m^2. Calculate the maximum vertical movement of the foundation due to the applied transient load.

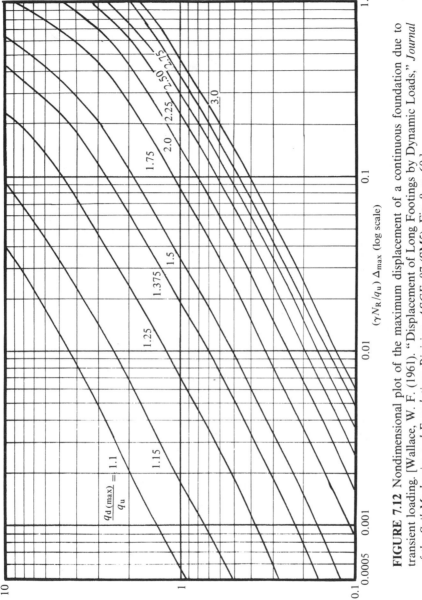

FIGURE 7.12 Nondimensional plot of the maximum displacement of a continuous foundation due to transient loading. [Wallace, W. F. (1961). "Displacement of Long Footings by Dynamic Loads," *Journal of the Soil Mechanics and Foundations Division, ASCE,* 87 (SM5), Fig. 8, p. 60.]

Solution

1. Calculate q_u:

$$q_u = cN_c + qN_q + \tfrac{1}{2}\gamma BN_\gamma$$

$$\phi = 20°, \qquad c = 40 \text{ kN/m}^2$$

$$q = \gamma D_f = 19 \times 1.5 = 28.5 \text{ kN/m}^2$$

Referring to Table 7.2 to determine the values of N_c, N_q, and N_γ for $\phi = 20°$, one needs to make several trials for various values of m (see Table 7.3).

The minimum value of q_u is obtained at $m = -0.05$, so

$$q_u = 1031 \text{ kN/m}^2 \qquad \text{at} \quad m = -0.05$$

2. For $m = -0.05$ and $\phi = 20°$,

$$N_R = 5.3068, \qquad \sqrt{N_R/N_I} = 5.7084$$

$$N_R\gamma/q_u = (5.3068 \times 19)/1031 = 0.0978$$

$$\zeta = \sqrt{2N_R/N_I B} = \sqrt{N_R/N_I}\sqrt{2/B} = (5.7084)\sqrt{2/2} = 5.7084$$

$$t_d\zeta = (0.25)(5.7084) = 1.4271$$

$$q_{d(max)}/q_u = 1420/1031 = 1.377$$

From Figure 7.12, for $t_d\zeta = 1.4271$ and $q_{d(max)}/q_u = 1.377$,

$$(N_R\gamma/q_u)\Delta_{max} = 0.04 \qquad \text{or} \qquad \Delta_{max} = 0.04/0.0978 = 0.409 \text{ m}$$

TABLE 7.3 Trials for Various Values of m (See Example 7.3)

$m = -0.35$	$q_u = (40)(23.9213) + (28.5)(9.7067) + \tfrac{1}{2}(19)(2)(3.8151)$ $= 1306$ kN/m^2
$m = -0.30$	$q_u = (40)(21.5876) + (28.5)(8.8572) + \tfrac{1}{2}(19)(2)(3.9952)$ $= 1192$ kN/m^2
$m = -0.25$	$q_u = (40)(20.0542) + (28.5)(8.2992) + \tfrac{1}{2}(19)(2)(4.2298)$ $= 1119$ kN/m^2
$m = -0.20$	$q_u = (40)(19.0369) + (28.5)(7.9289) + \tfrac{1}{2}(19)(2)(4.5161)$ $= 1073$ kN/m^2
$m = -0.15$	$q_u = (40)(18.3742) + (28.5)(7.6877) + \tfrac{1}{2}(19)(2)(4.8533)$ $= 1046$ kN/m^2
$m = -0.10$	$q_u = (40)(17.9678) + (28.5)(7.5398) + \tfrac{1}{2}(19)(2)(5.2413)$ $= 1033$ kN/m^2
$m = -0.05$	$q_u = (40)(17.7542) + (28.5)(7.4620) + \tfrac{1}{2}(19)(2)(5.6804)$ $= 1031$ kN/m^2
$m = 0.00$	$q_u = (40)(17.6903) + (28.5)(7.4388) + \tfrac{1}{2}(19)(2)(6.1717)$ $= 1037$ kN/m^2

7.3.4 Experimental Observation of Load–Settlement Relationship for Vertical Transient Loading

A limited number of laboratory tests for observation of load–settlement relationships of foundations under transient loading have so far been conducted (Cunny and Sloan, 1961; Shenkman and McKee, 1961; Jackson and Hadala, 1964: Carroll, 1963). The experimental evaluations of these tests are presented in this section.

Load–settlement observations of *square* model footings resting on sand and clay and subjected to transient loads have been presented by Cunny and Sloan (1961). The model footings were of varying sizes from 4.5 to 9-in. (114.3–228.6-mm) squares, and were placed on the surface of the compacted soil layers. The transient loads to which the footings were subjected were of the nature shown in Figure 7.13. The nature of the settlement of footings with time during the application of the dynamic load is also shown in the same figure. In general, during the *rise time* (t_r) of the dynamic load, the settlement of a footing increases rapidly. Once the peak load $[Q_{d(max)}]$ is reached, the rate of settlement with time decreases. However, the total settlement of a footing continues to increase during the *dwell time* of the load (t_{dw}) and reaches a maximum value (S_{max}) at the end of the dwell time. During the *decay period* of the load (t_{de}), the footing rebounds to some degree. The results of the model footing tests on sand obtained by Cunny and Sloan are given in Table 7.4. Also, the results of model tests for square

FIGURE 7.13 Nature of dynamic load applied to laboratory model footings.

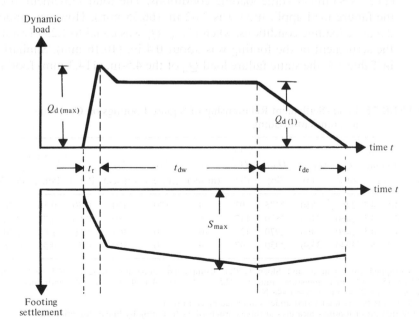

TABLE 7.4 Load–Settlement Relationship of Square Footings on Sand Due to Transient Loading[a]

Test No.	Size of footing (in.)	Q_u (lb)[b]	$Q_{d(max)}$ (lb)	$Q_{d(1)}$ (lb)	$\dfrac{Q_{d(1)}}{Q_u}$ (%)	t_r (m-sec)	t_{dw} (m-sec)	t_{de} (m-sec)	S_{max} (in.)[c] Pot. 1	Pot. 2	Pot. 3
1	6×6	770	800	800	104	18	122	110	0.28	0.05	0.11
2	8×8	1820	3140	2800	154	8	420	255	—	—	—
3	8×8	1820	2275	2175	120	90	280	290	0.83	0.93	0.95
4	9×9	2590	3500	3250	125	11	0	350	0.40	0.42	0.40

[a]Compiled from Cunny and Sloan (1961): Compacted dry unit weight of sand = 103.4 lb/ft³ (16.26 kN/m³); relative density of compaction of sand = 96%; triaxial angle of friction of sand = 32°. Note: 1 in. = 25.4 mm; 1 lb = 4.448 N.
[b]Ultimate failure load tested under static conditions.
[c]Settlement of footings measured at three corners of each footing by linear potentiometer.

surface footings on clay as reported by Cunny and Sloan are shown in Table 7.5. Based on these results, a few general observations may be made:

1. The settlement of foundations under transient loading is generally uniform. This can be seen by observing the settlements at the three corners of the model footings—both in sand and clay.
2. Footings under dynamic loading may fail by punching type of failure in soil, although general shear failure may be observed for the same footings tested under static conditions.
3. In Table 7.4, the 9-in. (228.6-mm) footing failed at a load of 2590 lb (11.52 kN) under static loading conditions. The total settlement after the failure load application was 2.62 in. (66.55 mm). However, under dynamic loading conditions, when $Q_{d(1)}/Q_u$ was equal to 1.25 (Test 4), the settlement of the footing was about 0.4 in. (10.16 mm). Similarly, in Table 7.5, the static failure load Q_u of the 4.5-in. (114.3-mm) footing

TABLE 7.5 Load–Settlement Relationship of Square Footings on Clay Due to Transient Loading[a]

Test No.	Size of footing (in.)	Q_u (lb)[b]	$Q_{d(max)}$ (lb)	$Q_{d(1)}$ (lb)	$\dfrac{Q_{d(1)}}{Q_u}$ (%)	t_r (m-sec)	t_{dw} (m-sec)	t_{de} (m-sec)	S_{max} (in.)[c] Pot. 1	Pot. 2	Pot. 3
1	4.5×4.5	2460	2850	2275	93	9	170	350	0.50	0.50	0.48
2	4.5×4.5	2460	3100	2820	117	9	0	380	0.66	0.72	0.70
3	4.5×4.5	2460	3460	2970	121	10	0	365	1.70	1.68	1.70
4	5×5	3040	3580	2950	97	9	0	360	0.58	0.55	0.55

[a]Compiled from Cunny and Sloan (1961): Compacted moist unit weight = 94.1–98.4 lb/ft³ (14.79–15.47 kN/m³); moisture content = 22.5 ± 1.7%; c = 2400 lb/ft²; ϕ = 4° (undrained test). Note: 1 in. = 25.4 mm; 1 lb = 4.448 N.
[b]Ultimate failure load tested under static loading conditions.
[c]Settlement of footings measures at three corners of each footing by linear potentiometer.

was 2460 lb (10.94 kN) with a settlement of 2 in. (50.8 mm). The same footing under dynamic loading with $Q_{d(1)}/Q_u = 1.17$ (Test 2) showed a total settlement of about 0.7 in. (17.78 mm).

The above facts show that, for a limiting settlement condition, a foundation can support higher load under dynamic loading conditions than those observed from static tests.

Dynamic Load vs Settlement Prediction in Clayey Soils

Jackson and Hadala (1964) reported several laboratory model tests on 4.5–8-in. (114.3–203.2-mm) square footings resting on highly saturated, compacted, plastic *Buckshop clay*. The tests were similar in nature to those described above in this section. Based on these results, Jackson and Hadala have shown that there is a unique nondimensional relation between $Q_{d(max)}/B^2 c_u$ and S_{max}/B (c_u = undrained shear strength). This is shown in Figure 7.14. Note that the tests on which Figure 7.14 are based have $t_{dw} = 0$. However, for dynamic loads with $t_{dw} > 0$, the results would not be too different.

The above finding is of great practical importance in estimation of the dynamic load–settlement relationships of foundations. Jackson and Hadala have recommended the following procedure for that purpose.

FIGURE 7.14 Nondimensional relationship of $Q_{d(max)}/B^2 c_u$ and S_{max}/B for model footing tests in Buckshot clay: $t_r = 2$–16 m-sec; $t_{de} = 240$–425 m-sec. [Jackson, J. G., Jr., and Hadala, P. F. (1964, Fig. 14, p. 36).]

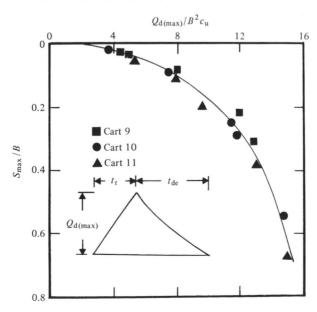

1. Determine the static load Q vs settlement S relationship for a foundation from plate bearing tests in the field.
2. Determine the unconfined compression strength of the soil q_{uc} in the
   laboratory.

$$q_{uc} = 2c_u$$

3. Plot a graph of $Q/B^2 c_u$ vs S_{stat}/B. (See Figure 7.15, curve a.)
4. For any given value of S_{stat}/B, multiply $Q/B^2 c_u$ by the strain rate
   factor (≈ 1.5) and plot in in the same graph. The resulting graph of
   S_{stat}/B vs $1.5Q/B^2 c_u$ will be the predicted relationship between
   $Q_{d(max)}/B^2 c_u$ and S_{max}/B. (See Figure 7.15, curve b.)

Example 7.4. The estimated static plate load bearing test results of a
foundation resting on stiff clay and 5 ft in diameter are given below.

Q (lb)	Settlement (in.)	Q (lb)	Settlement (in.)
0	0	6000	1.65
1000	0.25	8000	2.90
2000	0.48	9000	3.70
4000	1.10	10000	6.80

FIGURE 7.15 Predication of dynamic load–settlement relationship for foundations
on clay. Curve a: static loading; $Q/B^2 c_u$ vs S/B. Curve b: dynamic loading;
$Q_{d(max)}/B^2 c_u$ vs S_{max}/B.

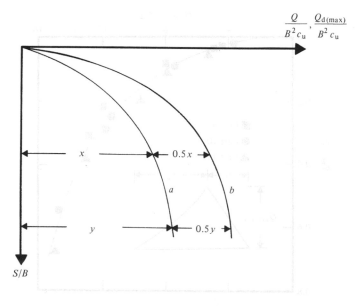

The unconfined compression strength of this clay was 3400 lb/ft^2.

   a. Plot a graph of estimated S_{max}/B vs $Q_{d(max)}/B^2c_u$ assuming a strain-rate factor of 1.5.

   b. Determine the magnitude of the maximum dynamic load $Q_{d(max)}$ which produces a maximum settlement S_{max} of 6 in.

Solution: Given $B = 60$ in. and $c_u = \frac{1}{2}(3400) = 1700$ lb/ft^2, the following table can be prepared.

Q (lb) (1)	S_{stat} (in) (2)	S/B (%) (3)	Q/B^2c_u (4)	$1.5Q/B^2c_u$ (5)
0	0	0	0	0
1000	0.25	0.417	0.0235	0.035
2000	0.48	0.8	0.0471	0.0707
4000	1.10	1.83	0.094	0.141
6000	1.65	2.75	0.141	0.212
8000	2.90	4.83	0.188	0.282
9000	3.70	6.17	0.212	0.318
10000	6.80	11.33	0.235	0.353

FIGURE 7.16

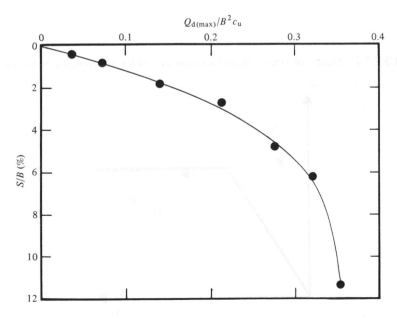

Assuming S/B (Col. 3) to be equal to S_{max}/B and $1.5Q/B^2c_u$ to be equal to $Q_{d(max)}/B^2c_u$, a graph can be plotted (Figure 7.16).

$$\text{For} \quad S_{max} = 6 \text{ in.}, \qquad S_{max}/B = 6/60 = 10\%$$

From Figure 7.16, the value of $Q_{d(max)}/B^2c_u$ corresponding to $S_{max}/B = 10\%$ is about 0.348. Hence

$$Q_{d(max)} = (0.348)(25)(1700) = 14,790 \text{ lb}$$

7.4 ELASTOPLASTIC ANALYSIS: FOUNDATION SUBJECTED TO TRANSIENT LOADING

In contrast to the rigid plastic type of analysis presented in the preceding sections, the assumed nature of variation of stress vs strain for soils in elastoplastic analysis is shown in Figure 7.17. This type of assumption of soil behavior is probably more realistic. Rao and Höeg (1967) have obtained numerical solutions for individual cases of strip foundations resting on saturated clay ($\phi = 0$ condition) soils and being subjected to triangular load pulses. Elastoplastic behavior of the supporting soil has been assumed for these analyses. The solution procedure is not elaborated upon here; readers are instead referred to the original source.

FIGURE 7.17 Nature of stress–strain variation in soil for elastoplastic behavior.

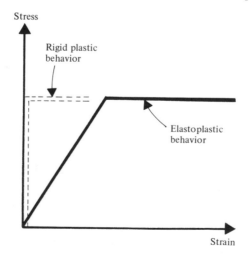

PROBLEMS

7.1. A 3-ft square foundation is supported by dense sand. The relative density of compaction, unit weight, and angle of friction (static) of this sand are 80%, 114 lb/ft^3, and 38°, respectively. Given the depth of foundation is 3 ft, estimate the minimum ultimate bearing capacity of this foundation that might be obtained if the vertical loading velocity on this foundation were varied from static to impact range.

7.2. Redo Problem 7.1 with the depth of foundation as 4 ft.

7.3. Redo Problem 7.1 with the following:
foundation width = 1.4 m
foundation depth = 0.6 m
angle of friction of sand = 36°
unit weight of compacted soil = 16.4 kN/m^3
relative density of compaction of sand = 75%

7.4. A rectangular foundation has a length L of 2.5 m. It is supported by a dense sand with a unit weight of 17 kN/m^3. The sand has an angle of friction of 35°. The depth of the foundation is 0.75 m. The foundation may be subjected to a dynamic load of 945 kN increasing at a moderate rate. Using a factor of safety equal to 2, determine the width of the foundation.

7.5. A foundation 2.25-m square is supported by saturated clay. The unit weight of this clay is 19.6 kN/m^3. The depth of the foundation is 1.5 m. Determine the ultimate bearing capacity of this foundation assuming that the load will be applied very rapidly and given the following for the clay [laboratory unconsolidated–undrained triaxial (static) test results]:

$$\sigma_3 = 70 \text{ kN/m}^2$$

$$\sigma_{1(\text{failure})} = 245 \text{ kN/m}^2$$

Assume a strain-rate factor of 1.4.

7.6. Redo Problem 7.5 with the following changes:
foundation width = 1.5 m
foundation length = 2.6 m
foundation depth = 1.75 m

7.7. Refer to Example 7.2.
   a. Determine the natural period (in seconds) of the foundation for rotational mode of failure due to transient loading.
   b. Using Eq. (7.36), trace the history of motion of the foundation, i.e. $\alpha =$ at 0.05, 1, 1.5, 2, 3, and 4 sec.

7.8. A 3-m-wide strip foundation on clay is subjected to a vertical transient stress pulse $q_d = 500\ e^{-\beta t}$ kN/m^2. Given $\beta = 5$ sec^{-1}, undrained cohesion of clay = 60 kN/m^2 and unit weight of soil = 19.5 kN/m^3, determine the maximum angular rotation that the footing might undergo.

7.9. Repeat Problem 7.8, given

$$q_d = 1000\,e^{-\beta t}\ \mathrm{kN/m^2}$$

$$\beta = 10\ \mathrm{sec}^{-1}$$

7.10. A strip foundation 5 ft wide is resting on a soil (as shown in Figure 7.7b) with $\phi = 15°$, $c = 1000$ lb/ft^2, and $\gamma = 115$ lb/ft^3 as given.
   a. Using a factor of safety of 2 and Eqs. (7.39)–(7.41), determine the allowable load the foundation can support per foot length.
   b. Let this foundation be subjected to a horizontal transient load as shown in Figure 7.7a. Given $t_d = 0.2$ sec and $H = 10$ ft (Figure 7.7b), determine the critical overload factor λ_{cr}.
   c. Using the critical overload factor determined in (b), trace the history of motion of the foundation, i.e., the angular rotation α at time $t = 0.2$, 0.3, 0.4, and 0.5 sec.

7.11. A continuous foundation 5 ft wide is subjected to a vertical transient load as shown in Figure 7.11b, with $q_{d(\max)} = 60{,}000$ lb/ft^2 and $t_d = 0.2$ sec. The foundation is resting on a soil with $c = 0$, $\phi = 35°$, and $\gamma = 115$ lb/ft^2. The depth of the foundation is 4 ft measured from the ground surface. Calculate the maximum vertical movement of the foundation based on the theory described in Section 7.3.3.

7.12. Repeat Problem 7.11 with the following as given:

   foundation width = 3 m
   foundation depth = 0.4 m

$$\gamma = 16\ \mathrm{kN/m^3}, \qquad c = 50\ \mathrm{kN/m^2}, \qquad \phi = 25°$$

$$q_{d(\max)} = 3000\ \mathrm{kN/m^2}, \qquad t_d = 0.4\ \mathrm{sec}$$

7.13. A clay deposit has an undrained cohesion (static test) of 90 kN/m^2. A static field plate load test was conducted with a plate having a diameter of 0.5 m. When the load per unit area q was 200 kN/m^2, the settlement was 20 mm.
   a. Assume that, for a given value of q, the settlement is proportional to the width of the foundation. Estimate the settlement of a prototype circular foundation in the same clay with a diameter of 3 m (static loading).
   b. The strain-rate factor of the clay is 1.4. If a vertical transient load pulse were applied to the foundation as given in part (a), what would be the maximum transient load (in kN) that will produce the same maximum settlement ($S_{\max}$) as calculated in part (a)?

REFERENCES

Brinch Hansen, J. (1970). "A Revised and Extended Formula for Bearing Capacity," *Bulletin No. 28*, Danish Geotechnical Institute, Copenhagen, Denmark.

Carroll, W. F. (1963). "Dynamic Bearing Capacity of Soils. Vertical Displacements of Spread Footings on Clay: Static and Impulsive Loadings," *Technical Report*

No. 3-599, Report 5, U.S. Army Corps of Engineers, Waterways Experiment Station, Vicksburg, Mississippi.

Caquot, A., and Kerisel, J. (1953). "Sur le Terme de Surface Dans le Calcul des Foundations en Milieu Pulve'rulent," *Proceedings, 3rd International Conference on Soil Mechanics and Foundation Engineering*, Zurich, Switzerland, Vol. I, pp. 336–337.

Cunny, R. W., and Sloan, R. C. (1961). "Dynamic Loading Machine and Results of Preliminary Small-Scale Footing Tests," *Special Technical Publication No. 305*, American Society for Testing and Materials, pp. 65–77.

DeBeer, E. E. (1970). "Experimental Determination of the Shape Factors and the Bearing Capacity Factors of Sand,"*Geotechnique* Vol. 20 (4), 387–411.

Heller, L. W. (1964). "Failure Modes of Impact-Loaded Footings on Dense Sand," *Technical Report R-281*, U.S. Naval Civil Engineering Laboratory, Port Hueneme, California.

Jackson, J. G., Jr., and Hadala, P. F. (1964). "Dynamic Bearing Capacity of Soils. Report 3: The Application of Similitude to Small-Scale Footing Tests," U.S. Army Corps of Engineers, Waterways Experiment Station, Vicksburg, Mississippi.

Prakash, S., and Chummar, A. V. (1967). "Response of Footings to Lateral Loads," *Proceedings, International Symposium on Wave Propagation and Dynamic Properties of Earth Materials*, University of New Mexico Press, Albuquerque, New Mexico, pp. 679–691.

Rao, H. A. B., and Höeg, K. (1967). "Two-Dimensional Analysis of Stress and Strain in Soils, Dynamic Response of Strip Footing on Elastic Plastic Soil," *Contract Report 3-129*, Report No. 4, U.S. Army Corps of Engineers, Waterways Experiment Station, Vicksburg, Mississippi.

Reissner, H. (1924). "Zum Erddrukproblem," *Proceedings, First International Conference on Applied Mechanics*, Delft, Netherlands, pp. 294–311.

Shenkman, S., and McKee, K. E. (1961). "Bearing Capacity of Dynamically Loaded Footings," *Special Technical Publication No. 305*, American Society for Testing and Materials, pp. 78–90.

Terzaghi, K. (1943). *Theoretical Soil Mechanics*, Wiley, New York.

Triandafilidis, G. E. (1965). "The Dynamic Response of Continuous Footings Supported on Cohesive Soils," *Proceedings, 6th International Conference on Soil Mechanics and Foundation Engineering*, Vol. II, pp. 205–208.

Vesic, A. S. (1973). "Analysis of Ultimate Loads of Shallow Foundations," *Journal of the Soil Mechanics and Foundations Division, ASCE* 99 (SM1), 45–73.

Vesic, A. S., Banks, D. C., and Woodard, J. M. (1965). "An Experimental Study of Dynamic Bearing Capacity of Footings on Sand," *Proceedings, 6th International Conference on Soil Mechanics and Foundation Engineering*, Montreal, Canada, Vol. II, pp. 209–213.

Wallace, W. F. (1961). "Displacement of Long Footings by Dynamic Loads," *Journal of the Soil Mechanics and Foundations Division, ASCE* 87 (SM5), 45–68.

Whitman, R. V., and Healy, K. A. (1962). "Shear Strength of Sands During Rapid Loading," *Journal of the Soil Mechanics and Foundations Division, ASCE* 88 (SM2), 99–132.

8
EARTHQUAKE
AND GROUND VIBRATION

The ground vibrations due to earthquakes have resulted in several major structural damages in the past. In the North American continent, earthquakes are believed to originate from the rupture of faults. The ground vibration resulting from an earthquake is due to the upward transmission of the stress waves from rock to the softer soil layer(s). In recent times, several major studies have been performed to study the nature of occurrence of earthquakes and the associated amount of energy released. Also, modern techniques have been developed to analyze and estimate the physical properties of soils under earthquake conditions and to predict the ground motion. These developments are the subjects of discussion in this chapter.

8.1 DEFINITION OF SOME EARTHQUAKE-RELATED TERMS

Focus. The focus of an earthquake is a point below the ground surface where the rupture of a fault first occurs (point F in Figure 8.1a).

Focal Depth. The vertical distance from the ground surface to the focus (EF in Figure 8.1a). The maximum focal depth of all earthquakes recorded so far does not exceed 700 km (437.5 miles). Based on focal depth, earthquakes may be divided into the three following categories:

1. *Deep-focus earthquakes*: These have focal depths of 300–700 km (187.5–437.5 miles). They constitute about 3% of all earthquakes recorded around the world and are mostly located in the Circum-Pacific belt.
2. *Intermediate-focus earthquakes*: These have focal depths of 70–300 km (43.75–187.5 miles).
3. *Shallow-focus earthquakes*: The focal depth for these is less than 70 km

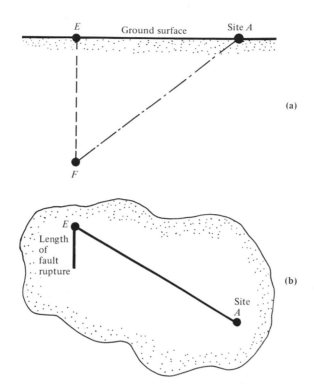

FIGURE 8.1 Definition of focus and epicenter: **(a)** section, **(b)** plane.

(43.75 miles). About 75% of all the earthquakes around the world belong to this category. The California earthquakes have focal depths of about 10–15 km (6.25–9.38 miles).

Epicenter. The point vertically above the focus located on the ground surface (point E in Figure 8.1).

Epicentric Distance. The horizontal distance between the epicenter and a given site (line EA in Figure 8.1).

Hypocentric Distance. The distance between a given site and the focus (line FA in Figure 8.1a).

Effective Distance to Causative Fault. The distance from a fault to a given site for calculation of ground motion (Figure 8.2).

This distance is commonly presumed to be the epicentric distance. This type of assumption, under certain circumstances, may lead to gross errors. It can be explained with reference to Figure 8.2, which shows the plans of

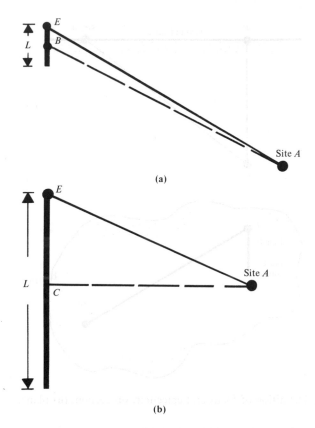

(a)

(b)

FIGURE 8.2 Effective distance from a site to the causative fault. L is the length of the fault rupture.

two cases of fault rupture. In Figure 8.2a, the length of the fault rupture L is small as compared to the epicentric distance EA. In this case, the effective distance could be taken to be equal to the epicentric distance. However, a better estimate of the effective distance is BA (B is the midpoint of the ruptured fault). Figure 8.2b shows the case where the length of the fault rupture is large. In such circumstances, the length AC is the effective distance, which is the perpendicular distance from the site of the line of fault rupture in the plan.

Intensity. An arbitrary scale developed to measure its destructiveness. The *Modified Mercalli Scale* is presently in use in the United States for that purpose, divided into 12 degrees of intensity. An abridged version of the *Modified Mercalli Scale* is given in Table 8.1.

TABLE 8.1 Abridged Modified-Mercalli Intensity Scale[a]

Intensity	Description
I	Detected only by sensitive instruments
II	Felt by a few persons at rest, especially on upper floors; delicate suspended objects may swing
III	Felt noticeably indoors, but not always recognized as a quake; standing autos rock slightly, vibration like passing trucks
IV	Felt indoors by many, outdoors by a few; at night some awaken; dishes, windows, doors disturbed; motor cars rock noticeably
V	Felt by most people; some breakage of dishes, windows and plaster; disturbance of tall objects
VI	Felt by all; many are frightened and run outdoors; falling plaster and chimneys; damage small
VII	Everybody runs outdoors; damage to building varies, depending on quality of construction; noticed by drivers of autos
VIII	Panel walls thrown out of frames; fall of walls, monuments, chimneys; sand and mud ejected; drivers of autos disturbed
IX	Buildings shifted off foundations, cracked, thrown of out plumb; ground cracked underground pipes broken
X	Most masonary and frame structures destroyed; ground cracked; rails bent; landslides
XI	New structures remain standing; bridges destroyed; fissures in ground, pipes broken; landslides; rails bent
XII	Damage total; waves seen on ground surface; lines of sight and level distorted; objects thrown up into air

[a]Wiegel, R. W. (1970, p. 90).

8.2 EARTHQUAKE MAGNITUDE

Magnitude. Is a measure of the size of an earthquake, based on the amplitude of elastic waves it generates. The *magnitude scale* presently in use was first developed by C. F. Richter. The historical developments of the magnitude scale have been summarized by Richter himself (1958).

Richter's earthquake magnitude is defined by the equation

$$\log_{10} E = 11.4 + 1.5M \tag{8.1}$$

where E is the energy released (in ergs) and M is magnitude. Båth (1966) has slightly modified the constants given in Eq. (8.1) and presented it in the

TABLE 8.2 Comparison of the Richter Scale Magnitude with the Modified Mercalli Scale

Richter scale magnitude M	Maximum intensity, Modified Mercalli Scale
1	—
2	I, II
3	III
4	IV
5	VI, VII
6	VIII
7	IX, X
8	XI

form

$$\log_{10} E = 12.24 + 1.44M \qquad (8.2)$$

From the above equation, it can be seen that the increase of M by one unit will generally correspond to about a 30-fold increase of the energy released (E) due to the earthquake. A comparison of the magnitude M of an earthquake with the *maximum intensity* of the Modified Mercalli Scale is given in Table 8.2. Table 8.3 gives a list of some of the past major earthquakes around the world with their magnitudes.

As mentioned previously, the main cause of earthquakes is the rupture of faults. In general, the greater the length of fault rupture, the greater the

TABLE 8.3 Some Past Major Earthquakes

Name	Epicenter Location	Date	Magnitude
Alaska	61.1° N, 147.5° W	March 27, 1964	8.4
Chile (South America)	38° S, 73.5° W	May 22, 1960	8.4
Colombia (South America)	1° N, 82° W	January 31, 1906	8.6
Peru (South America)	9.2° S, 78.8° W	May 31, 1970	7.8
San Francisco, California	38° N, 123° W	April 18, 1906	8.3
Kern County, California	35° N, 119° W	July 21, 1952	7.7
Dixie Valley, Nevada	39.8° N, 118.1° W	December 16, 1954	6.8
Hebgen Lake, Montana	44.8° N, 111.1° W	August 17, 1959	7.1

magnitude of an earthquake. Several relations for the magnitude of the earthquake and the length of fault rupture have been presented by various investigators (Tocher, 1958; Bonilla, 1967; Housner, 1969). Tocher (1958), based on observations of some earthquakes in the area of California and Nevada, suggested the relationship

$$\log L = 1.02 M - 5.77 \qquad (8.3)$$

where L is the length of fault rupture (kilometers).

Based on Eq. (8.3), it can be seen that for an earthquake of magnitude 6, the length of fault rupture is about 2.3 kilometers (1.44 miles). However, when the magnitude is increased to 8, the length of fault rupture associated is about 250 kilometers (153.3 miles).

8.3 AREAS OF INFLUENCE OF AN EARTHQUAKE IN THE UNITED STATES

Because of the differences in geologic conditions, the area over which an earthquake of a given magnitude is felt is different in different areas. Housner (1970) has given the following general approximations for the areas of influence in the United States:

Western United States

$$A = e^{(M+5.1)} - 3000 \qquad (8.4)$$

Rocky Mountain and Central Region

$$A = e^{(M+6.6)} - 14{,}000 \qquad (8.5)$$

Eastern Region

$$A = e^{(M+7.55)} - 34{,}000 \qquad (8.6)$$

where A is the area over which earthquake is felt (in square miles).

8.4 CHARACTERISTICS OF ROCK MOTION DURING AN EARTHQUAKE

The ground motion near the surface of a soil deposit is mostly attributed to the upward propagation of shear waves from the underlying rock or "rocklike" layers. The term "rocklike" implies that the shear wave velocity in the material is similar to that associated with soft rocks. The typical range of shear wave velocities in hard rocks such as granite is about 10,000–12,000 ft/sec ($\approx$ 3050–3660 m/sec). Shear wave velocities associated with soft rocks can be in the low range of 2500–3000 ft/sec (762–915 m/sec). However, the rocklike material may not exhibit the characteristics associated with hard base rocks (Seed et al. 1969). Hence, for arriving at a solution of the nature of ground motion at or near the ground surface, one needs to know some aspects of the earthquake-induced motion in the rock or rocklike materials. The most important of these are

duration of the earthquake

predominant *period* of acceleration

maximum *amplitude* of motion

Each of the above factors have been well summarized by Seed et al. (1969).

Duration of Earthquake

In general, it can be assumed that the duration of an earthquake will be somewhat similar to that of the fault rupture. The rate of propagation of fault rupture has been estimated by Housner (1965) to be about 2 miles/sec (3.2 km/sec). Based on this, Housner has estimated the following variation of the duration of fault rupture with the magnitude of an earthquake.

Magnitude of earthquake (Richter scale)	Duration of fault break (sec)
5	5
6	15
7	25–30

It may be noted that the approximate duration of fault rupture can be estimated from Eq. (8.3). Once the length of rupture L for a given magnitude of earthquake is estimated, the duration can be given by $L/$(velocity of rupture).

Predominant Period of Rock Acceleration

Gutenberg and Richter (1956) have given an estimate of the predominant periods of *accelerations* developed in rock for *California* earthquakes. Similar results for earthquakes of magnitude $M > 7$ have been reported by Figueroa (1960). Using these results, Seed et al. (1969) developed a chart for the *average predominant periods of accelerations* for various earthquake magnitudes. This is shown in Figure 8.3. Note that in this figure the predominant periods are plotted against the distance from the causative fault. This distance is approximately the epicentric distance for the case

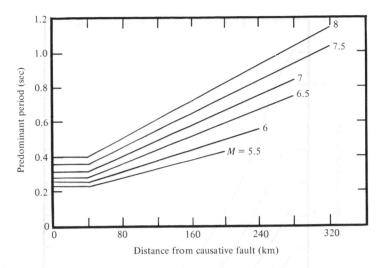

FIGURE 8.3 Predominant period for maximum rock acceleration. Note: 1.6 km = 1 mile. [Seed, H. B., Idriss, I. M., and Kiefer, F. W. (1969). "Characteristics of Rock Motion During Earthquakes," *Journal of the Soil Mechanics and Foundations Division*, *ASCE*, 9 (SM5), Fig. 7, p. 1204.]

where the length of fault rupture is small. However, where the length of fault rupture is large, it is the perpendicular distance from the site to the ruptured fault line in the plan (Figure 8.2).

Maximum Amplitude of Acceleration

The maximum amplitude of acceleration in rock in the epicentric region for shallow earthquakes [focal depth about 16 km (10 miles)] can be approximated as (Gutenberg and Richter, 1956)

$$\log a_0 = -2.1 + 0.81M - 0.027M^2 \tag{8.7}$$

where a_0 is the maximum amplitude of acceleration.

At any other point away from the epicenter, the magnitude of the maximum amplitude of acceleration decreases. Relations for the attenuation factor of the maximum acceleration have been given by Gutenberg and Richter (1956), Banioff (1962), Esteva and Rosenblueth (1963), Kanai (1966) and Blume (1965). Based on these studies, Seed et al. (1969) have given the average values of maximum acceleration for various magnitudes of earthquakes and distances from the causative faults. These are given in Figure 8.4.

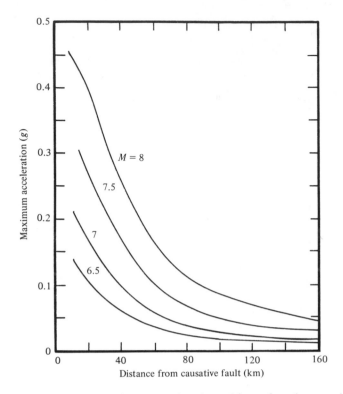

FIGURE 8.4 Variation of maximum acceleration with earthquake magnitude and distance from causative fault. Note: 1.6 km = 1 mile. [Seed, H. B., Idriss, I. M., and Kiefer, F. W. (1969). "Characteristics of Rock Motion During Earthquakes," *Journal of the Soil Mechanics and Foundations Division*, ASCE, 95 (SM5), Fig. 14, p. 1209.]

8.5 VIBRATION OF HORIZONTAL SOIL LAYERS WITH LINEARLY ELASTIC PROPERTIES

As stated before, the vibration of the soil layers due to an earthquake is due to the upward propagation of shear waves from the underlying rock or rocklike layer. The response of a horizontal soil layer with *linearly elastic properties*, developed by Idriss and Seed (1968), is presented in this section.

Homogeneous Soil Layer

Figure 8.5 shows a horizontal soil layer of thickness H underlain by a rock or rocklike material. Let the underlying rock layer be subjected to a seismic motion u_g which is a function of time t. Considering a soil column of unit

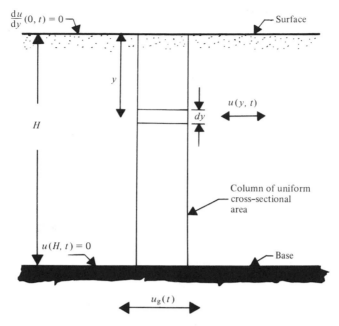

FIGURE 8.5 Cross-section and boundary conditions of a semi-infinite soil layer subjected to a horizontal seismic motion at its base. [Idriss, I. M., and Seed, H. B. (1968). "Seismic Response of Horizontal Soil Layers," *Journal of the Soil Mechanics and Foundations Division*, *ASCE*, 94 (SM4), Fig. 1, p. 1005.]

cross-sectional area, the equation of motion an be written as

$$\rho(y)\frac{\partial^2 u}{\partial t^2} + c(y)\frac{\partial u}{\partial t} - \frac{\partial}{\partial y}\left[G(y)\frac{\partial u}{\partial y}\right] = -\rho(y)\frac{\partial^2 u_g}{\partial t^2} \qquad (8.8)$$

where

u = relative displacement at depth y and time t

$G(y)$ = shear modulus at depth y

$c(y)$ = viscous damping coefficient at depth y

$\rho(y)$ = density of soil at depth y

The shear modulus can be given by the equation (see Section 8.7)

$$G(y) = Ay^B \qquad (8.9)$$

where A and B are constant depending on the nature of the soil.

Substituting Eq. (8.9) into Eq. (8.8), we obtain

$$\rho\frac{\partial^2 u}{\partial t^2} + c\frac{\partial u}{\partial t} - \frac{\partial}{\partial y}\left[Ay^B\frac{\partial u}{\partial y}\right] = -\rho\frac{\partial^2 u_g}{\partial t^2} \qquad (8.10)$$

For the case of $B \neq 0$ (but < 0.5), using the method of separation of variables, the solution to Eq. (8.10) can be given in the form

$$u(y,t) = \sum_{n=1}^{n=\infty} Y_n(y) X_n(t) \tag{8.11}$$

where

$$Y_n(y) = \left(\tfrac{1}{2}\beta_n\right)^b \Gamma(1-b)(y/H)^{b/\theta} J_{-b}\left[\beta_n(y/H)^{1/\theta}\right] \tag{8.12}$$

and

$$\ddot{X}_n + 2D_n\omega_n \dot{X}_n + \omega_n^2 X_n = -R_n \ddot{u}_g \tag{8.13}$$

J_{-b} is the Bessel function of first kind of order $-b$, β_n represents the roots of $J_{-b}(\beta_n) = 0$, $n = 1, 2, 3 \ldots$, and the circular natural frequency of nth mode of vibration

$$\omega_n = \beta_n \sqrt{A/\rho} / \theta H^{1/\theta} \tag{8.14}$$

damping ratio in the nth mode

$$D_n = \tfrac{1}{2}c/\rho\omega_n \tag{8.15}$$

and Γ is the gamma function,

$$R_n = \left[\left(\tfrac{1}{2}\beta_n\right)^{1+b} \Gamma(1-b) J_{1-b}(\beta_n)\right]^{-1} \tag{8.16}$$

The terms b and θ are related as follows:

$$B\theta - \theta + 2b = 0 \tag{8.17}$$

and

$$B\theta - 2\theta + 2 = 0 \tag{8.18}$$

For detailed derivations, see Idriss and Seed (1967).

For obtaining the relative displacement at a depth y, the general procedure is as follows:

1. Determine the system shape $Y_n(y)$ during the nth mode of vibration [Eq. (8.12)].
2. Determine $X_n(t)$ from Eq. (8.13). This can be done by direct numerical step-by-step procedure (Berg and Housner, 1961; Wilson and Clough, 1962) or the iterative procedure as proposed by Newmark (1962).
3. Determine $u(y, t)$ from Eq. (8.11).
4. The relative velocity $[\dot{u}(y, t)]$, relative acceleration $[\ddot{u}(y, t)]$, and strain $\partial u / \partial y$ can be obtained by differentiation of Eq. (8.11).
5. The values of total acceleration, velocity, and displacement can be

obtained as

$$\text{total acceleration} = \ddot{u} + \ddot{u}_g$$
$$\text{total velocity} = \dot{u} + \dot{u}_g$$
$$\text{total displacement} = u + u_g$$

The values of $\dot{u}_g$ and u_g can be obtained by integration of the acceleration record $[\ddot{u}_g(t)]$.

Special Cases

Cohesionless Soils

In the case of cohesionless soils, the shear modulus [Eq. (8.9)] can be approximated to be (see Section 8.7)

$$G(y) = Ay^{1/2} \quad \text{or} \quad G(y) = Ay^{1/3}$$

Assuming the latter to be representative (i.e., $B = \frac{1}{3}$), Eqs. (8.17) and (8.18) can be solved, yielding

$$b = 0.4 \quad \text{and} \quad \theta = 1.2$$

Hence, Eqs. (8.12)–(8.14) take the following form:

$$Y_n(y) = \left(\tfrac{1}{2}\beta_n\right)^{0.4}\Gamma(0.6)(y/H)^{1/3}J_{-0.4}\left[\beta_n(y/H)^{5/6}\right] \quad (8.19)$$

$$\ddot{X}_n + 2D_n\omega_n\dot{X}_n + \omega_n^2 X_n = -\ddot{u}_g\left[\left(\frac{\beta_n}{2}\right)^{1.4}\Gamma(0.6)J_{0.6}(\beta_n)\right]^{-1} \quad (8.20)$$

and

$$\omega_n = \beta_n\sqrt{A/\rho}/1.2H^{5/6} \quad (8.21)$$

(*Note*: $\beta_1 = 1.7510$, $\beta_2 = 4.8785$, $\beta_3 = 8.0166$, $\beta_4 = 11.1570\ldots$.

Cohesive Soils

In cohesive soils, the shear modulus may be considered to be approximately constant with depth; so, in Eq. (8.9), $B = 0$ and

$$G(y) = A \quad (8.22)$$

With this assumption, Eqs. (8.12)–(8.14) are simplified as

$$Y_n(y) = \cos\left[\tfrac{1}{2}(2n-1)(y/H)\right] \quad (8.23)$$

$$\ddot{X}_n + 2D_n\omega_n\dot{X}_n + \omega_n^2 X_n = (-1)^n[4/(2n-1)\pi]\ddot{u}_g \quad (8.24)$$

and

$$\omega_n = [(2n-1)\pi/2H]\sqrt{G/\rho} \quad (8.25)$$

Computer programs for determination of acceleration, velocity, and displacement of soil profiles for these two special cases can be found in Idriss and Seed (1967, Appendix C).

An example of a solution for cohesionless (granular) soil is given in Figure 8.6. Figure 8.7 shows the variation of shear modulus, maximum shear strain, and maximum shear stress with depth for the same soil layer shown

FIGURE 8.6 Surface response of layer with modulus proportional to cube root of depth. Note: 1 in./sec = 25.4 mm/sec; 1 ft = 0.3048 m. [Idriss, I. M., and Seed, H. B. (1968). "Seismic Response of Horizontal Soil Layers," *Journal of the Soil Mechanics and Foundations Division*, *ASCE*, 94 (SM4), Fig. 5, p. 1009.]

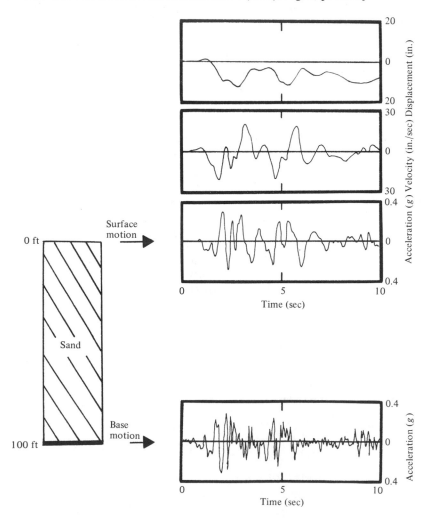

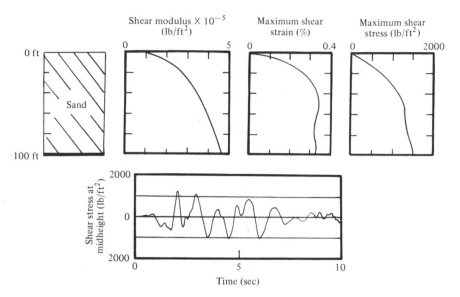

FIGURE 8.7 Stresses and strains developed within the layer of soil shown in Figure 8.6. Note: 1 lb/ft² = 47.88 N/m²; 1 ft = 0.3048 m. [Idriss, I. M., and Seed, H. B. (1968). "Seismic Response of Horizontal Soil Layers," *Journal of the Soil Mechanics and Foundations Division, ASCE*, 94 (SM4), Fig. 6, p. 1010.]

in Figure 8.6. For this example

$$H = 100 \text{ ft } (30.49 \text{ m})$$

$$\text{Total unit weight of soil} = \gamma = 125 \text{ lb/ft}^3 \ (19.65 \text{ kN/m}^3)$$

$$\text{Effective unit weight of soil} = \gamma' = 60 \text{ lb/ft}^3 \ (9.43 \text{ kN/m}^3)$$

$$\text{Shear modulus of soil} = 1 \times 10^5 y^{1/3} \text{ lb/ft}^2$$

$$D = 0.2 \ (\text{for all modes, } n = 1, 2, \ldots, \infty)$$

For a discussion on the damping coefficient of soil under earthquake conditions, see Section 8.7.

Layered Soils

If a soil profile consists of several layers of varying properties which are linearly elastic, a lumped mass type of approach can be taken (Idriss and Seed, 1968). These lumped masses $(m_1, m_2, \ldots, m_N)$ are shown in Figure 8.8. Note

$$m_1 = \gamma_1 h_1 / g \tag{8.26}$$

where m_1 is a lumped mass placed at the top of soil layer No. 1, γ_1 is the

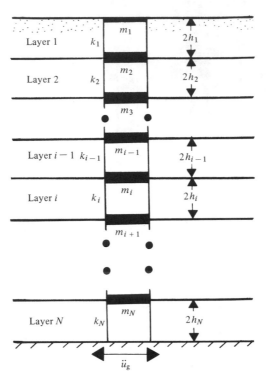

FIGURE 8.8 Lumped mass idealization of horizontal soil layers.

unit weight of soil in layer No. 1, h_1 is the half thickness of soil layer No. 1, and

$$m_i = (\gamma_{i-1} h_{i-1} + \gamma_i h_i)/g, \qquad i = 2, 3, \dots, N \qquad (8.27)$$

These masses are connected by springs which resist lateral deformation. The spring constants can be given by

$$k_i = G_i/2h_i, \qquad i = 1, 2, \dots, N \qquad (8.28)$$

where k_i is the spring constant of the spring connecting the masses m_i and m_{i+1}, and G_i is the shear modulus of layer i.

The equation of motion of the system can be given by the expression

$$[M]\{\ddot{u}\} + [C]\{\dot{u}\} + [K]\{u\} = \{R(t)\} \qquad (8.29)$$

where $[M]$ is a matrix for mass, $[C]$ is a matrix for viscous damping, $[K]$ is the stiffness matrix, and $\{u\}$, $\{\dot{u}\}$, and $\{\ddot{u}\}$ are relative displacement, relative velocity, and relative acceleration vectors, respectively. The matrices $[M]$, $[C]$, and $[K]$ are of the order N (the number of layers considered). The matrix $[M]$ is a diagonal matrix such that

$$\text{diag}[M] = (m_1, m_2, m_3, \dots, m_N) \qquad (8.30)$$

The matrix $[K]$ is tridiagonal and symmetric and

$$K_{11} = k_1$$

$$K_{ij} = k_{i-1} + k_i \qquad \text{for} \quad i = j$$

$$K_{ij} = -k_i \qquad \text{for} \quad i = j - 1$$

$$K_{ij} = -k_j \qquad \text{for} \quad i = j + 1$$

All other K_{ij} are equal to zero.

The load vector $\{R(t)\}$ is

$$\{R(t)\} = -\text{col}(m_1, m_2, \ldots, m_N) \ddot{u}_g \tag{8.31}$$

A computer program for solution of Eq. (8.29) is given in Idriss and Seed (1967, Appendix C). The general outline of the solution is as follows:

1. The number of layers of soil (N) and the mass and stiffness matrices are first obtained.
2. The mode shapes and frequencies are obtained from the characteristic value problem as

$$[K]\{\phi^n\} = \omega_n^2 [M]\{\phi^n\} \tag{8.32}$$

   where ϕ_i^n is the mode shape at the ith level during the nth mode of vibration and ω_n is the circular frequency at the nth mode of vibration.
3. Eq. (8.29) is then reduced to a set of uncoupled normal equations. The normal equations are solved for the response of each mode at each instant of time. The relative displacement at level i can then be expressed as

$$u_i(t) = \sum_{n=1}^{N} \phi_i^n X_n(t) \tag{8.33}$$

   where $X_n(t)$ is the normal coordinate for the nth mode and $u_i(t)$ is the relative displacement at the ith level at time t.
4. The relative velocity $[\dot{u}_i(t)]$ and the relative acceleration $[\ddot{u}_i(t)]$ can be obtained by differentiation of Eq. (8.33), or

$$\dot{u}_i(t) = \sum_{n=1}^{N} \phi_i^n \dot{X}_n(t) \tag{8.34}$$

$$\ddot{u}_i(t) = \sum_{n=1}^{N} \phi_i^n \ddot{X}_n(t) \tag{8.35}$$

5. The total acceleration, velocity, and displacement at level i and time t can be given as follows:

$$\text{total acceleration} = \ddot{u}_i(t) + \ddot{u}_g$$

$$\text{total velocity} = \dot{u}_i(t) + \dot{u}_g$$

$$\text{total displacement} = u_i(t) + u_g$$

6. The shear strain between level i and $i+1$ can be expressed as

$$\text{shear strain} = [u_i(t) - u_{i+1}(t)]/2h_i \qquad (8.36)$$

7. The shear stress between level i and $i+1$ can now be obtained as

$$\tau_i(t) = (\text{shear strain})G \qquad (8.37)$$

Degree of Accuracy and Stability of the Analysis

The degree of accuracy of the lumped mass solution depends on the number of layers of soils used in an analysis. (*Note:* The value of the shear modulus for each layer is assumed to be constant.) In order to select a reasonable number of layers N with a tolerable degree of accuracy, Idriss and Seed (1968) have prepared the graph shown in Figure 8.9, where ERS means the percentage of error in the lumped mass representation. The use of this figure can be explained as follows.

Let the height, shear modulus, and unit weight of the ith layer of soil be H_i, G_i, and γ_i, respectively. The fundamental frequency of this layer can be obtained from Eq. (8.25) as

$$\omega_{n(i)} = [(2n-1)\pi/2H_i]\sqrt{G_i/\rho_i}$$
$$= (\pi/2H_i)\sqrt{G_i/\rho_i} \qquad \text{for} \quad n=1$$

FIGURE 8.9 Plot of N vs T_1 for equal values of ERS. [Idriss, I. M., and Seed, H. B. (1968). "Seismic Response of Horizontal Soil Layers," *Journal of the Soil Mechanics and Foundations Division, ASCE*, 94 (SM4), Fig. 10, p. 1015.]

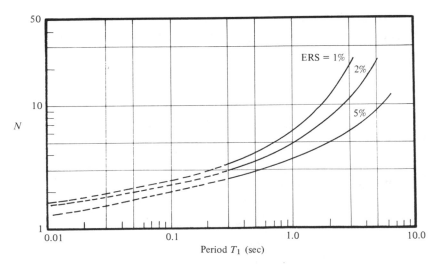

Hence, the fundamental period can be given by

$$T_{1(i)} = 2\pi/\omega_{1(i)} = 4H_i/\sqrt{G_i g/\gamma_i} \tag{8.38}$$

where g is acceleration due to gravity.

Using the value of this $T_{1(i)}$ and a given value of ERS, the value of N_i can be obtained from Figure 8.9; this is the number of layers into which the ith layer has to be divided for the analysis of the ground vibration. Since this needs to be done for each layer of soil,

$$N = \sum N_i \tag{8.39}$$

For the stability of the lumped mass solution, Idriss and Seed (1978) have suggested the following condition:

For the step-by-step solution (Berg and Housner, 1961; Wilson and Clough, 1962):

$$T_{NN} \geqslant 2\Delta t \tag{8.40}$$

For Newmark's iterative solution (1962)

$$T_{NN} \geqslant 5\Delta t \tag{8.41}$$

where Δt is the time interval used for integrating the normal equations and T_{NN} is the lowest period included in the analysis. Note that this corresponds to the highest mode of vibration.

General Remarks for Ground Vibration Analysis

First of all, it should be kept in mind that soil deposits, in general, tend to amplify the underlying rock motion to some degree.

Secondly, for appropriate analysis of ground motion due to an earthquake, it is necessary that an earthquake acceleration–time record be available at the level of the *bedrock* or *bedrocklike material* for a given site. The design accelerogram can be obtained by selecting an actual motion, which has been recorded in the past, of a somewhat similar magnitude and fault distance as the design conditions. This accelerogram is then modified by taking into account the differences between the recorded and design conditions. This modification can be better explained by the following example.

Let the design earthquake be of magnitude 7 and the site be located at a distance of 80 km. Hence, its predominant period at *bedrock* or *bedrocklike material* is 0.4 sec (Figure 8.3) and the maximum acceleration is of the order of $0.04g$ (Figure 8.4). The estimated duration of this earthquake is about 30 sec (equal to the duration of the fault break; Section 8.4). Also, let the recorded earthquake have a predominant period of 0.45 sec, maximum acceleration of $0.05g$, and a duration of 40 sec. The recorded earthquake may now be modified by reducing the ordinates (i.e., magnitudes of acceleration) by $0.04/0.05 = \frac{4}{5}$ and by compressing the time scale by $0.40/0.45 = \frac{8}{9}$. This results in a maximum acceleration of $0.04g$ with a predominant period of 0.4 sec and a duration of 35.5 sec. The first 30 sec of this accelerogram can now be taken for the analysis of ground motion.

Appropriate parts of an accelerogram could be repeated to obtain the desired period of predicted significant motion.

Example 8.1 In a soil deposit, a clay layer has a thickness of 16 m. The unit weight and the shear modulus of the clay soil deposit are 17.8 kN/m^3 and $24,000 \text{ kN/m}^2$, respectively. Determine the number of layers into which this should be divided so that the ERS in the lumped mass solution does not exceed 5%.

Solution: Given that $H_i = 16$ m and $G_i = 24,000 \text{ kN/m}^2$,

$$T_{1(i)} = \frac{4H_i}{\sqrt{G_i g / \gamma_i}} = \frac{(4)(16)}{\sqrt{(24,000 \times 9.81)/17.8}} = 0.556 \text{ sec}$$

From Figure 8.9, with $T_{1(i)} = 0.556$ sec and ERS = 5%, the value of N_i is equal to 3. Thus, this clay layer should be subdivided into at least 3 layers with thicknesses of 5.33 m each.

FIGURE 8.10 Shear stress–strain characteristics of soil: **(a)** stress–strain curve; **(b)** bilinear idealization. [Idriss, I. M., and Seed, H. B. (1968). "Seismic Response of Horizontal Soil Layers," *Journal of the Soil Mechanics and Foundations Division*, *ASCE*, 94 (SM4), Fig. 11, p. 1018.]

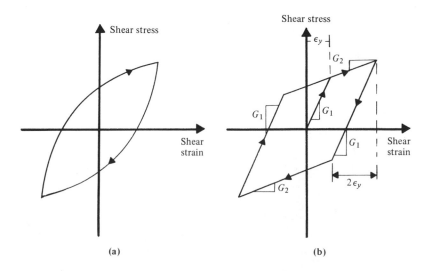

(a)   (b)

8.6 OTHER STUDIES FOR VIBRATION OF SOIL LAYERS DUE TO EARTHQUAKES

In the preceding section, for the evaluation of the ground vibration, it was assumed that

the soil layer(s) possess linearly elastic properties and

the soil layer(s) are horizontal

Under strong ground-shaking conditions, the stress–strain relationships may be of the nature shown in Figure 8.10a, and not linearly elastic. This type of stress–strain relationship can be approximated to a bilinear system as shown in Figure 8.10b and the analysis of ground vibration can then be carried out. The lumped mass type of solution using bilinear stress–strain relationships of horizontally layered soils (Figure 8.11) have been presented

FIGURE 8.11 Lumped mass solution of a semi-infinite layer: bilinear solution. [Idriss, I. M. and Seed, H. B. (1968). "Seismic Response of Horizontal Soil Layers," *Journal of the Soil Mechanics and Foundations Division*, *ASCE*, 94 (SM4), Fig. 12, p. 1019.]

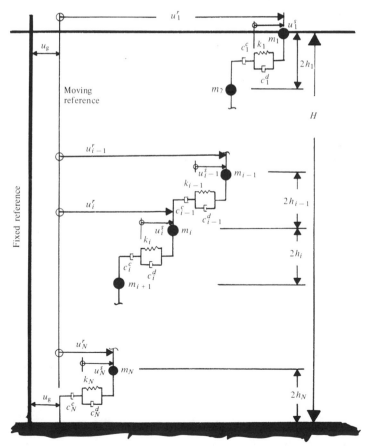

by Parmelee et al. (1964) and Idriss and Seed (1967, 1968), whose works may be examined for further details.

Studies of the vibration of soils with sloping boundaries have also been made by Idriss et al. (1969) and Dezfulian and Seed (1970). This involves a finite element method of analysis. For a computer program of such an analysis, refer to Idriss et al. (1969).

8.7 SHEAR MODULUS AND DAMPING IN SOILS (FOR RESPONSE CALCULATION)

The procedures for the determination of earthquake-induced ground vibration have been outlined in Sections 8.5 and 8.6. They involve use of the shear moduli and damping ratios of soil layers involved. In Chapter 2 (Sections 2.5–2.7), the procedure for experimental determination of these parameters and the empirical correlations therefrom were discussed. The parameters, however, are valid for *low* amplitudes of strain. In earthquake-related problems, the shear strains to which soil layers are subjected are considerably higher. The level of shear strain has considerable effects on the shear moduli and damping of soils: As the magnitude of shear strain increases, the value of shear modulus G of a soil decreases (Figure 8.12), and the damping increases. From Figure 8.12, it may be seen that the value of

FIGURE 8.12 Nature of variation of shear modulus with strain. [Hardin, B. O., and Drnevich, V. P. (1972). "Shear Modulus and Damping Soils: Design Equation and Curves," *Journal of the Soil Mechanics and Foundations Division*, *ASCE*, 98 (SM7), Fig. 2, p. 670.

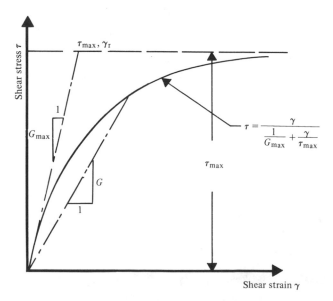

$$\tau = \frac{\gamma}{\dfrac{1}{G_{max}} + \dfrac{\gamma}{\tau_{max}}}$$

TABLE 8.4 Test Procedures for Measuring Moduli and Damping Characteristics[a]

General procedure	Test condition	Approximate strain range	Properties determined
Determination of hysteretic stress–strain relationships	Triaxial compression	10^{-2} to 5%	Modulus; damping
	Simple shear	10^{-2} to 5%	Modulus; damping
	Torsional shear	10^{-2} to 5%	Modulus; damping
Forced vibration	Longitudinal vibrations	10^{-4} to 10^{-2}%	Modulus; damping
	Torsional vibration	10^{-4} to 10^{-2}%	Modulus; damping
	Shear vibrations—lab	10^{-4} to 10^{-2}%	Modulus; damping
	Shear vibrations—field		Modulus
Free vibration tests	Longitudinal vibrations	10^{-3} to 1%	Modulus; damping
	Torsional vibrations	10^{-3} to 1%	Modulus; damping
	Shear vibration—lab	10^{-3} to 1%	Modulus; damping
	Shear vibrations—field	10^{-3} to 1%	Modulus
Field wave velocity measurements	Compression waves	$\approx 5 \times 10^{-4}$%	Modulus
	Shear waves	$\approx 5 \times 10^{-4}$%	Modulus
	Rayleigh waves	$\approx 5 \times 10^{-4}$%	Modulus
Field seismic response	Measurement of motions at different levels in deposit		Modulus; damping

[a] After Seed, H. B., and Idriss, I. M. (1970, Table 1).

the maximum shear modulus $G = G_{max}$ is for a very low strain range and corresponds to those cases discussed in Chapters 2 and 3 (i.e., measurement of field wave velocity). In this section, the procedure for the determination of the shear modulus and the damping at large strain levels is discussed. Table 8.4 presents a summary of the test procedures available and the corresponding strain ranges for evaluation of moduli and damping characteristics of soils.

Estimation of Shear Modulus

Based on several experimental observations, Hardin and Drnevich (1972) proposed a generalized method according to which the variation of shear stress vs strain of all soils can be approximated by a hyperbolic relation as (Figure 8.12)

$$\tau = \frac{\gamma}{1/G_{max} + \gamma/\tau_{max}} \tag{8.42}$$

where τ is shear stress, γ is shear strain,

$$G_{max} = \tau_{max}/\gamma_r \tag{8.43}$$

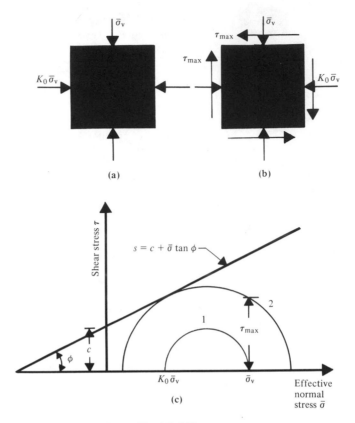

FIGURE 8.13 Definition of τ_{max} Eq. [(8.44)].

γ_r is the reference strain, τ_{max} is the maximum shear stress at failure and

$$G_{max} = 1230 \frac{(2.973 - e)^2}{1 + e} (OCR)^K \overline{\sigma}_0^{1/2} \qquad (2.78)$$

An expression for τ_{max} can be obtained in the following manner. Figure 8.13a shows a soil element at a given depth being subjected to vertical and horizontal effective stresses of $\overline{\sigma}_v$ and $K_0\overline{\sigma}_v$, respectively (K_0 is the earth pressure coefficient at rest). The corresponding Mohr's circle is show as circle 1 in Figure 8.13c. From this Mohr's circle, the relation for τ_{max} can be derived as

$$\tau_{max} = \left\{ \left[\tfrac{1}{2}(1 + K_0)\overline{\sigma}_v \sin\phi + c\cos\phi \right]^2 - \left[\tfrac{1}{2}(1 - K_0)\overline{\sigma}_v \right]^2 \right\}^{1/2} \qquad (8.44)$$

where c is cohesion and ϕ is the effective angle of friction of the soil.

Combining Eqs. (8.43), (2.78), and (8.44) and other empirical correlations, the variation of $\gamma_r/(\overline{\sigma}_v)^{1/2}$ with the angle of friction for sands and the variation of $\gamma_r/(\overline{\sigma}_v)^{1/2}$ with the plasticity indices for saturated cohesive soils can be determined and are shown in Figure 8.14.

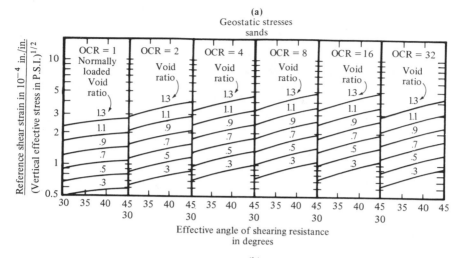

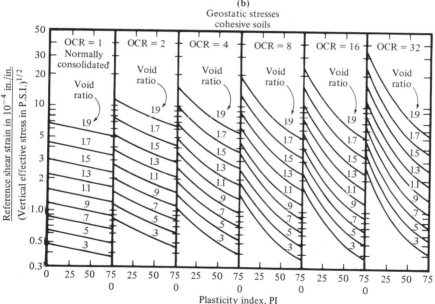

FIGURE 8.14 Reference strain for geostatic stress conditions. [Hardin, B. O., and Drnevich, V. P. (1972). "Shear Modulus and Damping Soils: Design Equation and Curves," *Journal of the Soil Mechanics and Foundations Division, ASCE,* 98 (SM7), Fig. 6, p. 672.]

Knowing that $G = \tau/\gamma$ and $G_{max} = \tau_{max}/\gamma_r$, Eq. (8.42) can be rewritten in the form

$$G = G_{max}/(1 + \gamma/\gamma_r) \qquad (8.45)$$

At this point it needs to be mentioned that, for real soils, the stress–strain relationship somewhat deviated from that given by Eq. (8.42). Hence, Eq. (8.45) has to be modified slightly as

$$G = G_{max}/(1+\gamma_h) \tag{8.46}$$

where hyperbolic strain

$$\gamma_h = (\gamma/\gamma_r)[1 + ae^{-b(\gamma/\gamma_r)}] \tag{8.47}$$

and a and b are constants for soils, representative values of which are given in Table 8.5.

Equation (8.46) can now be used to determine the shear modulus of a soil at any given strain level (see Example 8.2).

Seed and Idriss (1970) have made a detailed parametric evaluation of the works of Hardin and Drnevich described above, and have suggested a simplified relation for the shear modulus of sand. This can be expressed as

$$G = 1000K_2(\overline{\sigma_0})^{1/2} \tag{8.48}$$

where the shear modulus G and the mean effective stress $\overline{\sigma_0}$ are in lb/ft^2 and the values of K_2 are shown in Figure 8.15.

In order to estimate the shear modulus of gravelly soils, Eq. (8.48) may also be used; however, the values of K_2 will change. The variation of some representative values of K_2 with shear strain is shown in Figure 8.16.

For the variation of the *in situ* shear modulus of saturated clays, Seed and Idriss (1970) have collected the experimental results from various sources.

TABLE 8.5 Values of Hyperbolic Strain Constants[a]

Soil type	Modulus or damping	Hyperbolic strain constant[b]	
		Value of a, a_1	Value of b, b_1
Clean dry sand	Modulus	$a = -0.5$	$b = 0.16$
	Damping[c]	$a_1 = 0.6(N^{-1/6})-1$	$b_1 = 1 - N^{-1/12}$
Clean saturated sands	Modulus[c]	$a = -0.2\log N$	$b = 0.16$
	Damping[c]	$a_1 = 0.54(N^{-1/6})-0.9$	$b_1 = 0.65 - 0.65N^{-1/12}$
Saturated cohesive soils	Modulus	$a = 1+0.25(\log N)$	$b = 1.3$
	Damping[d]	$a_1 = 1+0.2(f^{1/2})$	$b_1 = 0.2f(e^{-\overline{\sigma_0}})+2.25\overline{\sigma_0}+0.3(\log N)$

[a] Hardin, B. O., and Drnevich, V. P. (1972). "Shear Modulus and Damping Soils: Design Equation and Curves," *Journal of the Soil Mechanics and Foundations Division*, ASCE, 98 (SM7), Table 2, p. 675.
[b] Number of cycles of loading = N.
[c] These values for modulus and damping of clean sands are for less than 50,000 cycles of loading. Beyond 50,000 cycles, the damping begins to increase with number of cycles, possibly due to fatigue effects. The behavior of modulus for saturated sands beyond 50,000 cycles is not yet established.
[d] Frequency f is in cps and $\overline{\sigma_0}$ is in kg/cm^2.

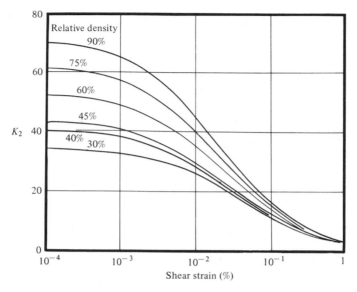

FIGURE 8.15 Values of K_2 [Eq. (8.48)] for sand at different relative densities. [Seed, H. B., and Idriss, I. M. (1970, Fig. 5).]

FIGURE 8.16 Values of K_2 for gravelly soil. [Seed, H. B., and Idriss, I. M. (1970, Fig. 16).]

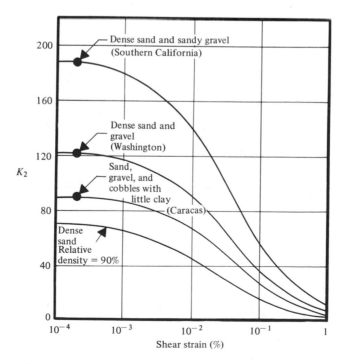

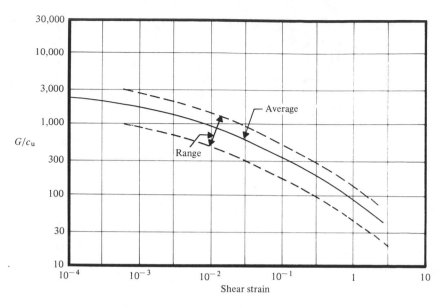

FIGURE 8.17 *In situ* shear modulus for saturated clays. [Seed, H. B., and Idriss, I. M. (1970, Fig. 13).]

FIGURE 8.18 Definition of damping ratio.

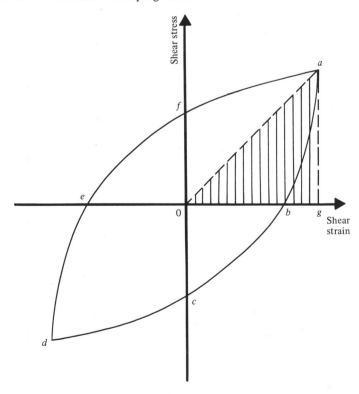

Based on these results, the variation of G/c_u (where c_u is undrained shear strength) with shear strain is shown in Figure 8.17.

Estimation of Damping Ratio

Figure 8.18 shows a shear stress–strain loop for a soil. From this figure, the damping ratio D can be defined as

$$D = \tfrac{1}{4}A_1/\pi A_2 \tag{8.49}$$

where A_1 is the area of the loop $abcdefa$ and A_2 is the area of the triangle $0ag$.

Hardin and Drnevich (1972) presented the relation between the damping ratio and the shear modulus as

$$D = D_{max}(1 - G/G_{max}) \tag{8.50}$$

where D_{max} is the maximum damping ratio, which occurs when $G = 0$. Representative values of D_{max} are given in Table 8.6.

Combining Eqs. (8.45) and (8.50),

$$D/D_{max} = (\gamma/\gamma_r)/(1 + \gamma/\gamma_r) \tag{8.51}$$

In a manner similar to that for the shear modulus, a hyperbolic strain can be defined:

$$\gamma_h = (\gamma/\gamma_r)\left[1 + a_1 e^{-b_1(\gamma/\gamma_r)}\right] \tag{8.52}$$

where a_1 and b_1 are constants for soil. These are given in Table 8.5 for various soils.

Modification of Eq. (8.51) to conform to the real soil behavior gives

$$D/D_{max} = \gamma_h/(1 + \gamma_h) \tag{8.53}$$

TABLE 8.6 Values[a][b] of D_{max}

Soil type	Value of D_{max} (%)
Clean dry sands	$33 - 1.5(\log N)$
Clean saturated sands	$28 - 1.5(\log N)$
Saturated Lick Creek silt	$26 - 4\bar{\sigma}_0^{-1/2} + 0.7f^{1/2} - 1.5(\log N)$
Various saturated cohesive soils including Rhodes Creek Clay	$31 - (3 + 0.03f)\bar{\sigma}_0^{-1/2} + 1.5f^{1/2} - 1.5(\log N)$

[a] Hardin, B. O., and Drnevich, V. P. (1972). "Shear Modulus and Damping Soils: Design Equation and Curves," *Journal of the Soil Mechanics and Foundations Division, ASCE*, 98 (SM7), Table 3, p. 677.

[b] Frequency f is in cps and $\bar{\sigma}_0$ is in kg/cm^2.

The above equation may now be used to obtain the damping ratio for a soil at any strain level.

Seed and Idriss (1970) combined the experimental results for damping from various sources. The upper limit, average, and lower limit for the damping ratio at various strain levels as obtained from this study for sands and saturated clays are shown in Figures 8.19 and 8.20.

Example 8.2 The ground water table in a normally loaded sand layer is located at a depth of 10 ft below the ground surface. The unit weight of sand above the ground water table is 100 lb/ft^3. Below the ground water table, the saturated unit weight of sand is 120 lb/ft^3. Assuming that the void ratio and effective angle of friction of sand below the ground water table are 0.6 and 36°, respectively, determine the damping ratio and the *shear modulus* of this sand at a depth of 25 ft below the ground surface in the fifth cycle if the strain is expected to be about 0.12%. Assume $K_0 = 1 - \sin \phi$. Use the equations proposed by Hardin and Drnevich.

Solution

Calculation of Reference Strain, γ_r. Vertical effective stress at a depth of 25 ft below the ground surface equals

$$\bar{\sigma}_v = (10)(100) + 15(120 - 62.4) = 1864 \text{ lb/ft}^2 = 12.95 \text{ lb/in.}^2$$

FIGURE 8.19 Damping ratio for sand. [Seed, H. B., and Idriss, I. M. (1970, Fig. 10).]

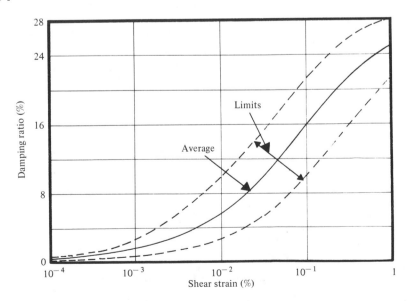

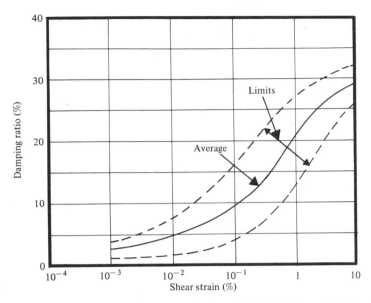

FIGURE 8.20 Damping ratio for saturated clays. [Seed, H. B., and Idriss, I. M. (1970, Fig. 15).]

and the earth pressure coefficient

$$K_0 = 1 - \sin \phi = 1 - \sin 36° = 0.41$$

From Eq. (2.78), for normally consolidated sand

$$G_{max} = \frac{1230(2.973 - e)^2}{1 + e} \bar{\sigma}_0^{-1/2}$$

$$\bar{\sigma}_0 = \tfrac{1}{3}(\bar{\sigma}_v + 2K_0\bar{\sigma}_v) = \tfrac{1}{3}\{12.95[1 + 2(0.41)]\} = 7.86 \text{ lb/in.}^2$$

Thus,

$$G_{max} = \frac{1230(2.973 - 0.6)^2}{1 + 0.6}(7.86)^{1/2} = 12{,}136 \text{ lb/in.}^2$$

Again, from Eq. (8.44) for sand ($c = 0$),

$$\tau_{max} = \left\{\left[\tfrac{1}{2}(1 + K_0)\bar{\sigma}_v\sin\phi\right]^2 - \left[\tfrac{1}{2}(1 - K_0)\bar{\sigma}_v\right]^2\right\}^{1/2}$$

$$= \left\{\left[\tfrac{1}{2}(1 + 0.41)(12.95)\sin 36°\right]^2 - \left[\tfrac{1}{2}(1 - 0.41)(2.95)\right]^2\right\}^{1/2}$$

$$= (28.8 - 14.6)^{1/2} = 3.77 \text{ lb/in.}^2$$

Therefore [Eq. (8.43)],

$$\gamma_r = \tau_{max}/G_{max} = 3.77/12{,}136 = 3.1 \times 10^{-4}$$

Note. If Figure 8.14 is used for $\phi = 36°$ and OCR $= 1$,

$$\gamma_r / \bar{\sigma}_v^{1/2} \approx 0.85 \times 10^{-4}$$

Thus

$$\gamma_r = 0.85\sqrt{12.95} \times 10^{-4} = 3.06 \times 10^{-4}$$

which checks with the above calculation.

Calculation of Shear Modulus, G. From Eq. (8.47)

$$\gamma_h = (\gamma/\gamma_r)[1 + ae^{-b(\gamma/\gamma_r)}]$$

From Table 8.5

$$a = -0.2\log N = -0.2\log 5 = -0.14$$
$$b = 0.16$$

Thus

$$\gamma_h = \frac{12 \times 10^{-4}}{3.06 \times 10^{-4}}\left\{1 - 0.14\exp\left[-0.16\left(\frac{12 \times 10^{-4}}{3.06 \times 10^{-4}}\right)\right]\right\}$$

$$= 3.92(1 - 0.075) = 3.626$$

From Eq. (8.46),

$$G = G_{max}/(1 + \gamma_h) = 12,136/(1 + 3.626) = 2623 \text{ lb/in.}^2$$

This can be approximately checked with the equation given by Seed and Idriss. When $\phi = 36°$, the relative density of sand is about 40%–50%. Using Figure 8.15, K_2 (for 45% relative density and shear strain about 0.12%) is about 12; so

$$G = 1000K_2(\bar{\sigma}_0)^{1/2} \tag{8.48}$$

or

$$G = (1000)(12)(7.86 \times 144)^{1/2} = 403,714 \text{ lb/ft}^2$$

$$= 2804 \text{ lb/in.}^2$$

which checks with less than 10% difference.

Calculation of Damping Ratio. From Eq. (8.53)

$$D = D_{max}\gamma_h/(1 + \gamma_h)$$

$$D_{max} = 28 - 1.5\log N \quad \text{(Table 8.6)}$$

or

$$= 28 - 1.5\log 5 = 26.95$$

and [Eq. (8.52)]

$$\gamma_h = (\gamma/\gamma_r)[1 + a_1 e^{-b_1(\gamma/\gamma_r)}]$$

Table 8.5 gives the equations for a_1 and b_1.

$$a_1 = 0.54(N^{-1/6}) - 0.9 = 0.54(5^{-1/6}) - 0.9 = -0.487$$
$$b_1 = 0.65 - 0.65N^{-1/12} = 0.65 - 0.65(5)^{-0.083} = 0.0813$$

Therefore,

$$\gamma_h = 3.92[1 - 0.487e^{-0.0813(3.92)}] = 2.53$$

and

$$D = (26.95)(2.53)/(1 + 2.53) = 19.32\%$$

Referring to the average curve for damping given in Figure 8.19 (for strain = 0.12%), $D \approx 17\%$. Note that there is a difference of 13.6% between the two values of D determined above. Considering the type of problem, they are in good agreement.

8.8 EQUIVALENT NUMBER OF SIGNIFICANT UNIFORM STRESS CYCLES FOR EARTHQUAKES

In the study of soil liquefaction of granular soils (Chapter 11), it becomes necessary to determine the equivalent number of significant uniform stress cycles for an earthquake which has *irregular stress-time history*. This is explained with the aid of Figure 8.21. Figure 8.21a shows the irregular pattern of shear stress on a soil deposit with time for an earthquake. The maximum shear stress induced is τ_{max}. This irregular stress-time history may be *equivalent* to uniformly intense N number of cyclic shear stresses of maximum magnitude equal to $\beta\tau_{max}$ (Figure 8.21b). The term "equivalent" means that the effect of the stress history shown in Figure 8.21a on a given soil deposit should be same as the uniform stress cycles as shown in Figure 8.21b. From the point of view of soil liquefaction, this fact has been studied by Lee and Chan (1972), Seed et al. (1975), Seed (1976, 1979), and Valera and Donovan (1977).

The basic procedure involved in developing the equivalent stress cycle is fairly simple and has been described by Seed et al. (1975). This is done by using the results of the soil liquefaction study by simple shear tests obtained by DeAlba et al. (1975). Figure 8.22 shows a plot of τ/τ_{max} against the equivalent number of uniform cyclic stresses N at a maximum stress magnitude of $0.65\tau_{max}$. This means, for example, that *one cycle* of shear stress of maximum magnitude τ_{max} is equivalent to three cycles of shear stress of maximum magnitude $0.65\tau_{max}$. Similarly, *one cycle* of shear stress with maximum magnitude of $0.75\tau_{max}$ is equivalent to 1.4 cycles of shear stress with a maximum magnitude of $0.65\tau_{max}$. Figure 8.22 can be used to evaluate the values of N for various earthquakes for a maximum magnitude

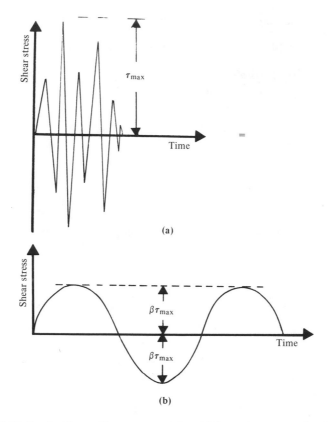

(a)

(b)

FIGURE 8.21 Equivalent uniform stress cycles: **(a)** Irregular stress–time history; **(b)** equivalent uniform stress–time history.

FIGURE 8.22 Plot of τ/τ_{max} vs N at $\tau = 0.65\tau_{max}$. [Seed, H. B., Idriss, I. M., Makdisi, F., and Banerjee, N. 1975, Fig. 7).]

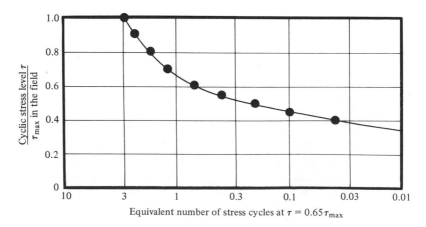

of uniform cyclic shear stress level equaling $0.65\tau_{max}$. (*Note*: $\beta = 0.65$.) This can be most effectively explained by a numerical example. While doing this, one must recognize that, within the top 20 ft ($\approx 6–7$ m) of a given soil deposit, the cyclic shear stress–time history of an earthquake is similar in form to the acceleration–time history at the ground surface. The acceleration–time history for the San Jose earthquake (1955) is shown in Figure 8.23. Note that the maximum acceleration in this case is $0.106g$. Hence, τ_{max} is proportional to $0.106g$. In order to determine N, one needs to prepare Table 8.7. This can be done in the following manner.

1. Looking at Figure 8.23, determine the number of stress cycles at various stress levels such as τ_{max}, $0.95\tau_{max}$, $0.9\tau_{max}$,... above the horizontal axis (col. 2) and below the horizontal axis (col. 5).
2. Determine the conversion factors from Figure 8.22 (cols. 3 and 6).
3. Determine the equivalent number of uniform cycles at a maximum stress level of $0.65\gamma_{max}$ (cols. 4 and 7).

$$\text{col. } 2 \times \text{col. } 3 = \text{col. } 4$$

and

$$\text{col. } 5 \times \text{col. } 6 = \text{col. } 7$$

4. Determine the total number of equivalent stress cycles at $0.65\tau_{max}$ above and below the horizontal axis.
5. $N = \frac{1}{2}$(Equivalent No. of cycles above the horizontal + Equivalent No. of cycles below the horizontal)

Equivalent numbers of uniform stress cycles (at a maximum level of $0.65\tau_{max}$) for several earthquakes with magnitudes of 5.3–7.7 analyzed in

FIGURE 8.23 San Jose earthquake record, 1955. [Seed, H. B., Idriss, I. M. Makdisi, F., and Banerjee, N. (1975, Fig. 8).]

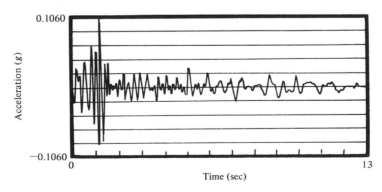

TABLE 8.7 Example of Determination of Equivalent Uniform Cyclic Stress Series from Figure 8.23

	Above horizontal axis			Below horizontal axis		
Stress level $(\otimes\tau_{max})$ (1)	No. of stress cycles (2)	Conversion factor (3)	Equivalent No. of cycles at $0.65\tau_{max}$ (4)	No. of stress cycles (5)	Conversion factor (6)	Equivalent No. of cycles at $0.65\tau_{max}$ (7)
1.00	1	3.00	3.00			
0.95						
0.90	—	—	—			
0.85	—	—	—	1	2.05	2.05
0.80	—	—	—	1	1.70	1.70
0.75	—	—	—			
0.70	—	—	—			
0.65	—	—	—			
0.60	1	0.70	0.70			
0.55	1	0.40	0.40	1	0.40	0.40
0.50						
0.45						
0.40	1	0.04	0.04	1	0.04	0.04
0.35	2	0.02	0.02	1	0.02	0.02
		Total	4.2		Total	4.2

Average number of cycles at $0.65\tau_{max} \approx 4.2$

[a]Seed, H. B., Idriss, I. M., Makdisi, F., and Banerjee, N. (1975, Fig. 8).

the above manner are shown in Figure 8.24. These are for the strongest component of the ground motion recorded. The mean and the mean ± 1 standard deviation (i.e., 16, 50, and 84 percentile) are also shown. This helps the designer choose the proper value of the equivalent uniform stress cycles depending on the degree of conservation required.

Using a similar procedure, Lee and Chan (1972) have given the variation of N with the earthquake magnitude for maximum uniform cyclic stress levels of $0.65\tau_{max}$, $0.75\tau_{max}$, and $0.85\tau_{max}$.

A cumulative damage approach has also been described by Valera and Donovan (1977) for determination of N. This approach is based on Miner's law and involves the natural period of the soil deposit and the duration of earthquake shaking.

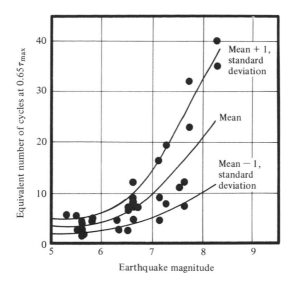

FIGURE 8.24 Equivalent numbers of uniform stress cycles based on strong component of ground motion. [Seed, H. B., Idriss, I. M.,Makdisi, F., and Banerjee, N. (1975, Fig. 12).]

PROBLEMS

8.1. A sand layer 8 m thick is underlain by a 6-m-thick clay layer. Under the clay layer, there is bedrock. Given:

	Unit weight (kN/m^3)	Average shear modulus (kN/m^2)
Sand	16.8	28,000
Clay	18.2	36,000

Determine the number of sublayers into which the sand and clay layers have to be divided so that the ERS in the lumped mass solution does not exceed 2%.

8.2. For a deposit of sand layer, at a certain depth in the field the effective vertical pressure is 120 kN/m^2. The void ratio and the angle of friction are 0.72 and 30°, respectively. Determine the shear modulus and damping ratio for a shear strain level of 0.05% in the 10th cycle. Use the Hardin–Drnevich equations.

8.3. A layer of clay deposit extends to a depth of 50 ft below the ground surface. The ground water table coincides with the ground surface. Given for the clay
void ratio $= 1.0$
specific gravity of soil solids $= 2.78$

plasticity index = 25%
overconsolidation ratio = 2
Determine the shear modulus and damping ratio of this clay at a depth of 25
ft for the fifth cycle at a strain level of 0.1%. Assuming that the frequency of
the earthquake is about 1 cps. (*Note:* Use Figure 8.14.)

$$K_{0(\text{overconsol})} \approx K_{0(\text{norm consol})}(\sqrt{\text{OCR}})$$

$$K_{0(\text{norm consol})} \approx 0.4 + 0.007(\text{PI}) \qquad \text{for} \quad 0 < \text{PI} < 40$$

8.4. The unit weight of a sand layer is 108 lb/ft^3 at a relative density of 60%.
Assume that, for this sand,

$$\phi = 30 + 0.15 R_D$$

where ϕ is the friction angle and R_D is the relative density (in percent). At a
depth of 20 ft below the ground surface, estimate its shear modulus and
damping ratio at a shear strain level of 0.01%. Use the equation proposed by
Seed and Idriss (1970).

8.5. The results of a standard unconsolidated undrained triaxial test on an undisturbed saturated clay specimen are as follows:

$$\text{confining pressure} = 70 \text{ kN/m}^2$$

$$\text{total axial stress at failure} = 166.6 \text{ kN/m}^2$$

Using the method proposed by Seed and Idriss (1970), determine and plot the
variation of shear modulus and damping ratio with shear strain (strain range
$10^{-3}\%{-}1\%$).

8.6. For the determination of the equivalent uniform cyclic shear stress at a
maximum stress level of $0.65\tau_{\text{max}}$, the acceleration–time record of an earthquake was analyzed. The results are tabulated here. Determine the equivalent
number of uniform cyclic shear stress.

Stress level ($\otimes \tau_{\text{max}}$)	No. of stress cycles		Stress level ($\otimes \tau_{\text{max}}$)	No. of stress cycles	
	Above horiz. axis	Below horiz. axis		Above horiz. axis	Below horiz. axis
1.00	1	—	0.60	—	—
0.95	2	1	0.55	—	—
0.90	—	—	0.50	2	1
0.85	—	—	0.45	1	0
0.80	2	2	0.40	—	2
0.75	—	—	0.35	1	1
0.70	1	1	0.30	—	1
0.65	1	1			

REFERENCES

Banioff, H. (1962). Unpublished report to A. R. Golze, Chief Engineer, Department of Water Resources, Sacramento, California, by Department of Water Resources Consulting Board for Earthquake Analysis.

Båth, M. (1966). "Earthquake Seismology," *Earth Science Reviews* 1, 69.

Berg, G. V., and Housner, G. W. (1961). "Integrated Velocity and Displacement of Strong Earthquake Ground Motion," *Bulletin, Seismological Society of America* 51 (2), 175–189.

Blume, J. A. (1965). "Earthquake Ground Motion and Engineering Procedures for Important Installations Near Active Faults," *Proceedings, 3rd World Conference on Earthquake Engineering*, New Zealand, Vol. III.

Bonilla, M. G. (1967). "Historic Surface Faulting in Continental United States and Adjacent Parts of Mexico," Interagency Report, U.S. Department of the Interior, Geological Survey.

DeAlba, P., Chan, C., and Seed, H. B. (1975). "Determination of Soil Liquefaction Characteristics by Large-Scale Laboratory Tests," Earthquake Engineering Research Center, *Report No. EERC 75-14*, University of California, Berkeley.

Dezfulian, H., and Seed, H. B. (1970). "Seismic Response of Soil Deposits Underlain by Sloping Rock Boundaries," *Journal of the Soil Mechanics and Foundations Division, ASCE* 96 (SM6), 1893–1916.

Esteva, L., and Rosenblueth, E. (1963). "Espectros de Temblores a Distancias Moderadas y Grandes," *Proceedings, Chilean Conference on Seismology and Earthquake Engineering*, Vol. 1, University of Chile.

Figueroa, J. J. (1960). "Some Considerations About the Effect of Mexican Earthquakes," *Proceedings, 2nd World Conference on Earthquake Engineering*, Japan, Vol. III.

Gutenberg, B., and Richter, C. F. (1956). "Earthquake Magnitude, Intensity, Energy and Acceleration," *Bulletin of the Seismological Society of America* 46 (2), 105–146.

Hardin, B. O., and Drnevich, V. P. (1972). "Shear Modulus and Damping in Soils: Design Equations and Curves," *Journal of the Soil Mechanics and Foundations Division, ASCE* 98 (SM7), 667–692.

Housner, G. W. (1965). "Intensity of Earthquake Ground Shaking Near the Causative Fault," *Proceedings, 3rd World Conference on Earthquake Engineering*, New Zealand, Vol. 1.

Housner, G. W. (1969). "Engineering Estimate of Ground Shaking and Maximum Earthquake Magnitude," *Proceedings, 4th World Conference on Earthquake Engineering*, Santiago, Chile.

Housner, G. W. (1970). "Design Spectrum," in *Earthquake Engineering* (R. W. Wiegel, Ed.), Prentice-Hall, Englewood Cliffs, New Jersey, pp. 97–106.

Idriss, I. M., Dezfulian, H., and Seed, H. B. (1969). "Computer Programs for Evaluating the Seismic Response of Soil Deposits with Nonlinear Characteristics Using Equivalent Linear Procedures," *Research Report*, Earthquake Engineering Research Center, College of Engineering, University of California, Berkeley.

Idriss, I. M., and Seed, H. B. (1967). "Response of Horizontal Soil Layers During Earthquakes," *Research Report*, Soil Mechanics and Bituminous Materials Laboratory, University of California, Berkeley.

Idriss, I. M., and Seed, H. B. (1968). "Seismic Response of Horizontal Soil Layers," *Journal of the Soil Mechanics and Foundations Division, ASCE* 94 (SM4), 1003–1031.

Kanai, K. (1966). "Improved Empirical Formula for the Characteristics of Strong Earthquake Motions," *Proceedings, Japan Earthquake Engineering Symposium*, Tokyo, pp. 1–4.

Lee, K. L., and Chan, K. (1972). "Number of Equivalent Significant Cycles in Strong Motion Earthquakes," *Proceedings, International Conference on Microzonation*, Seattle, Washington, Vol. II. pp. 609–627.

Newmark, N. M. (1962). "A Method of Computations for Structural Dynamics," *Transactions, ASCE* 127 (Part I), 1406–1435.

Parmellee, R., Penzien, J., Scheffey, C. F., Seed, H. B., and Thiers, G. R. (1964). "Seismic Effects on Structures Supported on Piles Extending Through Deep Sensitive Clays," *Report No. 64-2*, Institute of Engineering Research, University of California, Berkeley.

Richter, C. F. (1958). *Elementary Seismology*, W. H. Freeman, San Francisco, California.

Seed, H. B. (1976). "Evaluation of Soil Liquefaction Effects on Level Ground During Earthquakes," *Preprint No. 2752, ASCE National Convention*, Sept. 27–Oct. 1, pp. 1–104.

Seed, H. B. (1979). "Soil Liquefaction and Cyclic Mobility Evaluation for Level Ground During Earthquakes," *Journal of the Geotechnical Engineering Division, ASCE* (GT2), 201–255.

Seed, H. B., Idriss, I. M., and Kiefer, F. W. (1969). "Characteristics of Rock Motion During Earthquakes," *Journal of the Soil Mechanics and Foundations Division, ASCE* 95 (SM5), 1199–1218.

Seed, H. B., and Idriss, I. M. (1970). "Soil Moduli and Damping Factors for Dynamic Response Analysis," *Report No. EERC 70-10*, Earthquake Engineering Research Center, University of California, Berkeley.

Seed, H. B., Idriss, I. M., Makdisi, F., and Banerjee, N. (1975). "Representation of Irregular Stress–Time Histories by Equivalent Uniform Stress Series in Liquefaction Analyses," *Report No. EERC 75-29*, Earthquake Engineering Research Center, University of California, Berkeley.

Tocher, D. (1958). "Earthquake Energy and Ground Breakage," *Bulletin of the Seismological Society of America* 48 (2), 147–153.

Valera, J. E., and Donovan, N. C. (1977). "Soil Liquefaction Procedures—A Review," *Journal of the Geotechnical Engineering Division, ASCE* 103 (GT6), 607–625.

Wiegel, R. W. (ed.) (1970). *Earthquake Engineering*, Prentice-Hall, Englewood Cliffs, New Jersey.

Wilson, E. L., and Clough, R. W. (1962). "Dynamic Response by Step-by-Step Matrix Analysis," *Proceedings, Symposium on the Use of Computers in Civil Engineering*, Lisbon, Portugal.

9

LATERAL EARTH PRESSURE ON RETAINING WALLS

Excessive dynamic lateral earth pressure on retaining structures resulting from earthquakes has caused several major damages in the past. The increase of lateral earth pressure during earthquakes induces sliding and/or tilting to the retaining structures. The majority of case histories of failures reported in the literature until now concern waterfront structures such as quay walls and bridge abutments. Some of the examples of failures and lateral movements of quay walls due to earthquakes are given in Table 9.1. Sccd and Whitman (1970) have suggested that some of these failures may have been due to several reasons, such as

increase of lateral earth pressure behind the wall

reduction of water pressure at the front of the wall

liquefaction of the backfill material (see Chapter 11)

Nazarian and Hadjan (1979) have given a comprehensive review of the dynamic lateral earth pressure studies advanced so far. Based on this study, the theories can be divided into three broad categories, such as

1. fully plastic (static) solution
2. solutions based on elastic wave theory
3. solutions based on elastoplastic and nonlinear theory

In this chapter, the lateral earth pressure theory, based on the fully plastic solution which is widely used by most of the design engineers, is developed.

9.1 MONONOBE–OKABE ACTIVE EARTH PRESSURE THEORY

In 1776, Coulomb derived an equation for active earth pressure on a retaining wall due to a dry cohesionless backfill (Figure 9.1), which is of the form

$$P_A = \tfrac{1}{2}\gamma H^2 K_A \qquad (9.1)$$

302

TABLE 9.1 Failures and Movements of Quay Walls[a]

Earthquake	Date	Magnitude	Harbor	Distance from Epicenter
Kitaizu	25 November 1930	7.1	Shimizu	30 miles (48 km)
Shizuoka	11 July 1935		Shimizu	
Tonankai	7 December 1944	8.2	Shimizu	110 miles (175 km)
			Nagoya	80 miles (128 im)
			Yokkaichi	90 miles (144 km)
Nankai	21 December 1946	8.1	Nagoya	
			Osaka	125–190 miles
			Yokkaichi Uno	(200–304 km)
Tokachioki	4 March 1952	7.8	Kushiro	90 miles (144 km)
Chile	22 May 1960	8.4	Puerto Montt	70 miles (112 km)
Niigata	16 June 1964	7.5	Niigata	32 miles (51.2 km)

[a]Seed, H. B., and Whitman, R. V. (1970). "Design of Earth Retaining Structures for Dynamic Loads," *Proceedings, Specialty Conference on Lateral Stresses in the Ground and Design of Earth Retaining Structures, ASCE*, Table 1, p. 112.

FIGURE 9.1 Coulomb's active earth pressure: BC is the failure plane; W = weight of the wedge ABC; S and N = shear and normal forces on the plane BC; and F = resultant of S and N.

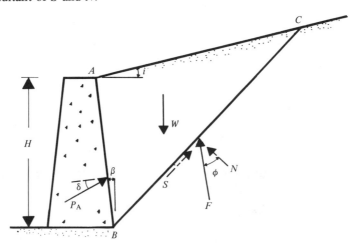

TABLE 9.1 (*continued*)

Damage	Approximate Movement
Failure of gravity walls[b]	26 ft (7.93 m)
Retaining wall collapse[b]	16 ft (4.88 m)
Sliding of retaining wall[b] Outward movement of bulkhead with relieving platform[b] Outward movement of pile-supported deck[b]	10–13 ft (3.05–3.96 m) 12 ft (3.66 m)
Outward movement of bulkhead with relieving platform[b] Failure of retaining wall above relieving platform[b] Outward movement of pile-supported deck[b] Outward movement of gravity wall[b]	13 ft (3.96 m) 14 ft (4.27 m) 12 ft (3.66 m) 2 ft (0.61 m)
Outward movement of gravity wall[b]	18 ft (5.49 m)
Complete overturning of gravity walls[c]	> 15 ft (4.57 m)
Outward movement of anchored bulkheads[c]	2–3 ft (0.61–0.915 m)
Tilting of gravity wall[d] Outward movement of anchored bulkheads[d]	10 ft (3.05 m) 1–7 ft (0.305–2.13 m)

[b] Reported by Amano, Azuma and Ishii (1956).
[c] Reported by Duke and Leeds (1963).
[d] Reported by Hayashi, Kubo and Nakase (1966).

where P_A is the active force/unit length of the wall, γ is the unit weight of soil, H is the height of the retaining wall, and K_A is the active earth pressure coefficient

$$K_A = \frac{\cos^2(\phi - \beta)}{\cos^2\beta \cos(\delta + \beta)\left[1 + \left\{\frac{\sin(\delta + \phi)\sin(\phi - i)}{\cos(\delta + \beta)\cos(\beta - i)}\right\}^{1/2}\right]^2} \quad (9.2)$$

where ϕ is the soil friction angle, δ is the angle of wall friction, β is the slope of the back of the wall with respect to the vertical, and i is the slope of the backfill with respect to the horizontal. The values of K_A for various i, β, δ, and ϕ values are given in Table 9.2.

Coulomb's active earth pressure equation can be modified to take into account the vertical and horizontal coefficients of acceleration induced by

TABLE 9.2 Values[a] of K_A

δ	$\phi = 26$	28	30	32	34	36	38	40
				$\beta = 0, \quad i = 0$				
0	0.390	0.361	0.333	0.307	0.283	0.260	0.238	0.217
16	0.349	0.324	0.300	0.278	0.257	0.237	0.218	0.201
17	0.348	0.323	0.299	0.277	0.256	0.237	0.218	0.200
20	0.345	0.320	0.297	0.276	0.255	0.235	0.217	0.199
22	0.343	0.319	0.296	0.275	0.254	0.235	0.217	0.199
				$\beta = 0, \quad i = 5$				
0	0.414	0.382	0.352	0.323	0.297	0.272	0.249	0.227
16	0.373	0.345	0.319	0.295	0.272	0.250	0.229	0.210
17	0.372	0.344	0.318	0.294	0.271	0.249	0.229	0.210
20	0.370	0.342	0.316	0.292	0.270	0.248	0.228	0.209
22	0.369	0.341	0.316	0.292	0.269	0.248	0.228	0.209
				$\beta = 0, \quad i = 10$				
0	0.443	0.407	0.374	0.343	0.314	0.286	0.261	0.238
16	0.404	0.372	0.342	0.315	0.289	0.265	0.242	0.221
17	0.404	0.371	0.342	0.314	0.288	0.264	0.242	0.221
20	0.402	0.370	0.340	0.313	0.287	0.263	0.241	0.220
22	0.401	0.369	0.340	0.312	0.387	0.263	0.241	0.220

[a]From *Foundation Analysis and Design*, Second edition, by J. E. Bowles. Copyright © 1978 McGraw-Hill. Used with the permission of McGraw-Hill Book Company.

an earthquake. This is generally referred to as the *Mononobe – Okabe analysis* (Mononobe, 1929; Okabe, 1926). The Mononobe–Okabe solution is based on the following assumptions:

1. The failure in soil takes place along a plane such as *BC* shown in Figure 9.2.
2. The movement of the wall is sufficient to produce minimum active pressure.
3. The shear strength of the dry cohesionless soil can be given by the equation

$$s = \sigma' \tan \phi \tag{9.3}$$

where σ' is the effective stress and s is shear strength.
4. At failure, full shear strength along the failure plane (plane *BC*, Figure 9.2) is mobilized.
5. The soil behind the retaining wall behaves as a rigid body.

Figure 9.2 shows the forces considered in the Mononobe–Okabe solution. Line *AB* is the back face of the retaining wall and *ABC* is the soil wedge which will fail. The forces on the failure wedge per unit length of the wall

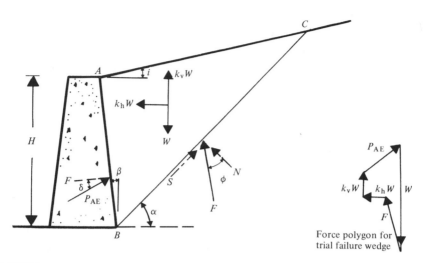

FIGURE 9.2 Derivation of Mononabe–Okabe equation: unit weight of soil $= \gamma$; friction angle $= \phi$.

are:

weight of wedge W

active force P_{AE}

resultant of shear and normal forces along the failure plane F

$k_h W$ and $k_v W$, the inertia forces in the horizontal and vertical directions, respectively, where

$$k_h = \frac{\text{horiz. component of earthquake accel.}}{g}$$

$$k_v = \frac{\text{vert. component of earthquake accel.}}{g}$$

and g is acceleration due to gravity.

The active force determined by the wedge analysis described above may be expressed as

$$P_{AE} = \tfrac{1}{2}\gamma H^2 (1 - k_v) K_{AE} \qquad (9.4)$$

where K_{AE} is the active earth pressure coefficient with earthquake effect,

$$K_{AE} = \frac{\cos^2(\phi - \theta - \beta)}{\cos\theta\cos^2\beta\cos(\delta + \beta + \theta)\left[1 + \sqrt{\dfrac{\sin(\phi + \delta)\sin(\phi - \theta - i)}{\cos(\delta + \beta + \theta)\cos(i - \beta)}}\right]^2}$$

$$(9.5)$$

$$\theta = \tan^{-1}\left[k_h / (1 - k_v)\right] \qquad (9.6)$$

Equation (9.4) is generally referred to as the *Mononobe – Okabe active earth pressure equation.*

Since the values of K_A (Table 9.2) are available in most standard handbooks and textbooks, I. Arango (1969) developed a simple procedure for obtaining the values of K_{AE} from the standard charts of K_A. This procedure has been described by Seed and Whitman (1970). Referring to Eq. (9.1),

$$P_A = \tfrac{1}{2}\gamma H^2 K_A = \tfrac{1}{2}\gamma H^2 A_c \left(\cos^2\beta\right)^{-1} \tag{9.7}$$

where

$$A_c = K_A \cos^2\beta$$

$$= \frac{\cos^2(\phi - \beta)}{\cos(\delta + \beta)\left[1 + \left\{\dfrac{\sin(\delta + \phi)\sin(\phi - i)}{\cos(\delta + \beta)\cos(\beta - i)}\right\}^{1/2}\right]^2} \tag{9.8}$$

In a similar manner, from Eq. (9.4)

$$P_{AE} = \tfrac{1}{2}\gamma H^2 (1 - k_v) K_{AE} = \tfrac{1}{2}\gamma H^2 (1 - k_v)\left(\cos\theta\cos^2\beta\right)^{-1}(A_m) \tag{9.9}$$

where

$$A_m = K_{AE}\cos\theta\cos^2\beta$$

$$= \frac{\cos^2(\phi - \beta - \theta)}{\cos(\delta + \beta + \theta)\left[1 + \left\{\dfrac{\sin(\phi + \delta)\sin(\phi - i - \theta)}{\cos(\delta + \beta + \theta)\cos(i - \beta)}\right\}^{1/2}\right]^2} \tag{9.10}$$

Now let

$$i' = i + \theta \tag{9.11}$$

and

$$\beta' = \beta + \theta \tag{9.12}$$

Substitution of Eqs. (9.11) and (9.12) into Eq. (9.10) yields

$$A_m = \frac{\cos^2(\phi - \beta')}{\cos(\delta + \beta')\left[1 + \left\{\dfrac{\sin(\phi + \delta)\sin(\phi - i')}{\cos(\delta + \beta')\cos(\beta' - i')}\right\}^{1/2}\right]^2} \tag{9.13}$$

The preceding equation is similar to Eq. (9.8) except for the fact that i' and β' are used in place of i and β. Thus, it can be said that

$$A_m = A_c(i', \beta') = K_A(i', \beta')\cos^2\beta'$$

The active earth pressure P_{AE} can now be expressed as

$$P_{AE} = \tfrac{1}{2}\gamma H^2 (1 - k_v)\left[\cos^2\beta'/(\cos\theta\cos^2\beta)\right] K_A(i',\beta')$$
$$= P_A(i',\beta')(1 - k_v)(\overset{*}{p}) \tag{9.14}$$

where

$$\overset{*}{p} = \left[\cos^2\beta'/(\cos\theta\cos^2\beta)\right] \tag{9.15}$$

In order to calculate P_{AE} by using Eq. (9.14), one needs to follow these steps:

1. Calculate i' [Eq. (9.11)].
2. Calculate β' [Eq. (9.12)].
3. With known values of ϕ, δ, i', and β', calculate K_A (from Table 8.2 or other available charts).
4. Calculate P_A as equal to $\tfrac{1}{2}\gamma H^2 K_A$ (K_A from Step 3).
5. Calculate $(1 - k_v)$.
6. Calculate $\overset{*}{p}$ [Eq. (9.15)].
7. Calculate $P_{AE} = \underset{\text{(step 4)}}{P_A(i',\beta')} \ \underset{\text{(step 5)}}{(1 - k_v)} \ \underset{\text{(step 6)}}{(\overset{*}{p})}$

For convenience, some typical values of $\overset{*}{p}$ are plotted in Figure 9.3.

Some Comments on the Active Force Equation [Eq. (9.4)]

Considering the active force relation given by Eqs. (9.4)–(9.6), the term $\sin(\phi - \theta - i)$ in Eq. (9.5) has some important implications:

First, if $\phi - \theta - i < 0$ (i.e., negative), no real solution of K_{AE} is possible. Physically it implies that an *equilibrium condition will not exist*. Hence, for stability, the limiting slope of the backfill may be given by

$$i \leqslant \phi - \theta \tag{9.16}$$

For no earthquake condition, $\theta = 0$; for stability, Eq. (9.16) gives the familiar relation

$$i \leqslant \phi \tag{9.17}$$

Secondly, for horizontal backfill, $i = 0$; for stability,

$$\theta \leqslant \phi \tag{9.18}$$

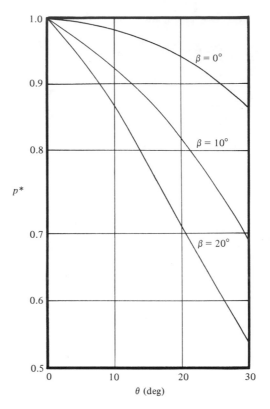

$$\theta \ (\text{deg})$$

FIGURE 9.3 Variation of $\overset{*}{p}$ with θ.

Since $\theta = \tan^{-1}[k_h/(1-k_v)]$, for stability, combining Eqs. (9.6) and (9.18) results as

$$k_h \leqslant (1-k_v)\tan\phi \qquad (9.19)$$

Hence, the critical value of the horizontal acceleration can be defined as

$$k_{h(cr)} = (1-k_v)\tan\phi \qquad (9.20)$$

where $k_{h(cr)}$ is the critical value of horizontal acceleration (Figure 9.4).

Example 9.1 Refer to Figure 9.2. If $\beta = 0°$, $i = 0°$, $\phi = 36°$, $\delta = 18°$, $H = 15$ ft, $\gamma = 110$ lb/ft^3, $k_v = 0.2$, and $k_h = 0.3$, determine the active force per unit length of the wall.

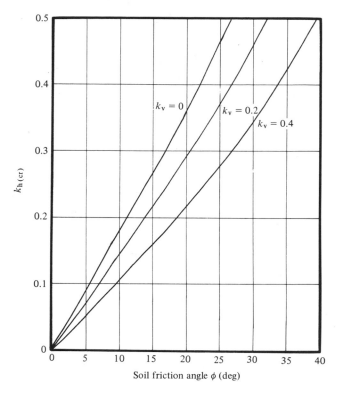

FIGURE 9.4 Critical values of horizontal acceleration.

Solution

$$\theta = \tan^{-1}[k_h/(1-k_v)] = \tan^{-1}[0.3/(1-0.2)] = 20.56°$$
$$i' = i + \theta = 0 + 20.56° = 20.56°$$
$$\beta' = \beta + \theta = 0 + 20.56° = 20.56°$$

$$K_A(i', \beta') = \frac{\cos^2(\phi - \beta')}{\cos^2\beta' \cos(\delta + \beta')\left[1 + \left\{\dfrac{\sin(\delta + \phi)\sin(\phi - i')}{\cos(\delta + \beta')\cos(\beta' - i')}\right\}^{1/2}\right]^2}$$

$$= \frac{\cos^2(15.44)}{(\cos^2 20.56)(\cos 38.56)\left[1 + \left\{\dfrac{(\sin 54)(\sin 15.44)}{(\cos 38.56)(\cos 0)}\right\}^{1/2}\right]^2}$$

$$= 0.583$$

$$P_A(i', \beta') = \tfrac{1}{2}\gamma H^2 K_A(i', \beta') = \tfrac{1}{2}(110)(15)^2(0.583) = 7214.62 \text{ lb/ft}$$

$$\overset{*}{p} = [\cos^2\beta'/(\cos\theta\cos^2\beta)] = \cos^2 20.56/[(\cos 20.56)(\cos 0)]$$

$$= 0.9363$$

310

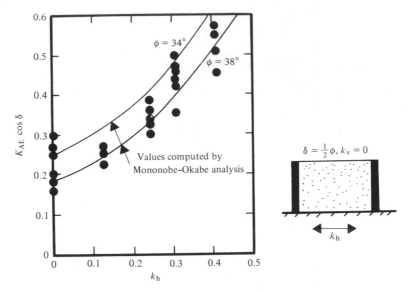

FIGURE 9.5 Results of model tests by Mononabe and Matsuo (1929). [Seed, H. B., and Whitman, R. V. (1970). "Design of Earth Retaining Structures for Dynamic Loads," *Proceedings, Specialty Conference on Lateral Stresses in the Ground and Design of Earth Retaining Structures, ASCE,* Fig. 19, p. 121.]

FIGURE 9.6 Results of model tests by Jacobson (1939). [Seed, H. B., and Whitman, R. V. (1970). "Design of Earth Retaining Structures for Dynamic Loads," *Proceedings, Specialty Conference on Lateral Stresses in the Ground and Design of Earth Retaining Structures, ASCE,* Fig. 20, p. 121.]

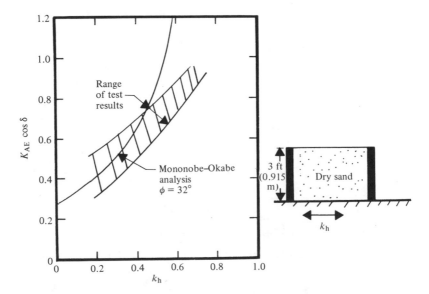

Hence, from Eq. (9.14),

$$P_{AE} = P_A(i', \beta')(1-k_v)\left(\overset{*}{p}\right) = (7214.62)(1-0.2)(0.9363) = 5404 \text{ lb/ft}$$

9.2 LABORATORY MODEL TEST RESULTS FOR DETERMINATION OF K_{AE}

Various small-scale laboratory model tests have been conducted in the past to compare the theoretical active earth pressure coefficient K_{AE} with experimental results. Mononobe and Matsuo (1929) conducted several tests using a box full of dry sand. The box was subjected to horizontal accelerations and measurements of the maximum pressures were made. The experimental values of $K_{AE}\cos\delta$ derived from these tests are shown in Figure 9.5. It can be seen from this figure that the experimental values are in good agreement with the Mononobe–Okabe theory.

Figure 9.6 also shows the experimental results of Jacobsen (1939) conducted on a 3-ft (0.915-m)-high wall with dry sand as backfill. The wall was

FIGURE 9.7 Influence of wall friction angle δ on K_{AE}: $k_v = 0$, $\beta = 0$, $i = 0$, $\phi = 35°$. [Seed, H. B., and Whitman, R. V. (1970). "Design of Earth Retaining Structures for Dynamic Loads," *Proceedings*, *Specialty Conference on Lateral Stress in the Ground and Design of Earth Retaining Structures*, *ASCE*, Fig. 12, p. 114.]

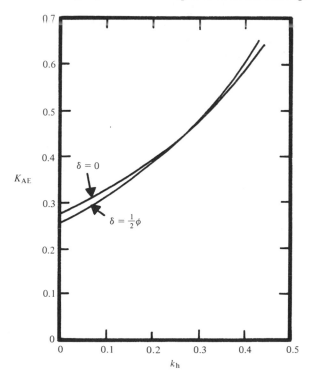

subjected to harmonic base excitation. The values of $K_{AE}\cos\delta$ derived from these experimental results are in good agreement with the Mononobe–Okabe theory up to a value of $k_h = 0.45$. For $k_h > 0.45$, the experimental values of $K_{AE}\cos\delta$ were considerably lower than predicted by the Mononobe–Okabe solution.

9.3 EFFECT OF VARIOUS PARAMETERS ON THE VALUE OF THE ACTIVE EARTH PRESSURE COEFFICIENT

Parameters such as the angle of wall friction, angle of friction of soil, and slope of the backfill influence the magnitude of the active earth pressure coefficient K_{AE} to varying degrees. The effect of each of these factors is briefly considered below.

Effect of the Wall Friction Angle δ

For all practical cases of design, one may usually adopt a value of δ varying from 0 to $\frac{1}{2}\phi$. For most practical cases, the effect of the wall friction angle $(0 < \delta < \frac{1}{2}\phi)$ on the value of K_{AE} is small. This is demonstrated in Figure 9.7 for a vertical wall with horizontal backfill for $\phi = 35°$.

FIGURE 9.8 Effect of soil friction angle on $K_{AE}\cos\delta$: $k_v = 0$, $\beta = 0$, $i = 0$, $\delta = \frac{1}{2}\phi$. [Seed, H. B., and Whitman, R. V. (1970). "Design of Earth Retaining Structures for Dynamic Loads," *Proceedings, Specialty Conference on Lateral Stresses in the Ground and Design of Earth Retaining Structures, ASCE*, Fig. 13, p. 117.]

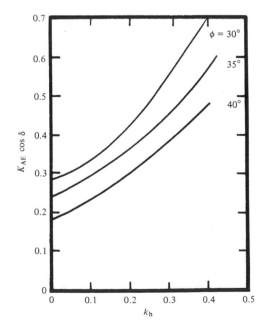

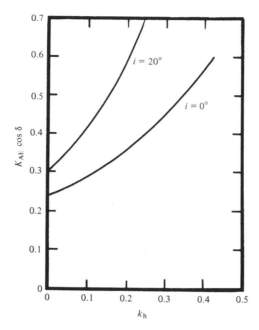

FIGURE 9.9 Effect of the inclination of the backfill on $K_{AE}\cos\delta$: $\phi = 35°$, $k_v = 0$, $\beta = 0$, $\delta = \frac{1}{2}\phi$. [Seed, H. B., and Whitman, R. V. (1970). "Design of Earth Retaining Structures for Dynamic Loads," *Proceedings, Specialty Conference on Lateral Stresses in the Ground and Design of Earth Retaining Structures, ASCE,* Fig. 14, p. 117.]

Effect of Soil Friction Angle ϕ

Figure 9.8 shows a plot of $K_{AE}\cos\delta$ for a vertical retaining wall with horizontal backfill ($i = 0°$, $\beta = 0°$). It may be seen that, for $k_v = 0$, $k_h = 0$, and $\delta = \frac{1}{2}\phi$, $K_{AE(\phi=30°)}$ is about 35% higher than $K_{AE(\phi=40°)}$. Similarly, for $k_v = 0$, $k_h = 0.4$, and $\delta = \frac{1}{2}\phi$, the value of $K_{AE(\phi=30°)}$ is about 30% higher as compared to $K_{AE(\phi=40°)}$. Hence, a small error in the assumption of the soil friction angle could lead to a large error in the estimation of P_{AE}.

Effect of the Slope of the Backfill i

Figure 9.9 shows the variation of the value of $K_{AE}\cos\delta$ with i for a wall with $\beta = 0$, $\delta = \frac{1}{2}\phi$, $\phi = 35°$, and $k_v = 0$. Note that the value of K_{AE} sharply increases with the increase of the slope of the backfill.

9.4 POINT OF APPLICATION OF THE RESULTANT EARTH PRESSURE

The original Mononobe–Okabe solution for the active force on retaining structures implied that the resultant force will act at a distance of $\frac{1}{3}H$ measured from the bottom of the wall (H = height of the wall) similar to that in the static case ($k_h = k_v = 0$). However, all the laboratory tests that have been conducted so far indicate that the resultant pressure P_{AE} acts at a distance $\overline{H}$ which is somewhat greater than $\frac{1}{3}H$ measured from the bottom of the wall. This is shown in Figure 9.10.

Prakash and Basavanna (1969) have made a theoretical evaluation for determination of $\overline{H}$. Based on the force-equilibrium analysis, their study shows that $\overline{H}$ increases from $\frac{1}{3}H$ for $k_h = 0$ to about $\frac{1}{2}H$ for $k_h = 0.3$ (for $\phi = 30°$, $\delta = 7.5°$, $k_v = 0$, $i = \beta = 0$). For similar conditions, the moment-equilibrium analysis gave a value of $\overline{H} = \frac{1}{3}H$ and $k_h = 0$, which increases to a value of $\overline{H} \approx H/1.9$ at $k_h = 0.3$.

For practical design considerations, Seed and Whitman (1969) have proposed the following procedure for determination of the line of action of P_{AE}.

1. Calculate P_A [Eq. (9.1)].
2. Calculate P_{AE} [Eq. (9.4)].
3. Calculate $\Delta P_{AE} = P_{AE} - P_A$. The term ΔP_{AE} is the incremental force due to earthquake condition.
4. Assume that P_A acts at a distance of $\frac{1}{3}H$ from the bottom of the wall (Figure 9.11).
5. Assume that ΔP_{AE} acts at a distance of $0.6H$ from the bottom of the wall (Figure 9.11); then

$$\overline{H} = \left[(P_A)(\tfrac{1}{3}H) + (\Delta P_{AE})(0.6H) \right]/P_{AE}$$

Example 9.2 Referring to Example 9.1, determine the location of the line of action for P_{AE}.

FIGURE 9.10 Point of application of resultant active earth pressure.

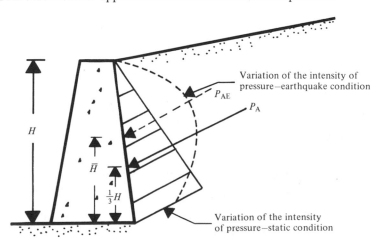

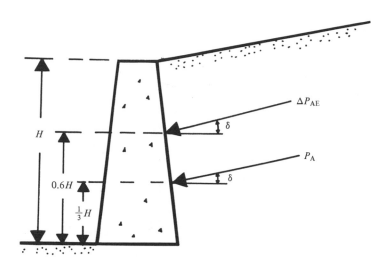

FIGURE 9.11

Solution: The value of P_{AE} in Example 9.1 has been determined to be 5404 lb/ft.

$$P_A = \tfrac{1}{2}\gamma H^2 K_A$$

For $\phi = 36°$, $\delta = 18°$, $K_A = 0.236$ (by interpolation from Table 9.2). Thus,

$$P_A = \tfrac{1}{2}(110)(15)^2(0.236) = 2920.5 \text{ lb/ft}$$

This acts at a distance equal to $\frac{15}{3} = 5$ ft from the bottom of the wall. Again,

$$\Delta P_{AE} = 5404 - 2920.5 = 2483.5 \text{ lb/ft}$$

The line of action of ΔP_{AE} intersects the wall at a distance of $0.6H = 9$ ft measured from the bottom, so

$$\bar{H} = \frac{(5)(2920.5) + (9)(2483.5)}{5404} = 6.83 \text{ ft}$$

9.5 GRAPHICAL CONSTRUCTION FOR DETERMINATION OF ACTIVE PRESSURE

A modified form of Culmann's graphical construction for determination of the active force P_{AE} per unit length of a retaining wall has been proposed by Kapila (1962). In order to understand this, consider the force polygon for the wedge ABC shown in Figure 9.2. For convenience, this has been replotted in Figure 9.12a. The force polygon can be reduced to a force triangle with forces P_{AE}, F, and $W\sqrt{(1-k_v)^2 + k_h^2}$. Note that, in Figures 9.12a, b, α is the angle that the failure wedge makes with the horizontal.

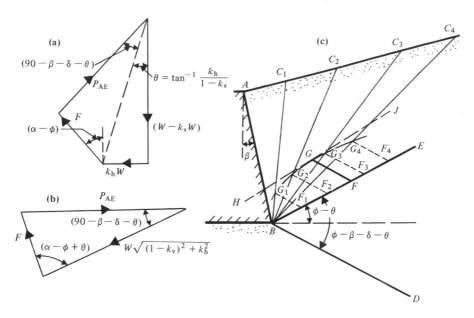

FIGURE 9.12 Modified Culmann construction.

The idea behind this graphical construction is to determine the *maximum* value of P_{AE} by considering several trial wedges. With reference to Figure 9.12c, following are steps for the graphical construction:

1. Draw line BE which makes an angle $\phi - \theta$ with the horizontal.
2. Draw a line BD which makes an angle $\phi - \beta - \delta - \theta$ with the line BE.
3. Draw $BC_1, BC_2, BC_3, \dots$, which are the trial failure surfaces.
4. Determine k_h and k_v and then $\sqrt{(1-k_v)^2 + k_h^2}$.
5. Determine the weights $W_1, W_2, W_3, \dots$ of trial failure wedges $ABC_1, ABC_2, ABC_3, \dots$, respectively (per unit length at right angle to the cross-section shown). Note

$$W_1 = (\text{area of } ABC_1) \cdot \gamma \cdot 1$$
$$W_2 = (\text{area of } ABC_2) \cdot \gamma \cdot 1$$

$$\vdots$$

6. Determine $W_1', W_2', W_3', \dots$ as

$$W_1' = \sqrt{(1-k_v)^2 + k_h^2}\, W_1$$

$$W_2' = \sqrt{(1-k_v)^2 + k_h^2}\, W_2$$

$$\vdots$$

7. Adopt a load scale.
8. Using the load scale adopted in Step 7, draw $BF_1 = W_1'$, $BF_2 = W_2'$, $BF_3 = W_3',\ldots$ on the line BE.
9. Draw F_1G_1, F_2G_2, $F_3G_3,\ldots$ parallel to line BD. Note that BF_1G_1 is the force triangle for the trial wedge ABC_1 similar to that shown in Figure 9.12b. Similarly, BF_2G_2, $BF_3G_3,\ldots$, are the force triangles for the trial wedges ABC_2, $ABC_3,\ldots$, respectively.
10. Join the points $G_1, G_2, G_3,\ldots$, by a smooth curve.
11. Draw a line HJ parallel to line BE. Let G be the point of tangency.
12. Draw line GH parallel to BD.
13. Determine active force P_{AE} as $GH \times$(load scale)

9.6 DESIGN OF GRAVITY RETAINING WALLS BASED ON LIMITED DISPLACEMENT

Richards and Elms (1979) have proposed a procedure for design of gravity retaining walls based on limited displacement. In their study, they have taken into consideration the wall inertia effect and concluded that there is some lateral movement of the wall even for mild earthquakes. In order to develop this procedure, consider a gravity retaining wall as shown in Figure 9.13, along with the forces acting on it during an earthquake. For stability,

FIGURE 9.13 Derivation of Eq. (9.25).

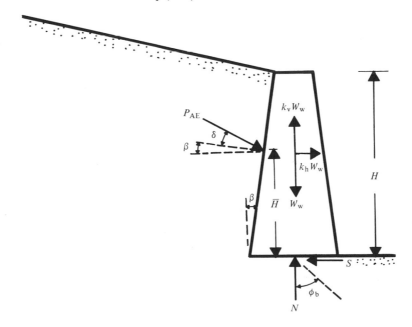

summing the forces in the vertical direction,

$$N = W_w - k_v W_w + P_{AE}\sin(\delta + \beta) \tag{9.21}$$

where N is the vertical component of the reaction at the base of the wall and W_w is the weight of the wall. Similarly, summing the forces in the horizontal direction,

$$S = k_h W_w + P_{AE}\cos(\delta + \beta) \tag{9.22}$$

where S is the horizontal component of the reaction at the base of the wall. At sliding,

$$S = N\tan\phi_b \tag{9.23}$$

where ϕ_b is the soil–wall friction angle at the base of the wall.

Substituting Eqs. (9.21) and (9.22) into Eq. (9.23), we get

$$k_h W_w + P_{AE}\cos(\delta + \beta) = \left[W_w(1 - k_v) + P_{AE}\sin(\delta + \beta)\right]\tan\phi_b$$

or

$$W_w\left[(1 - k_v)\tan\phi_b - k_h\right] = P_{AE}\left[\cos(\delta + \beta) - \sin(\delta + \beta)\tan\phi_b\right]$$

$$W_w = \frac{P_{AE}\left[\cos(\delta + \beta) - \sin(\delta + \beta)\tan\phi_b\right]]}{(1 - k_v)\tan\phi_b - k_h} \tag{9.24}$$

From Eq. (9.4), $P_{AE} = \frac{1}{2}\gamma H^2(1 - k_v)K_{AE}$. Substitution of this equation into Eq. (9.24) yields

$$W_w = \frac{\frac{1}{2}\gamma H^2 K_{AE}\left[\cos(\delta + \beta) - \sin(\delta + \beta)\tan\phi_b\right]}{(\tan\phi_b - \tan\theta)} \tag{9.25}$$

where $\tan\theta = k_h/(1 - k_v)$.

It may be noted that, in Eq. (9.25), W_w is equal to infinity if

$$\tan\phi_b = \tan\theta \tag{9.26}$$

This implies that infinite mass of the wall is required to prevent motion. The critical value of $k_h = k_{h(cr)}$ can thus be given by the relation

$$\tan\theta = k_{h(cr)}/(1 - k_v) = \tan\phi_b$$

or

$$k_{h(cr)} = (1 - k_v)\tan\phi_b \tag{9.27}$$

Equation (9.24) can also be written in the form

$$W_w = \left[\frac{1}{2}\gamma H^2(1 - k_v)K_{AE}\right]C_{IE} \tag{9.28}$$

where

$$C_{IE} = \frac{\cos(\delta + \beta) - \sin(\delta + \beta)\tan\phi_b}{(1 - k_v)(\tan\phi_b - \tan\theta)} \tag{9.29}$$

Figure 9.14 shows the variation of C_{IE} with k_h for various values of k_v ($\phi = \phi_b = 35°$, $\delta = \frac{1}{2}\phi$, $i = \beta = 0$). Also, Figure 9.15 shows the variation of C_{IE} with k_h for various values of wall friction angle, δ ($\phi = \phi_b = 35°$, $i = \beta = 0$, $k_v = 0$).

Note that Eq. (9.28) is for the limiting equilibrium condition for sliding with earthquake effects taken into consideration. For the static condition (i.e., $k_h = k_v = 0$), Eq. (9.28) becomes

$$W = \tfrac{1}{2}\gamma H^2 K_A C_I \tag{9.30}$$

where $W = W_w$ (for static condition) and

$$C_I = \left[\cos(\delta + \beta) - \sin(\delta + \beta)\tan\phi_b\right]/\tan\phi_b \tag{9.31}$$

Thus, comparing Eqs. (9.28) and (9.30), we can write that

$$W_w/W = F_T F_I = F_W \tag{9.32}$$

FIGURE 9.14 Effect of k_v on the value of C_{IE}: $\phi_b = \phi = 35°$; $\delta = \frac{1}{2}\phi$; $i = \beta = 0$. [Richards, R., and Elms, D. G. (1979). "Seismic Behavior of Gravity Retaining Walls," *Journal of the Geotechnical Division, ASCE*, 105 (GT4), Fig. 8, p. 455.]

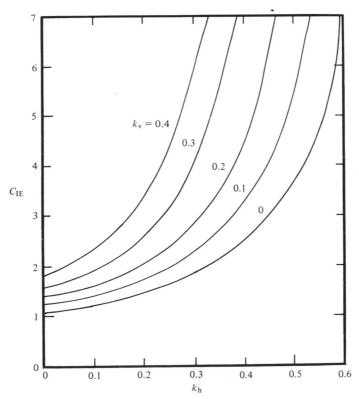

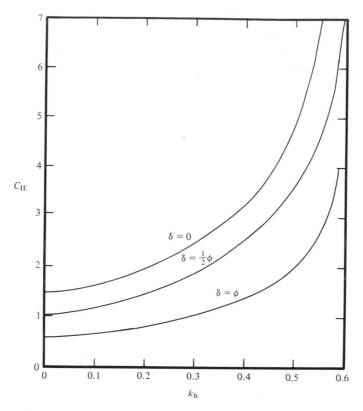

FIGURE 9.15 Effect of wall friction on C_{IE}: $\phi = \phi_b = 35°$; $i = \beta = 0$; $k_v = 0$. [Richards, R., and Elms, D. G. (1979). "Seismic Behavior of Gravity Retaining Walls," *Journal of the Geotechnical Division, ASCE*, 105 (GT4), Fig. 9, p. 455.]

where

$$F_T = K_{AE}(1 - k_v)/K_A = \text{soil thrust factor}$$

$$F_I = C_{IE}/C_I = \text{wall inertia factor}$$

and F_W is a factor of safety applied to the weight of the wall to take into account the effects of soil pressure and wall inertia.

Figure 9.16 shows a plot of F_T, F_I, and F_W for various values of k_h ($\phi = \phi_b = 35°$, $\delta = \frac{1}{2}\phi$, $k_v = 0$, $\beta = i = 0$). Richards and Elms (1969) have explained the importance of the inertia factors given in Eq. (9.32). Referring to Figure 9.16, suppose that one neglects the wall inertia factor (which is not considered in the design procedure outlined in Section 9.2; i.e., $F_I = 1$). In such a case,

$$F_W = F_T = W_w/W$$

For a value of $F_W = 1.5$, the critical horizontal acceleration is equal to 0.18.

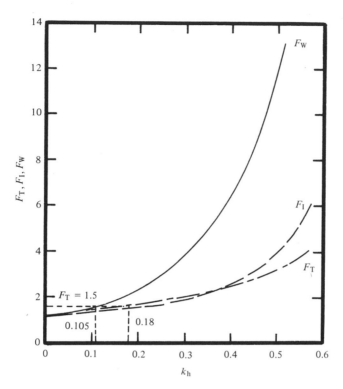

FIGURE 9.16 Variation of F_T, F_I, and F_W: $\phi = \phi_b = 35° = 2\delta$; $k_v = \beta = i = 0$. [Richards, R., and Elms, D. G. (1979). "Seismic Behavior of Gravity Retaining Walls," *Journal of the Geotechnical Division, ASCE,* 105 (GT4), Fig. 10, p. 457.]

However, if the wall inertial factor is considered, the critical horizontal acceleration corresponding to $F_W = 1.5$ is equal to 0.105. In other words, if a gravity retaining wall is designed such that $W_w = 1.5W$, the wall will start to move laterally at a value of $k_h = 0.105$. Based on the procedure described in Section 9.2, if $W_w = 1.5W$, it is assumed that the wall will not move laterally until a value of $k_h = 0.18$ is reached.

The above considerations show that, for no lateral movement, the weight of the wall has to be increased by a considerable amount over the static condition, which may prove to be very expensive. Thus, for actual design with reasonable cost, one has to assume some lateral displacement of the wall will take place during an earthquake; the procedure for determination of the wall weight (W_w) is then as follows:

1. Determine an acceptable displacement d of the wall.
2. Determine a design value of k_h from the equation

$$k_h = A_a \left(0.2 A_v^2 / A_a d \right)^{1/4} \tag{9.33}$$

where A_a and A_v are effective acceleration coefficients and displacement d is in inches. The values of A_a and A_v for a given region in the United States are given by the Applied Technology Council (1978).

Equation (9.33) has been suggested by Richards and Elms (1979), and is based on the study of Newmark (1965) and Franklin and Cheng (1977).

3. Using the above value of k_h, and assuming $k_v = 0$, determine the value of K_{AE}.
4. Determine the weight of the wall W_w from Eq. (9.28).
5. Apply a factor of safety to W_w obtained in Step 4.

Example 9.3 Determine the weight of a retaining wall 4 m high (given $\beta = 0$, $i = 0$, $\gamma = 17.29$ kN/m³, $\phi_b = \phi = 34°$, $\delta = \frac{1}{2}\phi$, $A_v = 0.2$, $A_a = 0.2$, factor of safety $= 1.5$)

(a) for static condition
(b) for zero displacement condition under earthquake loading
(c) for a displacement of 50.8 mm under earthquake loading

Solution:

(*a*) *For Static Condition.* From Eq. (9.30),

$$W = \tfrac{1}{2}\gamma H^2 K_A C_I$$

From Table 9.2, $K_A = 0.256$ (for $\phi = 34°$, $\delta = 17°$, $i = 0$, $\beta = 0$).

$$C_I = \frac{\cos(\delta + \beta) - \sin(\delta + \beta)\tan\phi_b}{\tan\phi_b} = \frac{\cos 17 - \sin 17(\tan 34)}{\tan 34}$$

$$= 1.125$$

Thus

$$W = \tfrac{1}{2}(17.29)(4)^2(0.256)(1.125) = 39.84 \text{ kN/m}$$

With a factor of safety of 1.5, the weight of the wall is equal to $(1.5)(39.84)$ $= 59.76$ kN/m.

(*b*) *For Zero Displacement Condition.* From Eq. (9.29),

$$W_w = \tfrac{1}{2}\gamma H^2(1 - k_v)K_{AE}C_{IE}$$

Assume $k_v = 0$.

$$C_{IE} = \frac{\cos(\delta + \beta) - \sin(\delta + \beta)\tan\phi_b}{(1 - k_v)(\tan\phi_b - \tan\theta)}$$

$$\tan\theta = k_h/(1 - k_v) = 0.2/1 = 0.2; \qquad \theta = 11.31°$$

$$C_{IE} = \frac{\cos 17 - \sin 17(\tan 34)}{\tan 34 - 0.2} = 1.6$$

Again, From Eq. (9.5),

$$K_{AE} = \frac{\cos^2(34-11.31)}{\cos(11.31)\left[\cos(17+11.31)\right]\left[1+\sqrt{\dfrac{\sin(34+17)\sin(34-11.31)}{\cos(17+11.31)}}\right]^2}$$

$$= 0.393$$

$$W_w = \tfrac{1}{2}(17.29)(4)^2(1-0)(0.393)(1.6) = 86.98 \text{ kN/m}$$

With a factor of safety of 1.5, the weight of wall $= (86.98)(1.5) = 130.47$ kN/m.

(c) *For Displacement Condition d = 50.8 mm (2 in.).* From Eq. (9.33),

$$k_h = A_a\left[0.2A_v^2/A_a d\right]^{1/4} = 0.2\left[(0.2)(0.2)^2/(0.2)(2)\right]^{1/4} = 0.075$$

$$\tan\theta = k_h/(1-k_v) = 0.075/(1-0) = 0.075$$

or

$$\theta = 4.29°$$

$$C_{IE} = \frac{\cos 17 - \sin 17(\tan 34)}{\tan 34 - 0.075} = \frac{0.7591}{0.5995} = 1.27$$

Using Eq. (9.5)

$$K_{AE} = \frac{\cos^2(34-4.29)}{\cos(4.29)\left[\cos(17+4.29)\right]\left[1+\sqrt{\dfrac{\sin(34+17)\sin(34-4.29)}{\cos(17+4.29)}}\right]^2}$$

$$= 0.3$$

Thus, with a factor of safety of 1.5,

$$W_w = (1.5)(\tfrac{1}{2})(17.29)(4)^2(0.3)(1.27) = 79.05 \text{ kN/m}$$

9.7 HYDRODYNAMIC EFFECTS OF PORE WATER

The lateral earth pressure theory developed in the preceding sections of this chapter has been for retaining walls with dry soil backfills. However, for quay walls (Figure 9.17), the hydrodynamic effect of the water also has to be taken into consideration. This is usually done according to the Westergaard theory (1933) which was derived to obtain the dynamic water pressure on the face of a concrete dam. Based on this theory, the water pressure due to an earthquake at a depth y (Figure 9.17) may be expressed as

$$p_1 = \tfrac{7}{8}k_h\gamma_w h^{1/2}y^{1/2} \tag{9.34}$$

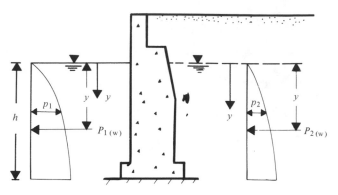

FIGURE 9.17 Hydrodynamic effects on a quay wall.

where p_1 is the intensity of pressure in the seaward side, γ_w is the unit weight of water, and h is the total depth of water. Hence, the total dynamic water force on the seaward side per unit length of the wall ($P_{1(w)}$ can be obtained by integration as

$$P_{1(w)} = \int p_1 \, dy = \int_0^h \tfrac{7}{8} k_h \gamma_w h^{1/2} y^{1/2} \, dy = \tfrac{7}{12} k_h \gamma_w h^2 \qquad (9.35)$$

The location of the resultant water pressure is

$$\bar{y} = \frac{1}{P_{1(w)}} \int_0^h (p \, dy) y = \frac{1}{P_{1(w)}} \left(\tfrac{7}{8} k_h \gamma_w h^{1/2} \right) \int_0^h (y^{1/2})(y) \, dy$$

$$= \frac{1}{P_{1(w)}} \left(\tfrac{7}{8} k_h \gamma_w h^{1/2} \right) (h^{5/2}) \tfrac{2}{5} = \frac{1}{P_{1(w)}} \left(\tfrac{7}{20} k_h \gamma_w h^3 \right)$$

or

$$\bar{y} = \left(\tfrac{7}{12} k_h \gamma_w h^2 \right)^{-1} \left(\tfrac{7}{20} k_h \gamma_w h^3 \right) = 0.6 h \qquad (9.36)$$

Matsuo and O'Hara (1960) have suggested that the increase of the pore water pressure in the landward side is approximately 70% of that in the seaward side. Thus

$$p_2 = 0.7 \left(\tfrac{7}{8} k_h \gamma_w h^{1/2} y^{1/2} \right) = 0.6125 k_h \gamma_w h^{1/2} y^{1/2} \qquad (9.37)$$

where p_2 is the dynamic pore water pressure on the landward side at a depth y. The total dynamic pore water force increase $[P_{2(w)}]$ per unit length of the wall is

$$P_{2(w)} = 0.7 \left(\tfrac{7}{12} k_h \gamma_w h^2 \right) = 0.4083 k_h \gamma_w h^2 \qquad (9.38)$$

During an earthquake, the force on the wall per unit length on the seaward side will be reduced by $P_{1(w)}$ and that on the landward side will be increased by $P_{2(w)}$. Thus, the total increase of the force per unit length of the wall is equal to

$$P_{\mathrm{w}} = P_{1(\mathrm{w})} + P_{2(\mathrm{w})} = 1.7\left(\tfrac{7}{12}k_{\mathrm{h}}\gamma_{\mathrm{w}}h^2\right) = 0.9917k_{\mathrm{h}}\gamma_{\mathrm{w}}h^2 \qquad (9.39)$$

Example 9.4 Refer to Figure 9.17. For the quay wall, $h = 10$ m. Determine the total dynamic force increase due to water for $k_{\mathrm{h}} = 0.2$.

  Solution: From Eq. (9.39)

$$P_{\mathrm{w}} = 0.9917k_{\mathrm{h}}\gamma_{\mathrm{w}}h^2 = 0.9917(0.2)(9.81)(10)^2 = 194.6 \text{ kN/m}$$

9.8 EQUATION FOR PASSIVE FORCE ON A RETAINING WALL

Using the basic assumptions for the soil given in Section 9.2, the passive force P_{PE} per unit length of a retaining wall (Figure 9.18) may also be derived (Kapila, 1962) as

$$P_{\mathrm{PE}} = \tfrac{1}{2}\gamma H^2(1 - k_{\mathrm{v}})K_{\mathrm{PE}} \qquad (9.40)$$

where

$$K_{\mathrm{PE}} = \frac{\cos^2(\phi + \beta - \theta)}{\cos\theta\cos^2\beta\cos(\delta - \beta + \theta)\left[1 - \left\{\dfrac{\sin(\phi - \delta)\sin(\phi + i - \theta)}{\cos(i - \beta)\cos(\delta - \beta + \theta)}\right\}^{1/2}\right]^2}$$

$$(9.41)$$

and $\theta = \tan^{-1}(k_{\mathrm{h}}/1 - k_{\mathrm{v}})$.

  Note that Eq. (9.41) has been derived for dry cohesionless backfill. Kapila has also developed a graphical procedure for determination of P_{PE}.

FIGURE 9.18 Passive force P_{PE} on a retaining wall.

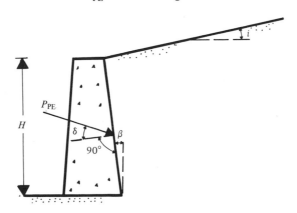

PROBLEMS

9.1. A retaining wall is 18 ft high with a vertical back ($\beta = 0$). It has a horizontal cohesionless soil (dry) as backfill. Given:

$$\text{unit weight of soil} = 95 \text{ lb/ft}^3$$
$$\text{angle of friction } \phi = 30°$$
$$k_h = 0.35, \qquad k_v = 0, \qquad \delta = 15°$$

Determine the active force P_{AE} per unit length of the retaining wall.

9.2. Refer to Problem 9.1. Determine the location of the point of intersection of the resultant force P_{AE} with the back face of the retaining wall.

9.3. Refer to Figure 9.2. Given

$$
\begin{array}{lll}
H = 3 \text{ m} & \phi = 40° & k_h = 0.3 \\
\beta = 10° & \delta = 13° & k_v = 0.1 \\
i = 10° & \gamma = 100 \text{ lb/ft}^3 &
\end{array}
$$

Determine the active force per unit length P_{AE} and the location of the resultant.

9.4. Redo Example 9.1 using the modified Culmann graphical solution procedure outlined in Section 9.5.

9.5. Redo Problem 9.1 by the modified Culmann graphical solution procedure.

9.6. Redo Problem 9.3 by the modified Culmann graphical solution procedure.

9.7. For the retaining wall and the backfill given in Problem 9.1, determine the passive force P_{PE} per unit length.

9.8. For the retaining wall and the backfill given in Problem 9.3, determine the passive force P_{PE} per unit length.

9.9. Refer to Figure 9.18. Develop a step-by-step procedure for a modified Culmann's graphical solution for obtaining P_{PE}.

9.10. Consider a 12-ft-high vertical retaining wall ($\beta = 0$) with a horizontal backfill ($i = 0$). Given for the soil, $\phi = 32°$, $c = 0$, $\gamma = 120 \text{ lb/ft}^3$, and $\delta = 0$:
   a. Calculate P_{AE} and the location of the resultant with $k_v = 0.1$ and $k_h = 0.15$.
   b. For the results of (a), what should be the weight of the wall per foot length for no lateral movement? The factor of safety against sliding is 1.4 (Section 9.6).
   c. What should be the weight of the wall for an allowable lateral displaement of one inch? Given: $A_v = A_a = 0.15$; factor of safety against sliding is 1.4 (Section 9.6).

REFERENCES

Amano, R., Azuma, H., and Ishii, Y. (1956). "Aseismic Design of Quay Walls in Japan," *Proceedings, 1st World Conference on Earthquake Engineering*, Berkeley, California.

Arango, I. (1969). Personal communication with Seed, H. B., and Whitman, R. V. (1970).

Applied Technology Council (1978). "Tentative Provisions for the Development of Seismic Regulations for Buildings," *Publication ATC 3-06*, Palo Alto, California.

Bowles, J. E. (1978). *Foundation Analysis and Design*, McGraw-Hill, New York, 2nd ed.

Duke, C. M., and Leeds, D. J. (1963). "Response of Soils, Foundations and Earth Structures," *Bulletin of the Seismological Society of America* 53 (2), 309–357.

Franklin, A. G., and Cheng, F. K. (1977). *"Earthquake Resistance of Earth and Rockfill Dams,"* Report 5, Miscellaneous Paper S-71-17, Soils and Pavement Laboratory, U.S. Army Engineer Waterways Experiment Station, Vicksburg, Mississippi.

Hayashi, S., Kubo, K., and Nakase, A. (1966). "Damage to Harbor Structures in the Nigata Earthquake," *Soils and Foundations* 6 (1), 26–32. (The Japanese Society of Soil Mechanics and Foundation Engineering).

Jacobsen, L. S. (1939). Described in Appendix D of "The Kentucky Project," Technical Report No. 13, Tennessee Valley Authority, 1951.

Kapila, J. P. (1962). "Earthquake Resistant Design of Retaining Walls," *Proceedings, 2nd Earthquake Symposium*, University of Roorkee, Roorkee, India.

Matsuo, H., and O'Hara, S. (1960). "Lateral Earth Pressures and Stability of Quay Walls During Earthquakes," *Proceedings, 2nd World Conference on Earthquake Engineering*, Japan, Vol. 1.

Mononobe, N. (1929). "Earthquake-Proof Construction of Masonry Dams," *Proceedings, World Engineering Conference* 9, 274–280.

Mononobe, N., and Matsuo, H. (1929). "On the Determination of Earth Pressures During Earthquakes," *Proceedings, World Engineering Conference* 9, 176–182.

Nazarian, H. N., and Hadjan, A. H. (1979). "Earthquake-Induced Lateral Soil Pressure on Structures," *Journal of the Geotechnical Engineering Division, ASCE* 105 (GT9), 1049–1066.

Newmark, N. M. (1965). "Effect of Earthquakes on Dams and Embankments," *Geotechnique* (London) 15 (2), 139–160.

Okabe, S. (1926). "General Theory of Earth Pressure," *Journal of the Japanese Society of Civil Engineers* (Tokyo) 12, (1).

Prakash, S., and Basavanna, B. M. (1969). "Earth Pressure Distribution Behind Retaining Wall During Earthquake," *Proceedings, 4th World Conference on Earthquake Engineering*, Santiago, Chile.

Richards, R., and Elms, D. G. (1979). "Seismic Behavior of Gravity Retaining Walls," *Journal of the Geotechnical Engineering Division, ASCE* 105 (GT4), 449–464.

Seed, H. B., and Whitman, R. V. (1970). "Design of Earth Retaining Structures for Dynamic Loads," *Proceedings, Specialty Conference on Lateral Stresses in the Ground and Design of Earth Retaining Structures, ASCE*, pp. 103–147.

Westergaard, H. M. (1933). "Water Pressures on Dams During Earthquakes," *Transactions, ASCE* 98, 418–433.

10

COMPRESSIBILITY OF SOILS
UNDER DYNAMIC LOADS

Permanent settlements under vibratory machine foundations can generally be placed under two categories:

1. elastic and consolidation settlement due to the static weight
2. settlement due to vibratory compaction of the foundation soil

Permanent settlement in soils can also be induced due to the vibration caused by an earthquake. The elastic and consolidation settlement due to static loads is not discussed here since the conventional methods of calculation can be found in most standard soil mechanics texts. In this chapter, the present available methods of evaluation of permanent settlement due to dynamic loading conditions are presented.

10.1 COMPACTION OF GRANULAR SOILS: EFFECT OF VERTICAL STRESS AND VERTICAL ACCELERATION

The fact that granular soils can be compacted by vibration is well known. Dry granular soils are likely to exhibit more compaction due to vibration as compared to moist soils. This is because of the surface tension effect in moist soils which offers a resistance for the soil particles to roll and slide and arrange themselves into a denser state.

Laboratory studies have been made in the past to evaluate the effect of cycling *controlled vertical stress* at low frequencies, i.e., at low acceleration levels on confined granular soils (D'Appolonia, 1970). Such laboratory tests can be performed by taking a granular soil specimen in a mold as shown in Figure 10.1a. A confining vertical air pressure σ_z is first applied to the specimen, after which a vertical dynamic stress of amplitude σ_d is applied repeatedly. The permanent compressions of the specimen are recorded after the elapse of several cycles of dynamic stress application.

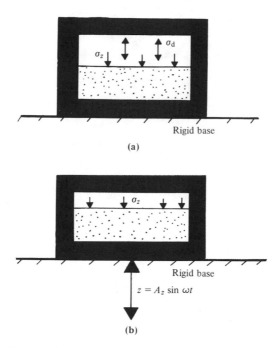

FIGURE 10.1 Compaction of granular soils by **(a)** controlled vertical stress and **(b)** controlled vertical acceleration.

Also, several investigations on confined dry granular soils have been conducted (e.g., see D'Appolonia and D'Appolonia, 1967; Ortigosa and Whitman, 1968) in which a *controlled vertical acceleration* is imposed on the specimen which produces small dynamic stress changes. For these tests, the specimen is placed in a mold fixed to a vibrating table (Figure 10.1b). Then a vertical confining air pressure σ_z is applied to the specimen. After that, the specimen is subjected to a vertical vibration for a period of time. Note that, for a vertical vibration,

$$z = A_z \sin \omega t$$

where A_z is the amplitude of the vertical vibration. The magnitude of the *peak acceleration* is equal to $A_z \omega^2 = A_z (2 \pi f)^2$. Thus, the peak acceleration is controlled by the *amplitude of displacement* and *the frequency of vibration*. For a *constant peak acceleration* of vibration, the drive mechanism has to be adjusted for A_z and f. The vertical compression of the specimen can be determined at the end of a test.

Thus the first type of test described above is run with repeated stresses with *negligible acceleration*; the second type is for repeated acceleration with *small dynamic stress on soils*.

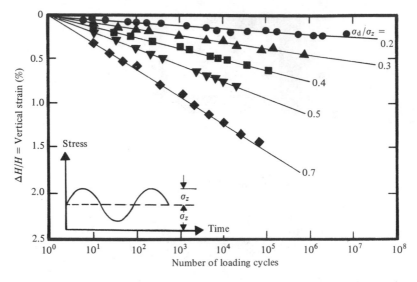

FIGURE 10.2 Compression of a dune sand under controlled vertical stress condition: relative density = 60% $\pm$; σ_z = 20 lb/in.^{2} (138 kN/m^2); frequency = 1.8–6 cps. [D'Appolonia, E. P. (1970). "Dynamic Loading," *Journal of the Soil Mechanics and Foundations Division*, *ASCE*, 96 (SM1) Fig. 9, p. 59.]

FIGURE 10.3 Nature of variation of void ratio or dry unit weight of dry sand in controlled vertical acceleration tests.

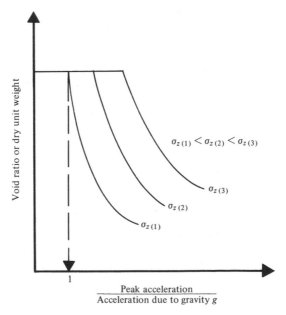

Figure 10.2 shows the results of a number of tests conducted on a dune sand for controlled vertical stress condition. For all tests, the sand specimens were compacted to an initial relative density of about 60%. The frequencies of load application were in a range of 1.8–6 cps. Along the ordinate are plotted the vertical strain, which is equal to $\Delta H/H$ (where H is the initial height of the specimen and ΔH is the vertical compression of the specimen after a given number of load cycles). It may be seen that, for a given value of σ_d/σ_z,

$$\frac{\Delta H}{H} \propto \log N$$

where N is the number of load cycle applications. Also note that, for a given number of load cycles, the vertical strain increases with increasing values of σ_d/σ_z.

Figure 10.3 shows the nature of the results obtained from controlled vertical acceleration tests on dry sand by Ortigosa and Whitman (1968). Note that, even at zero confining pressure, no vertical strain is induced up to a peak acceleration of about one g ($1 \times$ acceleration due to gravity). A similar test result of D'Appolonia (1970) is shown in Figure 10.4, for which

FIGURE 10.4 Correlation between terminal unit weight and peak vertical acceleration for a dune sand. Amplitude legend in in. (mm): ▼ 0.012 (0.305); ● 0.025 (0.635); ◆ 0.037 (0.940); ■ 0.050 (1.27); ▲ 0.075 (1.905). Tests at zero surcharge. [D'Appolonia, E. P. (1970). "Dynamic Loading," *Journal of the Soil Mechanics and Foundations Division, ASCE*, 96 (SM1), Fig. 7, p. 58.]

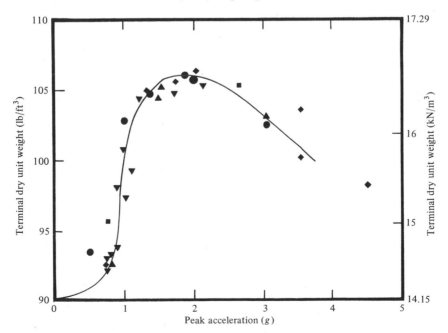

$\sigma_z = 0$. The terminal dry unit weight shown in Figure 10.4 is the unit weight of sand at the end of the test.

Krizek and Fernandez (1971) also conducted several laboratory tests with *controlled vertical acceleration* to study the densification of damp clayey sand. Tests were conducted with air-dry and damp specimens of Ottawa sand, grundite, and three mixtures of Ottawa sand and grundite: 90%–10%, 80%–20%, 70%–30%. Table 10.1 gives the details of the specimens used for the tests.

For conducting the tests, approximately 0.6 ft^3 (0.017 m^3) of soil samples—air dry and moist—were placed in a loose condition in a cylindrical mold 18 in. (0.457 m) high and 12 in. (0.305 m) in diameter. They were subjected to vertical vibrations for a period of time under various vertical pressures (σ_z). The range of time for vibratory compaction for the specimens was varied. Maximum vertical accelerations up to a value of about 6 g were used. Figure 10.5 shows the time rate of vibratory compaction of air-dry and moist sand–grundite mixtures. It needs to be pointed out that very few tests were conducted for $a_{max}/g < 1$ (a_{max} = peak acceleration). However, from this study the following general conclusion may be drawn.

1. Significant vibratory densification does not occur with peak acceleration levels of less than 1 g.
2. The terminal vibratory dry unit weight of *air-dry* soils slightly decreases for $a_{max}/g > 2$. This is only true for zero confining pressure ($\sigma_z = 0$).
3. An increase of the clay percentage in soils has a tendency to reduce $\gamma_{d(termin-vibrat)}/\gamma_{d(modif\ max\ Proctor)}$.
4. Increase of moisture content has a significant influence in reducing $\gamma_{d(termin-vibrat)}/\gamma_{d(modif\ max\ Proctor)}$.

TABLE 10.1 Details of the Specimens Used in the Tests by Krizek and Fernandez (1971)

Soil	Percent of mix	Moisture content air dry (%)	damp (%)	Modified Proctor dry unit weight (lb/ft^3)	(kN/m^3)	Optimum moisture content, modified Proctor test (%)
Ottawa sand	—	06	4.4 ± 0.5	107.6	16.92	11.0
Grundite	—	2.42	Not tested	101.8	16.00	18.5
Mix-10	90% Ottawa sand + 10% grundite	0.26	5 ± 0.5	114.4	17.99	8
Mix-20	80% Ottawa sand + 20% grundite	0.51	4.5 ± 0.5	120.6	18.96	9
Mix-30	70% Ottawa sand + 30% grundite	0.72	5 ± 0.3	124.0	19.49	9.5

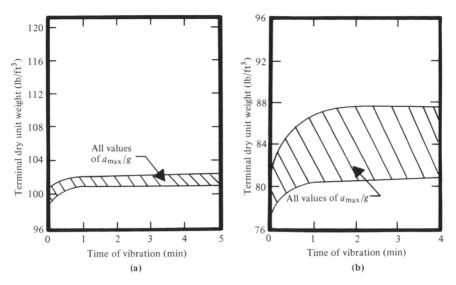

FIGURE 10.5 Time rate of vibratory compaction for air-dry and moist sand–grundite mixtures: **(a)** air-dry mix-10 (90% sand + 10% grundite); $\sigma_z = 3.26$ lb/in.^{2} (22.49 kN/m^2); **(b)** moist mix-10, moisture content = 4.5%; $\sigma_z = 3.26$ lb/in.^{2} (22.49 kN/m^2). Note: 1 lb/ft^3 = 0.1572 kN/m^3. [Krizek, R. J., and Fernandez, J. I. (1971). "Vibratory Densifications of Damp Clayey Sands," *Journal of the Soil Mechanics and Foundations Division, ASCE*, 97 (SM8), Fig. 3, p. 1073, and Fig. 5, p. 1076.]

10.2 SETTLEMENT OF STRIP FOUNDATION ON GRANULAR SOIL UNDER THE EFFECT OF CONTROLLED CYCLIC VERTICAL STRESS

In Section 10.1, some laboratory experimental observations of settlement of *laterally confined* sand specimens were presented. In these cases, the loads have been applied over the full surface area. However, in the field, the load covers only a small area and settlements in these cases include those caused by the induced shear strains. In the case of foundations, the shear strains increase with the increase of σ_d/q_u (where σ_d is the amplitude of dynamic load and q_u is the ultimate bearing capacity). In this section, some developments on settlements of strip footings under the effect of controlled cycling vertical stress applied *at low frequencies* (i.e., negligible acceleration) are discussed.

Raymond and Komos (1978) conducted laboratory model tests with strip footings with widths of 2.95 in. (75 mm) and 8.98 in. (228 mm) resting on 20–30 Ottawa sand in a large box. The cyclic loads on model strip footings were applied by a Bellofram loading piston activated by an air pressure system. The loadings approximated a rectangular wave form as shown in Figure 10.6 with a frequency of 1 cps. The settlements of the footings were measured by a dial gauge together with a DVDT activating a strip chart recorder. For conducting the tests, the ultimate static bearing capacities (q_u)

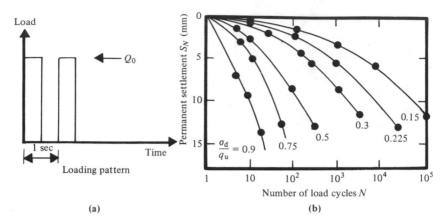

(a)   (b)

FIGURE 10.6 Plastic deformation due to repeated loading in plane strain tests. In (b), unit weight of compaction of sand = 16.97 kN/m³; width of foundation $B = 228$ mm. Note: 1 in. = 25.4 mm. [(b) is from Raymond, G. F., and Komos, F. E. (1978, Fig. 5, p. 193).]

were first experimentally determined. The footings were then subjected to various magnitudes of cyclic load ($\sigma_d/q_u = 13.5\%-90\%$, where $\sigma_d = Q_0/A$, and A is the area of the model footing). The load settlement relationships obtained from the tests for the 228-mm footing are shown in Figure 10.6. In this figure, S_N is the permanent settlement of the footing and N is the number of cycles of load application. Such plots may be given by an empirical relation as

$$S_N/\log N = a + bS_N \qquad (10.1)$$

where a and b are two constants.

The experimental values of a and b for these two footings may be approximated by the following equations.

For 2.95-in. (75-mm) wide footing

$$a = -0.0811 + 0.0115F \qquad (10.2)$$

$$b = 0.12420 + 0.00127F \qquad (10.3)$$

For 8.98-in. (228-mm) wide footing

$$a = -0.1053 + 0.0421F \qquad (10.4)$$

$$b = 0.0812 + 0.0031F \qquad (10.5)$$

where $F = \sigma_d/q_u$ and S_N is measured in millimeters. Equations (10.1)–(10.5) are valid up to a load cycle of $N = 10^5$.

Figure 10.7 shows the experimental results of the variation of σ_d with $\log N$ for various values of S_N. For a given value of S_N, the plot of σ_d vs $\log N$ is approximately linear up to a value of $\sigma_d \approx \frac{1}{4}q_u$. For $\sigma_d < \frac{1}{4}q_u$, the

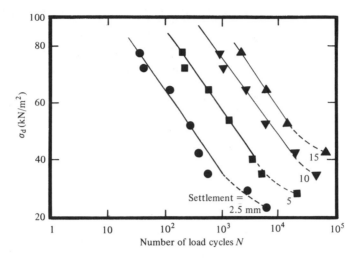

FIGURE 10.7 Variation of σ_d with $\log N$ for various values of permanent settlement: Ottawa sand; unit weight of compaction $= 16.97$ kN/m³, $B = 75$ mm. Note: 1 lb/in.² $= 6.9$ kN/m²; 1 in. $= 25.4$ mm. [Raymond, G. P., and Komos, F. E. (1978, Fig. 11, p. 196).]

slope of σ_d vs $\log N$ becomes smaller and the response tends toward elastic conditions.

From Eqs. (10.2)–(10.5), it may be seen that, *for a given soil*, the parameters a and b are functions of the width of the footing B. Thus, Eqs. (10.2) and (10.4) have been combined by Raymond and Komos to the form

$$a = -0.15125 + 0.0000693B^{1.18}(F + 6.09) \qquad (10.6)$$

where B is the width of the footing. Similarly, Eqs. (10.3) and (10.5) can be combined as,

$$b = 0.153579 + 0.0000363B^{0.821}(F - 23.1) \qquad (10.7)$$

Equations (10.6) and (10.7) are valid for only two different sizes of footing and for one soil. The general form of the equations for all sizes of footings and all soils can be written as,

$$a = a_1 + a_2 B^{1.18}F + a_3 B^{1.18} \qquad (10.8)$$

and

$$b = b_1 + b_2 B^{0.821}F - b_3 B^{0.821} \qquad (10.9)$$

where $a_1, a_2, a_3, b_1, b_2, b_3$ are parameters for a given soil. However,

$$F = \sigma_d / q_u$$

and

$$q_u = \tfrac{1}{2}\gamma BN_\gamma \quad \text{(for surface foundation)} \qquad (10.10)$$

where N_γ is the static bearing capacity factor and γ is the unit weight of soil. Thus,

$$F = \sigma_d / \tfrac{1}{2}\gamma B N_\gamma \tag{10.11}$$

Substitution of Eq. (10.11) into Eqs. (10.8) and (10.9) yields

$$a = a_1 + a_4 \sigma_d B^{0.18} + a_3 B^{1.18} \tag{10.12}$$

and

$$b = b_1 + b_4 \sigma_d B^{-0.18} - b_3 B^{0.82} \tag{10.13}$$

where

$$a_4 = \frac{a_2}{\tfrac{1}{2}\gamma N_\gamma} \tag{10.14}$$

$$b_4 = \frac{b_2}{\tfrac{1}{2}\gamma N_\gamma} \tag{10.15}$$

and B is in millimeters.

FIGURE 10.8 Plot of settlement vs number of load cycles (for a constant value of σ_d) for footings with $B = 75$, 114, 152, 190, and 228 mm: Ottawa sand; unit weight $= 16.97 \text{ kN/m}^3$, $\sigma_d = 64.6 \text{ kN/m}^2$. Note: 1 lb/in.$^2 = 6.9 \text{ kN/m}^2$, 1 in. $= 25.4$ mm. [Raymond, G. P., and Komos, F. E. (1978, Fig. 15, p. 198).]

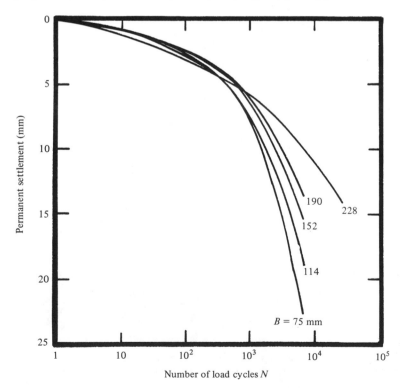

If the values of $a_1, a_3, a_4, b_1, b_3, b_4$, which are the plastic properties of a given soil at a given density of compaction, can be determined by laboratory testing, the settlement of a given strip footing can be determined by combining Eqs. (10.1), (10.12), and (10.13). It needs to be pointed out that, for given values of σ_d and N, the value of S_N *decreases* with the increase of the width of the footing. This fact is demonstrated in Figure 10.8 for five different footings.

Analysis of this type may be used in the estimation of the settlement of railroad ties subjected to dynamic loads due to the movement of trains.

10.3 SETTLEMENT OF MACHINE FOUNDATIONS ON GRANULAR SOILS SUBJECTED TO VERTICAL VIBRATION

For machine foundations subjected to vertical vibrations, many investigators believe that the *peak acceleration* is the main controlling parameter for the settlement of the foundation. Depending on the relative density of granular soils, the solid particles come to an equilibrium condition under a given peak acceleration level. This *threshold acceleration* level must be exceeded before additional densification can take place.

The general nature of the settlement–time relationship for a foundation is shown in Figure 10.9. Note that in Figure 10.9, A_z is the amplitude of the foundation vibration and W is the weight of the foundation. The foundation

FIGURE 10.9 Settlement–time relationship for a machine foundation subjected to vertical vibration.

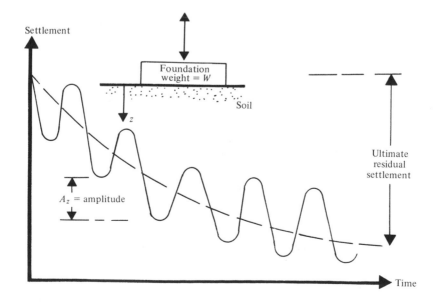

settlement gradually increases with time and reaches a maximum value S_u beyond which it remains constant.

Brumund and Leonards (1972) have studied the settlement of circular foundations resting on sand subjected to vertical excitation by means of laboratory model tests. According to them, the *energy per cycle of vibration* imparted to the soil by the foundation can be used as the parameter for determination of settlement of foundations.

The model tests of Brumund and Leonards were conducted in a 2-ft³ (0.057-m³) container. They used 20–30 Ottawa sand, compacted to a relative density of 70%. The model foundation used for the tests was 4 in. (101.6 mm) in diameter. The static ultimate bearing capacity was first experimentally determined before beginning the dynamic tests. The duration of vibration of the model foundation was chosen to be 20 min for all tests. Figure 10.10 shows the plot of the experimental results of settlement S against the peak acceleration for a *constant frequency of vibration*. For a

FIGURE 10.10 Plot of settlement vs peak acceleration for model foundation at a frequency of 20 Hz. Model foundation diameter = 4 in. (101.6 mm), 20–30 Ottawa sand: W [lb (N)] = ● 48.8 (217); ▲ 73.4 (326.5); ■ 98 (435.9). [Brumund, W. F., and Leonards, G. A. (1972). "Subsidence of Sand Due to Surface Vibration," *Journal of the Soil Mechanics and Foundations Division*, *ASCE*, 98 (SM1), Fig. 10, p. 35.]

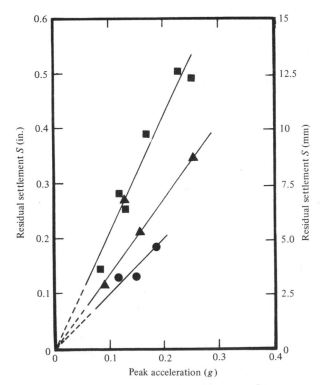

Peak acceleration (g)

given foundation weight W, the settlement increases linearly with the peak acceleration level. However, for a given frequency of vibration and peak acceleration level, the settlement increases with the increase of W. Figure 10.11 shows a plot of settlement S against the energy transmitted per cycle to the soil by the foundation. The data include

- a frequency range of 14–59.3 cps (both above and below the resonant frequency)
- a range in static pressure of 0.27–0.55 × static ultimate bearing capacity q_u (The static presure q can be defined as

$$q = W/A \tag{10.16}$$

  where A is the area of the foundation.)
- a range in maximum downward dynamic force of $0.3W-W$. The maximum downward dynamic force may theoretically be obtained from Eq. (1.90) as

$$F_{\text{dynam(max)}} = A_z \cdot \sqrt{k^2 + (c\omega)^2}$$

FIGURE 10.11 Plot of settlement vs transmitted energy per cycle. Model foundation diameter = 4 in. (101.6 mm), 20–30 Ottawa sand: ● below resonance; ■ above resonance; ▲ impact. Note: 1 in.-lb = 0.113 m-N. [Brummund, W. F., and Leonards, G. A. (1972). "Subsidence of Sand Due to Surface Vibrations," *Journal of the Soil Mechanics and Foundations Division, ASCE*, 98 (SM1), Fig. 9, p. 33.]

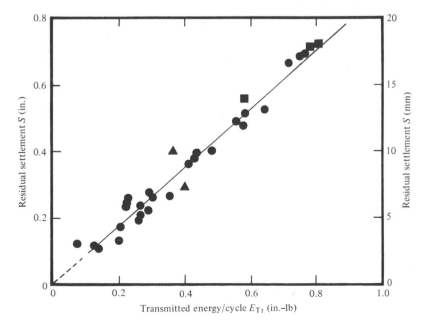

where A_z is the amplitude of foundation vibration, spring constant

$$k = 4Gr_0/(1-\mu) \qquad (6.40)$$

G is the shear modulus of soil, r_0 is the radius of the foundation, μ is Poisson's ratio of soil,

$$c = [3.4/(1-\mu)]r_0^2\sqrt{G\rho} \qquad (6.39)$$

ρ is the density of soil, $\omega = 2\pi f$ and f is oscillation frequency.

The equation for determination of *energy transmitted* to the soil/cycle E_{Tr} may be obtained as follows:

$$E_{Tr} = \int F\,dz = F_{av}A_z \qquad (10.17)$$

where F is the total contact force on the soil, and F_{av} is the average contact force on the soil/cycle; however,

$$F_{av} = \tfrac{1}{2}(F_{max} + F_{min}) \qquad (10.18)$$

$$F_{max} = W + F_{dynam(max)} \qquad (10.19)$$

FIGURE 10.12 Settlements vs peak acceleration for three levels of transmitted energy. Model foundation diameter = 4 in. (101.6 mm), 20–30 Ottawa sand. Frequency and E_{Tr} ranges shown below. Note: 1 in.-lb = 0.113 m-N. [Brumund, W. F., and Leonards, G. A. (1972). "Subsidence of Sand Due to Surface Vibrations," *Journal of the Soil Mechanics and Foundations Division*, ASCE, 98 (SM1), Fig. 11, p. 35.]

Symbol	Freq. (Hz)	E_{Tr} (in.-lb)
●	28.0–59.3	0.72–0.77
■	18.0–59.3	0.55–0.64
▲	18.7–37.3	0.23–0.28

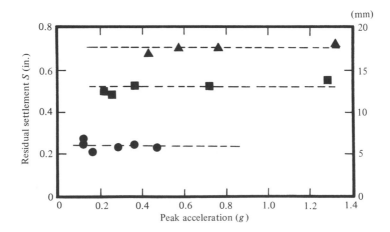

and

$$F_{min} = W - F_{dynam(max)}$$ (10.20)

Substituting Eqs. (10.19) and (10.20) into Eq. (10.18),

$$F_{av} = W$$ (10.21)

Thus, from Eqs. (10.17) and (10.21),

$$E_{Tr} = WA_z$$ (10.22)

Figure 10.11 shows that the transmitted energy per cycle of oscillation E_{Tr} varies linearly with the settlement. A plot of the experimental results of settlement against peak acceleration for different ranges of E_{Tr} is plotted in Figure 10.12. This clearly demonstrates that, if the value of the transmitted energy is constant, the residual settlement remains constant irrespective of the level of peak acceleration.

The above concept is very important for the analysis of settlement of foundations of machineries subjected to vertical oscillation. However, at this time, techniques of extrapolation of settlements of prototype foundations from laboratory model tests are not available. In any case, if the foundation soil is granular and loose, it is always advisable to take precautions to avoid possible problems in settlement. A specification of at least 70% relative density of compaction is often cited.

10.4 ONE-DIMENSIONAL CONSOLIDATION OF CLAY UNDER REPEATED LOADING

One-dimensional consolidation of a clay layer under repeated loading has been studied by Wilson and Elgohary (1974). For the development of the mathematical relationships, the assumptions incorporated in Terzaghi's consolidation theory due to static loading conditions (Terzaghi, 1943) were used. Those assumptions will not be repeated here. However, an additional assumption also needs to be made; namely, that the *coefficient of compressibility* $a_v = -de/d\bar{\sigma}$, is also equal to the *coefficient of expansion* (e = void ratio, $\bar{\sigma}$ = effective stress on soil).

Figure 10.13a shows a saturated clay layer of thickness H drained at the top only. Let a dynamic pressure σ_t acting on the clay layer be defined by Figure 10.13b. This means that the dynamic load $\sigma_t = \sigma$, is on for a time T_1 and off for a time $T - T_1$. The basic differential equation of Terzaghi for consolidation may be written as

$$c_v \partial^2 u_{(z,t)}/\partial z^2 = -\partial\bar{\sigma}_{(z,t)}/\partial t$$ (10.23)

where c_v is the coefficient of consolidation, $u_{(z,t)}$ is the excess pore water pressure at depth z, t = time, and $\partial\bar{\sigma}_{(z,t)}/\partial t$ is the rate of change of effective stress with time at a depth z.

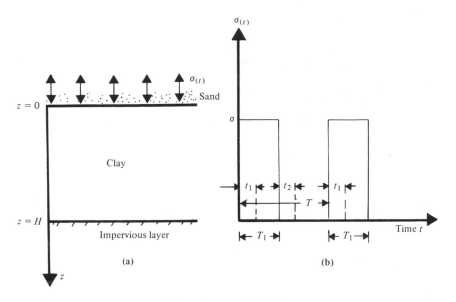

FIGURE 10.13 Definition of dynamic consolidation parameters.

The excess pore water pressure at a depth z and time t may be written as

$$u_{(z,t)} = \sigma_t - \bar{\sigma}_{(z,t)} \qquad (10.24)$$

where $\bar{\sigma}_{(z,t)}$ is that portion of the dynamic stress that has been transferred to the soil structure (i.e., effective stress). Since σ_t is a function of time, from Eq. (10.24) one can write

$$\partial^2 u_{(z,t)} / \partial z^2 = -\partial^2 \bar{\sigma}_{(z,t)} / \partial z^2 \qquad (10.25)$$

Combining Eqs. (10.23) and (10.25),

$$c_v \, \partial^2 \bar{\sigma}_{(z,t)} / \partial z^2 = \partial \bar{\sigma}_{(z,t)} / \partial t \qquad (10.26)$$

For solving the above equation, the boundary conditions are as follows:

At time $t = 0$, $\bar{\sigma}_{(z,t)} = 0$ for $0 \leqslant z \leqslant H$

At any time $t \geqslant 0$, $\partial \bar{\sigma}_{(z,t)} / \partial z = 0$ for $z = H$

$\bar{\sigma}_{(z,t)} = \sigma_{(t)}$ at $z = 0$ and $t \geqslant 0$

The value of $\sigma_{(t)}$ can be represented by the equations

$$\sigma_{(t)} = \sigma \qquad \text{for} \quad rT < t < rT + T_1 \qquad (10.27)$$

and

$$\sigma_{(t)} = 0 \qquad \text{for} \quad rT + T_1 < t < (r+1)T \qquad (10.28)$$

where $r = 0, 1, 2, \cdots$.

The solution of Eq. (10.26) with the above boundary conditions gives the following results.

During the loading period

$$u_{(z,t)} = \frac{4\sigma}{\pi} \sum_{m=0}^{\infty} \frac{1}{2m+1} \sin(MZ) \left[\frac{e^{\beta_m(T_1-t_1)} - e^{\beta_m(T-t_1)}}{1 - e^{\beta_m T}} \right.$$
$$\left. - \frac{e^{-\beta_m(t-T_1)} - e^{-\beta_m(t-T)}}{1 - e^{\beta_m T}} + e^{-\beta_m t} \right]$$

(10.29)

where

$$M = (2m+1)\tfrac{1}{2}\pi \qquad (10.30)$$
$$Z = z/H \qquad (10.31)$$
$$\beta_m = c_v M^2/H^2 \qquad (10.32)$$

t is the elapse of *total time* from the beginning of load application, and t_1 is the elapse of time from the beginning of each load cycle (see Figure 10.13).

The degree of consolidation at a depth z and time t can be given by

$$U_{(z,t)} = (\sigma - u_{(z,t)})/\sigma \qquad (10.33)$$

Substituting Eq. (10.29) into Eq. (10.33),

$$U_{(z,t)} = 1 - \frac{4}{\pi} \sum_{m=0}^{\infty} \frac{1}{2m+1} \sin(MZ) \left[\frac{e^{\beta_m(T_1-t_1)} - e^{\beta_m(T-t_1)}}{1 - e^{\beta_m T}} \right.$$
$$\left. - \frac{e^{-\beta_m(t-T_1)} - e^{-\beta_m(t-T)}}{1 - e^{\beta_m T}} + e^{-\beta_m t} \right]$$

(10.34)

The average degree of consolidation at any time t ($U_{av(t)}$) can also be evaluated as

$$U_{av(t)} = \left[\sigma - \frac{1}{H} \int_0^H u_{(z,t)} \right] / \sigma$$

$$= 1 - \frac{8}{\pi^2} \sum_{m=0}^{\infty} \frac{1}{(2m+1)^2} \left[\frac{e^{\beta_m(T_1-t_1)} - e^{\beta_m(T-t_1)}}{1 - e^{\beta_m T}} \right.$$
$$\left. - \frac{e^{-\beta_m(t-T_1)} - e^{-\beta_m(t-T)}}{1 - e^{\beta_m T}} + e^{-\beta_m t} \right]$$

(10.35)

where $U_{av(t)}$ = average degree of consolidation

During the period when the load $\sigma_t = 0$

$$u_{(z,t)} = -\frac{4\sigma}{\pi} \sum_{m=0}^{\infty} \frac{1}{2m+1} \sin(MZ) \left[\frac{e^{\beta_m(T-T_1-t_2)} - e^{\beta_m(T-t_2)}}{1 - e^{\beta_m T}} \right.$$

$$\left. + \frac{e^{-\beta_m(t-T_1)} - e^{-\beta_m(t-T)}}{1 - e^{\beta_m T}} - e^{-\beta_m t} \right] \qquad (10.36)$$

$$U_{(z,t)} = \frac{4}{\pi} \sum_{m=0}^{\infty} \frac{1}{2m+1} \sin(MZ) \left[\frac{e^{\beta_m(T-T_1-t_2)} - e^{\beta_m(T-t_2)}}{1 - e^{\beta_m T}} \right.$$

$$\left. + \frac{e^{-\beta_m(t-T_1)} - e^{-\beta_m(t-T)}}{1 - e^{\beta_m T}} - e^{-\beta_m t} \right] \qquad (10.37)$$

For definition of time t_2, see Figure 10.13b.

At time $t = \infty$, the degree of consolidation $U_{(z,t)}$ for loading and unloading periods can be obtained from Eqs. (10.34) and (10.37) as follows:

For period during which the load is on

$$U_{(z,t)} = 1 - \frac{4}{\pi} \sum_{m=0}^{\infty} \frac{1}{2m+1} \sin(MZ) \left[\frac{e^{\beta_m(T_1-t_1)} - e^{\beta_m(T-t_1)}}{1 - e^{\beta_m T}} \right] \qquad (10.38)$$

For period during which the load is off

$$U_{(z,t)} = \frac{4}{\pi} \sum_{m=0}^{\infty} \frac{1}{2m+1} \sin(MZ) \left[\frac{e^{\beta_m(T-T_1-t_2)} - e^{\beta_m(T-t_2)}}{1 - e^{\beta_m T}} \right] \qquad (10.39)$$

From Eqs. (10.34) and (10.37), it can be seen that, for given values of T, T_1, H, and c_v, they represent two sets of curves: one for loading and the second for unloading. The nature of the plot of $U_{(z,t)}$ vs z/H at time $t = rT + \frac{1}{2}T_1$ (i.e., at the middle of a loading period) and at time $t = rT + \frac{1}{2}T_1 + \frac{1}{2}T$ (i.e., at the middle of the following unloading period) is shown in Figure 10.14.

The physical process of consolidation under this type of cyclic loading can be explained in the following manner. During the time the load is "on" the soil, the excess pore water pressure $u_{(z,t)}$ is positive; but it is negative during the period the load is "off." With the increase of time t, the *positive excess pore water pressure* during the time the load is on decreases and the negative pressure during the period of load removal increases. At very large t, an equilibrium condition is reached; i.e., the water flowing out of the soil due to positive excess pore water pressure becomes equal to water flowing into the soil due to negative excess pore water pressure. It is not possible to reach 100% consolidation under cyclic loading.

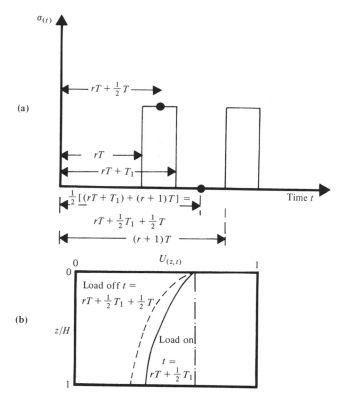

FIGURE 10.14 Nature of variation of the degree of consolidation with depth for dynamic loading.

10.5 SETTLEMENT OF SAND DUE TO CYCLIC SHEAR STRAIN

The laboratory experimental observations described in Section 10.1 have shown that, when a sand layer is subjected to controlled vertical acceleration, considerable settlement does not occur up to a peak acceleration level of $a_{max} = g$. However, in several instances, the cyclic shear strains induced in the soil layers due to ground-shaking of seismic events have caused considerable damage (Figure 10.15a). The controlling parameters for settlement in granular soils due to cyclic shear strain have been studied in detail by Silver and Seed (1971). Some of the results of this study are presented in this section.

The laboratory work of Silver and Seed was conducted on sand by using simple shear equipment developed by the Norwegian Geotechnical Institute. The frequency of the shear stress application to the sand specimens was 1 cps. Dry sand specimens were tested at various relative densities of compaction being subjected to varying normal stresses σ_z and amplitudes of shear

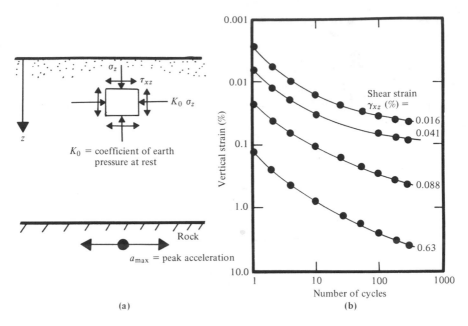

(a)   (b)

FIGURE 10.15 Settlement of sand due to cyclic shear strain. In (b): crystal silica No. 20 sand; $\sigma_z = 500$ lb/ft^2 (23.96 kN/m^2); relative density = 60%, frequency = 1 cps. [Silver, M. L., and Seed, H. B. (1971). "Volume Changes in Sand During Cyclic Loading," *Journal of the Soil Mechanics and Foundations Division, ASCE*, 97 (SM9), Fig. 2, p. 1174.]

strain γ_{xz}. An example of the nature of variation of the vertical strain $\epsilon_z = \Delta H/H$ (H = initial height of the specimen, ΔH = settlement) with number of cycles of shear strain application for a medium dense sand is shown in Figure 10.15b. For these tests, the initial relative density (R_D) of compaction was 60%. Based on Figure 10.15b, the following observations can be made.

For a given normal stress σ_z and amplitude of shear strain γ_{xz}, the vertical strain increases with the number of strain cycles. However, a large portion of the vertical strain occurs in the first few cycles. For example, in Figure 10.15b, the vertical strain occurring in the first 10 cycles is approximately equal to or more than that occurring in the next 40–50 cycles.

For a given value of the vertical stress and number of cycles N, the vertical strain increases with the increase of the shear strain amplitude.

However, one has to keep in mind that a small amount of compaction (i.e., increase in the relative density) could markedly reduce the settlement of a given soil. Silver and Seed (1971) also observed that at higher amplitudes of cyclic shear strain ($\gamma_{xz} > 0.05\%$, for a given value of N), the vertical strain is not significantly affected by the magnitude of the vertical stress. This may not be true where the shear strain is less than 0.05%.

The basic understanding of the laboratory test results for the settlement due to cyclic shear strain application may now be used for calculation of settlement of sand layers due to seismic effect. This is presented in the following section.

10.6 CALCULATION OF SETTLEMENT OF DRY SAND LAYERS SUBJECTED TO SEISMIC EFFECT

Seed and Silver (1972) have suggested a procedure to calculate the settlement of a sand layer subjected to seismic effect. This procedure is outlined below in a step-by-step manner.

1. Since the primary source of ground motion in a soil deposit during an earthquake is due to the upward propagation of motion from the underlying rock formation, adopt a representative history of horizontal acceleration for the base.
2. Divide the soil layer into n layers. They need not be of equal thickness.
3. Calculate the average value of the vertical effective stress σ_z for each layer (*Note*: In dry sand, total stress is equal to effective stress.)
4. Determine the representative relative densities for each layer.
5. Using the damping ratio and the shear moduli characteristics given in Section 8.7, calculate the history of shear strains at the middle of all n layers.
6. Convert the irregular strain histories obtained for each layer (Step 5) into average shear strains and equivalent number of uniform cycles (see Section 8.8).
7. Conduct laboratory tests with a simple shear equipment on representative soil specimens from each layer to obtain the vertical strains for the equivalent number of strain cycles calculated in Step 6. This has to be done for the average effective vertical stress levels σ_z calculated in Step 3 and the corresponding average shear strain levels calculated in Step 6.
8. Calculate the total settlement as

$$\Delta H = \epsilon_{z(1)} H_1 + \epsilon_{z(2)} H_2 + \cdots + \epsilon_{z(n)} H_n \qquad (10.40)$$

where $\epsilon_{z(1)}, \epsilon_{z(2)}, \ldots$ are average vertical strains determined in Step 7 for layers $1, 2, \ldots$ and $H_1, H_2, \ldots$ are layer thicknesses.

The applicability of the above procedure is explained in Example 10.1.

Example 10.1 A 20-m-thick sand layer is shown in Figure 10.16a. The unit weight of soil is 16.1 kN/m³. Using a design earthquake record, the average

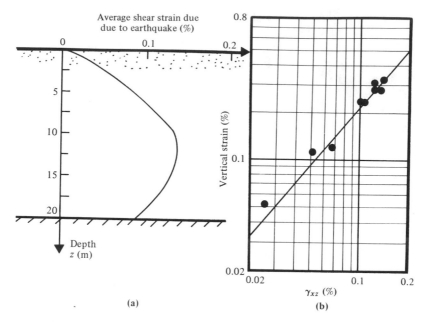

(a)   (b)

FIGURE 10.16 (a) Plot of average shear strain induced due to earthquake: sand unit weight $= 16.1$ kN/m³; relative density $= 50\%$. (b) Laboratory simple shear test results: number of cycles $= 10$.

shear strain in the soil layer has been evaluated and plotted in Figure 10.16a. (*Note*: It was assumed that $\gamma_{av} \approx 0.65\gamma_{max}$.) In this evaluation, the procedure outlined in Section 8.5 was followed with $G = 1000K_2\bar{\sigma}^{1/2}$ (lb/ft²) and damping $= 20\%$. The number of equivalent cycles of shear strain application was estimated to be 10. Cyclic simple shear tests on representative specimens of this sand were conducted with their *corresponding vertical*

TABLE 10.2[a]

Depth (m)	H (m)	Shear strain at the middle of layer γ_{xz} (%)	ϵ_z (%)	$H\epsilon_z \times 10^{-2}$ (m)
0 – 2.5	2.5	0.025	0.043	0.1075
2.5– 5	2.5	0.065	0.13	0.325
2 – 7.5	2.5	0.100	0.22	0.55
7.5–10	2.5	0.125	0.28	0.700
10 –12.5	2.5	0.140	0.31	0.775
12.5–15	2.5	0.135	0.3	0.750
15 –17.5	2.5	0.125	0.28	0.700
17.5–20	2.5	0.105	0.23	0.575

[a]$\Delta H = 4.4825 \times 10^{-2}$ m $= 44.8$ mm.

stresses as in the field. The results of these tests are shown in Figure 10.16b. Estimate the probable settlement of the sand layer.

Solution: From Figure 10.16b, it can be seen that, even though tests were conducted with different values of σ_z, the results of ϵ_z vs γ_{xz} are almost linear in a log–log plot. This shows that the magnitude of the effective overburden pressure has practically no influence on the vertical strain. Thus, for this calculation, the average line of the experimental results is used. For calculation of settlement, Table 10.2 can be prepared.

10.7 SETTLEMENT OF A DRY SAND LAYER DUE TO MULTIDIRECTIONAL SHAKING

Pyke et al. (1975) have made studies to calculate the settlement of a dry sand layer subjected to multidirectional shaking; i.e., shaking with accelerations in the x, y, and z directions as shown in Figure 10.17a. The conclusions of this study show that the settlements caused by combined *horizontal motions* are approximately equal to the sum of the settlements caused by the components acting separately. The effect of the vertical acceleration is again

FIGURE 10.17 Multidirectional shaking of a sand layer: **(a)** definition; **(b)** effect of vertical motion superimposed on horizontal motion—Monterey No. 0 sand, relative density = 60%. [Pyke, R., Seed, H. B., and Chan, C. K. (1975). "Settlement of Sands Under Multidirectional Shaking," *Journal of the Geotechnical Engineering Division*, *ASCE*, 101 (GT4), Fig. 15, p. 394.]

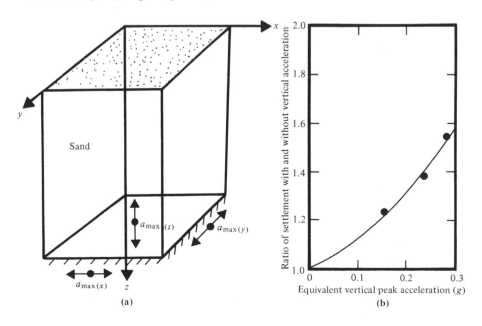

to increase the settlement. Figure 10.17b shows the effect of the vertical acceleration on settlement on Monterey No. 0 sand with an initial relative density of 60%. As an example, let us consider the problem of settlement given in Example 10.1. If the same sand layer is subjected to similar base accelerations in the x and y directions, and if the *average* vertical acceleration in the layer is about 0.2 g, the total settlement can be estimated as follows:

Settlement due to the component in the x direction = 44.8 mm

Settlement due to the component in the y direction = 44.8 mm

Total settlement due to horizontal motions = 89.6 mm

For $a_{z(max)} = 0.2$ g, from Figure 10.17b the ratio of settlement is about 1.3. Thus, the total settlement due to all three components is equal to 1.3(89.6) = 116.48 mm.

It needs to be pointed out that vertical acceleration acting alone without horizontal motion has practically no effect of settlement up to about 1 g (Section 10.1). However, when it acts in combination with the horizontal motions, it produces a marked increase of total settlement.

PROBLEMS

10.1. Explain the difference between the controlled vertical stress and controlled vertical acceleration types of test on laterally confined soil specimens for evaluation of compressibility.

10.2. The results of a set of laboratory simple shear tests on a dry sand are given below (vertical stress $\sigma_v = 400$ lb/ft^2; number of cycles = 12; frequency = 1 cps; initial relative density of specimens = 65%).

Peak shear strain γ_{xz} (%)	Vertical strain (%)	Peak shear strain γ_{xz} (%)	Vertical strain (%)
0.02	0.035	0.10	0.095
0.04	0.06	0.15	0.20
0.06	0.06	0.2	0.28
0.08	0.09		

Plot the results on log–log graph paper. Approximate the results in the form of an equation

$$\gamma_{xz} = m\epsilon_z^n$$

where m and n are constants and γ_{xz}, ϵ_z are percentages.

10.3. A dry sand deposit is 40 ft thick and its relative density is 65%. This layer of sand may be subjected to an earthquake. The number of equivalent cycles of shear stress application due to an earthquake is estimated to be 12. Following is the variation of the *average* expected shear strain with depth.

Depth (ft)	Average shear strain (%)	Depth (ft)	Average shear strain (%)
0	0	25	0.186
5	0.08	30	0.170
10	0.135	35	0.160
15	0.155	40	0.140
20	0.175		

Estimate the probable settlement of the sand layer using the laboratory test results given in Problem 10.2.

10.4. Repeat Problem 10.3 for the following (depth of sand layer $= 10$ m):

Depth (m)	Average shear strain (%)
0	0
2.5	0.1
5	0.14
7.5	0.135
10	0.117

REFERENCES

Brumund, W. F., and Leonards, G. A. (1972). "Subsidence of Sand Due to Surface Vibration," *Journal of the Soil Mechanics and Foundations Division, ASCE* 98 (SM1), 27–42.

D'Appolonia, D. J., and D'Appolonia, E. (1967). "Determination of the Maximum Density of Cohesionless Soils," *Proceedings, 3rd Asian Regional Conference on Soil Mechanics and Foundation Engineering*, Haifa, Israel, Vol. 1, p. 266.

D'Appolonia, E. (1970). "Dynamic Loadings," *Journal of the Soil Mechanics and Foundations Division, ASCE* 96 (SM1), 49–72.

Krizek, R. J., and Fernandez, J. I. (1971). "Vibratory Densification of Damp Clayey Sands," *Journal of the Soil Mechanics and Foundations Division, ASCE* 97 (SM8), 1069–1079.

Ortigosa, P., and Whitman, R. V. (1968). "Densification of Sand by Vertical Vibrations with Almost Constant Stress," *Publication No. 206*, Department of Civil Engineering, Massachusetts Institute of Technology, Cambridge.

Pyke, R., Seed, H. B., and Chan, C. K. (1975). "Settlement of Sands Under Multidirectional Shaking," *Journal of the Geotechnical Engineering Division, ASCE* 101 (GT4), 379–398.

Raymond, G. P., and Komos, F. E. (1978). "Repeated Load Testing of a Model Plane Strain Footing," *Canadian Geotechnical Journal* 15 (2), 190–201.

Seed, H. B., and Silver, M. L. (1972). "Settlement of Dry Sands During Earthquakes," *Journal of the Soil Mechanics and Foundations Division, ASCE* 98 (SM4), 381–397.

Silver, M. L., and Seed, H. B. (1971). "Volume Changes in Sands During Cyclic Loading," *Journal of the Soil Mechanics and Foundations Division, ASCE* 97 (SM9), 1171–1182.

Terzaghi, K. (1943). *Theoretical Soil Mechanics*, Wiley, New York.

Wilson, N. E., and Elgohary, M. M. (1974). "Consolidation of Soils Under Cyclic Loading," *Canadian Geotechnical Journal* 11 (3), 420–423.

11
LIQUEFACTION
OF SATURATED SAND

A loose saturated sand deposit, when subjected to vibration, tends to com-
pact and decrease in volume. If drainage is unable to occur, the pore water
pressure increases. If the pore water pressure in the sand deposit is allowed
to build up by continuous vibration, a condition will be reached at some
time where the overburden pressure will be equal to the pore water pressure.
Based on the effective stress principle,

$$\sigma' = \sigma - u \tag{11.1}$$

where σ' is the effective stress, σ is the total overburden pressure, and u is
the pore water pressure. If σ is equal to u, σ' is zero. Under this condition,
the sand does not possess any shear strength, and it develops into a
liquefied state.

Liquefaction of saturated sands during earthquakes has been the cause of
much damage to buildings, earth embankments, and retaining structures.
Several instances of damage of this type have been seen from the June 16,
1964 earthquake at Niigata (Japan) and the Alaskan earthquake of 1964.
This is a phenomenon which occurs mostly in medium- to fine-grained
sands.

One of the first attempts to explain the liquefaction phenomenon in
sandy soils was made by Casagrande (1936) and is based on the concept of
critical void ratio. Dense sand, when subjected to shear, tends to dilate;
loose sand, under similar conditions, tends to decrease in volume. The void
ratio at which sand does not change in volume when subjected to shear is
referred to as the *critical void ratio*. Casagrande explained that deposits of
sand which have a void ratio larger than the critical void ratio tend to
decrease in volume when subjected to a vibration by a *seismic effect*. If
drainage is unable to occur, the pore water pressure increases. This, in
effect, poses the possibility of soil liquefaction. However, one must keep in

mind the following facts which show that the critical void ratio concept may not be sufficient for a quantitative evaluation of soil liquefaction potential of sand deposits:

Critical void ratio is not a constant value, but changes with confining pressure.

Volume changes due to dynamic loading conditions are different than the one-directional static loading conditions realized in the laboratory by direct shear equipment and triaxial shear equipment.

For that reason, since the mid-1960s, intensive investigations have been carried out around the world to determine the soil parameters which control liquefaction. In this chapter, the findings of some of these studies are discussed.

11.1 LABORATORY STUDIES TO SIMULATE FIELD CONDITIONS FOR SOIL LIQUEFACTION

If one considers a soil element in the field as shown in Figure 11.1a, when earthquake effects are not present, the vertical effective stress on the element is equal to σ', which is equal to σ_v, and the horizontal effective

FIGURE 11.1 Application of cyclic shear stress on a soil element due to an earthquake.

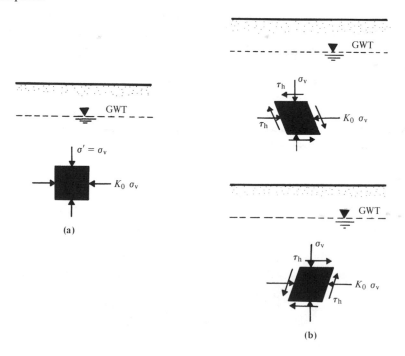

stress on the element equals $K_0\sigma_v$, where K_0 is the at-rest earth pressure coefficient.

Due to ground-shaking during an earthquake, a cyclic shear stress τ_h will be imposed on the soil element. This is shown in Figure 11.1b. Hence, any laboratory test to study the liquefaction problem must be designed in a manner so as to simulate the condition of a constant normal stress and a cyclic shear stress on a plane of the soil specimen. Various types of laboratory test procedure have been adopted in the past, such as the dynamic triaxial test (Seed and Lee, 1966; Lee and Seed, 1967), cyclic simple shear test (Peacock and Seed, 1968; Finn et al., 1970; Seed and Peacock, 1971), cyclic torsional shear test (Yoshimi and Oh-oka, 1970, 1973; Ishibashi and Sherif, 1974), and shaking table test (Prakash and Mathur, 1965). However, the most commonly used laboratory test procedures are the dynamic triaxial tests and the simple shear tests. These are discussed in detail in the following sections.

11.2 DYNAMIC TRIAXIAL TEST

11.2.1 General Concepts and Test Procedures

Consider a saturated soil specimen in a triaxial test as shown in Figure 11.2a, which* is consolidated under an all-around pressure of σ_3. The corresponding Mohr's circle is shown in Figure 11.2b. If the stresses on the specimen are changed such that the axial stress is equal to $\sigma_3 + \frac{1}{2}\sigma_d$ and the radial stress is $\sigma_3 - \frac{1}{2}\sigma_d$ (Figure 11.2c), and drainage into or out of the specimen is not allowed, then the corresponding total stress Mohr's circle is of the nature shown in Figure 11.2d. Note that the stresses on the plane $X-X$ are

$$\text{total normal stress} = \sigma_3, \qquad \text{shear stress} = +\tfrac{1}{2}\sigma_d$$

and the stresses on the plane $Y-Y$ are

$$\text{total normal stress} = \sigma_3, \qquad \text{shear stress} = -\tfrac{1}{2}\sigma_d$$

Similarly, if the specimen is subjected to a stress condition as shown in Figure 11.2e, the corresponding total stress Mohr's circle is as shown in Figure 11.2f. The stresses on the plane $X-X$ are

$$\text{total normal stress} = \sigma_3, \qquad \text{shear stress} = -\tfrac{1}{2}\sigma_d$$

The stresses on the plane $Y-Y$ are

$$\text{total normal stress} = \sigma_3, \qquad \text{shear stress} = +\tfrac{1}{2}\sigma_d$$

It can be seen that, if cyclic normal stresses of magnitude $\frac{1}{2}\sigma_d$ are applied simultaneously in the horizontal and vertical directions, one can achieve a stress condition along planes $X-X$ and $Y-Y$ that will be similar to the cyclic shear stress application shown in Figure 11.1b.

356

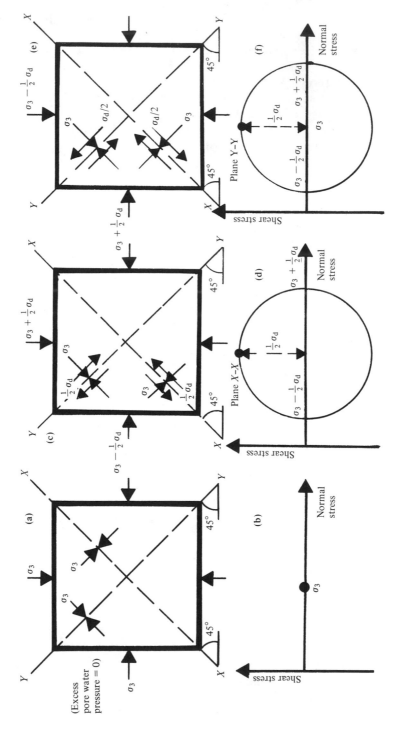

FIGURE 11.2 Simulation of cyclic shear stress on a plane for a triaxial test specimen.

However, for saturated sands, actual laboratory tests can be conducted by applying an all-around consolidation pressure of σ_3 and then applying a cyclic load having an amplitude of σ_d in the axial direction only without allowing drainage as shown in Figure 11.3a. The axial strain and the excess pore water pressure can be measured along with the number of cycles of load (σ_d) application. The question may now arise as to how the loading system shown in Figure 11.3a would produce stress conditions shown in Figures 11.2c and 11.2e. This can explained as follows. The stress condition shown in Figure 11.3b is the sum of the stress conditions shown in Figures 11.3c and 11.3d. The effect of the stress condition shown in Figure 11.3d is to reduce the excess pore water pressure of the specimen by an amount equal to $\frac{1}{2}\sigma_d$ without causing any change in the axial strain. Thus, the effect of the stress conditions shown in Figure 11.3b (which is the same as Figure 11.2c) can be achieved by only subtracting a pore water pressure $u = \frac{1}{2}\sigma_d$ from that observed from the loading condition shown in Figure 11.3c. Similarly, the loading condition shown in Figure 11.3e is the loading condition in Figure 11.3f plus the loading condition in Figure 11.3g. The effect of the stress condition shown in Figure 11.3g is only to increase the pore water pressure by an amount $\frac{1}{2}\sigma_d$. Thus the effect of the stress conditions shown in Figure 11.3e (which is the same as in Figure 11.2e) can be achieved by only adding $\frac{1}{2}\sigma_d$ to the pore water pressure observed from the loading condition in Figure 11.3f.

FIGURE 11.3

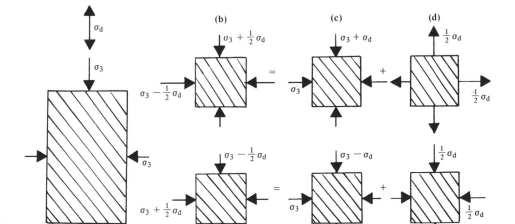

11.2.2 Typical Results from Cyclic Triaxial Tests

Several cyclic undrained triaxial tests on saturated soil specimens have been conducted by Seed and Lee (1966) on Sacramento River sand retained between No. 50 and No. 100 U.S. sieves. The results of a typical test in *loose sand* (void ratio, $e = 0.87$) is shown in Figure 11.4. For this test, the initial all-around pressure and the initial pore water pressure were 28.4 lb/in.^{2} (196.2 kN/m^2) and 14.2 lb/in.^{2} (98.2 kN/m^2), respectively. Thus, the all-around consolidation pressure σ_3 is equal to 14.2 lb/in.^{2} (98.1 kN/m^2). The cyclic deviator stress σ_d was applied with a frequency of 2 cps. Figure 11.5 is a plot of the axial strain, change in pore water pressure u, and the change in pore water pressure corrected to mean extreme principal stress conditions (i.e., subtracting or adding $\frac{1}{2}\sigma_d$ from or to the observed pore water pressure) against the number of cycles of load application. Figure

FIGURE 11.4 Typical pulsating load test on loose saturated Sacramento River sand: relative density = 38%; initial void ratio = 0.87; initial pore water pressure = 98.1 kN/m^2; initial confining pressure = 196.2 kN/m^2; $\sigma_d = 38.2$ kN/m^2. [Seed, H. B., and Lee, K. L. (1966). "Liquefaction of Saturated Sands During Cyclic Loading," *Journal of the Soil Mechanics and Foundations Division*, ASCE, 92 (SM6), Fig. 4, p. 113.]

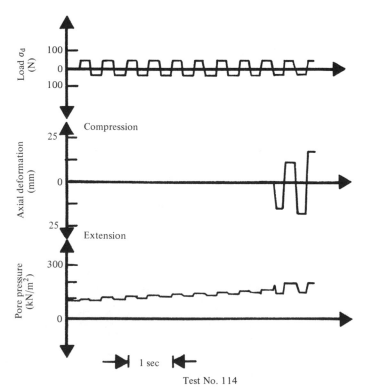

Test No. 114

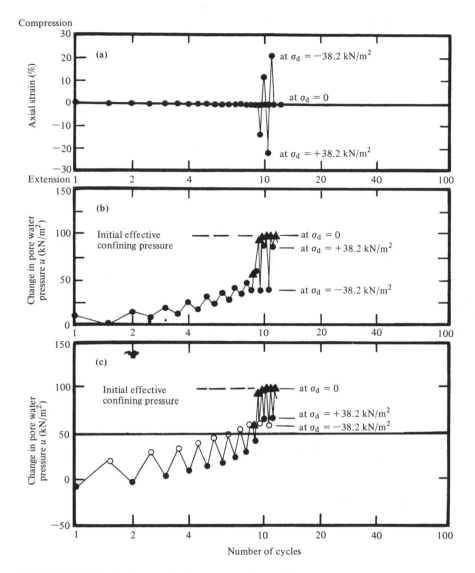

FIGURE 11.5 Typical pulsating load test on loose Sacramento River sand—plot of axial strain and pore water pressure vs number of cycles of load application: **(a)** axial strain vs number of cycles, test No. 114, relative density = 38%, $\sigma_3 = 98.1$ kN/m², $\sigma_d = \pm 38.2$ kN/m²; **(b)** observed change in pore water pressure vs number of cycles; **(c)** change in pore water pressure vs number of cycles (corrected to mean principal stress conditions). [Seed, H. B., and Lee, K. L. (1966). "Liquefaction of Saturated Sands During Cyclic Loading," *Journal of the Soil Mechanics and Foundations Division*, *ASCE*, 92 (SM6), Fig. 5, p. 114.]

11.5c shows that the change in pore water pressure becomes equal to σ_3 during the ninth cycle, indicating that the effective confining pressure is equal to zero.

During the tenth cycle, the axial strain exceeded 20% and the soil liquefied. The relationship between the magnitude of σ_d against the number of cycles of pulsating load applications for the liquefaction of the same loose sand [$e = 0.87$, $\sigma_3 = 14.2$ lb/in.^{2} (98.1 kN/m^2)] is shown in Figure 11.6. Note that the number of cycles of pulsating load application increases with the decrease of the value of σ_d.

The nature of variation of the axial strain and the *corrected* pore water pressure for a pulsating load test in a dense Sacramento River sand is shown in Figure 11.7. After about 13 cycles, the change in pore water pressure becomes equal to σ_3; however, the axial strain amplitude did not exceed 10% after even 30 cycles of load application. This is a condition with a peak cyclic pore pressure ratio of 100%, with limited strain potential due to the remaining resistance of the soil to deformation, or due to the fact that the soil dilates. Dilation of the soil reduces the pore water pressure and helps stabilization of soil under load. This may be referred to as *cyclic mobility* (Seed, 1979).

A summary of axial strain, number of cycles for liquefaction, and the relative density for Sacramento River sand are given in Figure 11.8 [for $\sigma_3 = 14.2$ lb/in.^{2} (98.1 kN/m^2)]. However, a different relationship may be obtained if the confining pressure σ_3 is changed.

FIGURE 11.6 Relationship between pulsating deviator stress and number of cycles required to cause failure in Sacramento River sand: initial void ratio = 0.87; $\sigma_3 = 14.2$ lb/in.^{2} (98.1 kN/m^2). [Seed, H. B., and Lee, K. L. (1966). "Liquefaction of Saturated Sands During Cyclic Loading," *Journal of the Soil Mechanics and Foundations Division, ASCE*, 92 (SM6), Fig. 6, p. 115.]

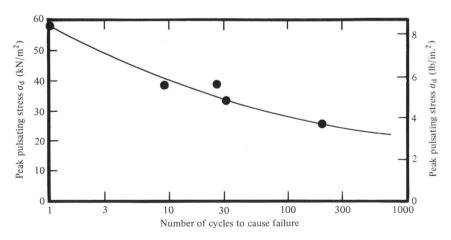

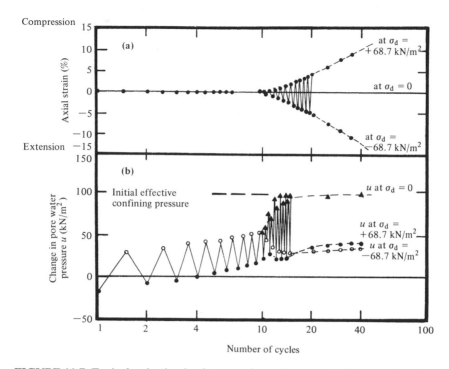

FIGURE 11.7 Typical pulsating load test on dense Sacramento River sand—plot of axial strain and corrected pore water pressure vs number of cycles of load application: **(a)** axial strain vs number of cycles, test No. 119, relative density = 78%, $\sigma_3 = 98.1$ kN/m², $\sigma_d = \pm 68.7$ kN/m²; **(b)** corrected change in pore water pressure. [Seed, H. B. and Lee, K. L. (1966). "Liquefaction of Saturated Sands During Cyclic Loading," *Journal of the Soil Mechanics and Foundations Division, ASCE*, 92 (SM6), Fig. 8, p. 118.]

It has been mentioned earlier that the critical void ratio of a sand cannot be used as a unique criterion for a quantitative evaluation of the liquefaction potential. This can now be seen from Figure 11.9, which shows the critical void ratio line for Sacramento River sand. Based on the initial concept of critical void ratio, one would assume that a soil specimen represented by a point to the left of the critical void ratio line would *not* be susceptible to liquefaction; likewise, a specimen which plots to the right of the critical void ratio line *would* be vulnerable to liquefaction. In order to test this concept, the cyclic load test results on two specimens are shown as *A* and *B* in Figure 11.9. Under a similar pulsating stress $\sigma_d = \pm 17.06$ lb/in.² (117.72 kN/m²), specimen *A* liquefied in 57 cycles, whereas specimen *B* did not fail even in 10,000 cycles. This is contrary to the

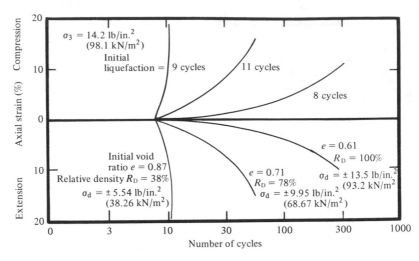

FIGURE 11.8 Axial strain after initial liquefaction for pulsating load tests at three densities for Sacramento River sand. [Seed, H. B., and Lee, K. L. (1966). "Liquefaction of Saturated Sands During Cyclic Loading," *Journal of the Soil Mechanics and Foundations Division*, *ASCE*, 92 (SM6), Fig. 9, p. 118.]

FIGURE 11.9 Critical confining pressure–void ratio relationship for Sacramento River sand: specimen A—sample liquefied in 57 cycles, $\sigma_d = \pm 17.06$ lb/in.^{2} (117.72 kN/m^2); specimen B—sample did not fail in 10,000 cycles, $\sigma_d = \pm 17.06$ lb/in.^{2}. [Seed, H. B., and Lee, K. L. (1966). "Liquefaction of Saturated Sands During Cyclic Loading," *Journal of the Soil Mechanics and Foundations Division*, *ASCE*, 92 (SM6), Fig. 15, p. 126.]

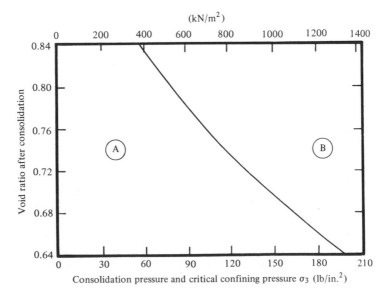

aforementioned assumptions. Thus, the liquefaction potential depends on five important factors:

1. relative density R_D
2. continuing pressure σ_3
3. peak pulsating stress σ_d
4. number of cycles of pulsating stress application
5. overconsolidation ratio

The importance of the first four factors is discussed in the following section. The overconsolidation ratio is discussed in Section 11.3.2. Soil grain characteristics are also known to have some effects on the liquefaction potential.

FIGURE 11.10 Influence of initial relative density on liquefaction for Sacramento River sand: **(a)** initial void ratio = 0.87, relative density = 38%, $\sigma_3 = 14.2 \text{ lb/in.}^2$ (98.1 kN/m^2); **(b)** initial void ratio = 0.61; relative density = 100%, $\sigma_3 = 14.2 \text{ lb/in.}^2$. [Lee, K. L., and Seed, H. B. (1967). "Cyclic Stress Conditions Causing Liquefaction of Sand," *Journal of the Soil Mechanics and Foundations Division*, *ASCE*, 93 (SM1), Fig. 4, p. 52.]

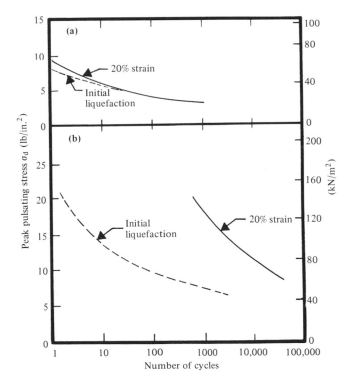

364

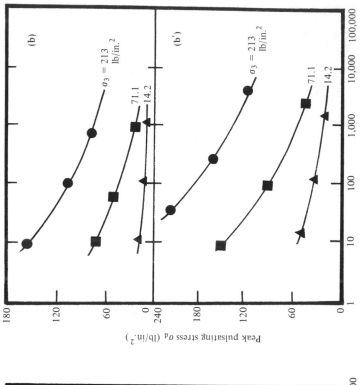

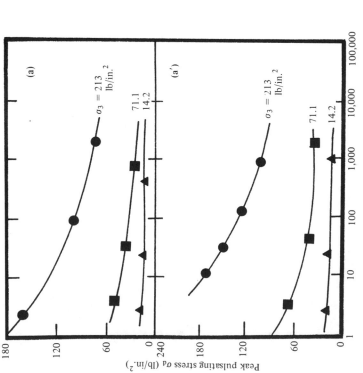

FIGURE 11.11 Influence of confining pressure on liquefaction of Sacramento River sand: **(a, a')** initial liquefaction; **(b, b')** 20% strain; **(a, b)** initial void ratio = 0.71, relative density = 78% **(a', b')** initial void ratio = 0.61, relative density = 100%. *Note:* 1 lb/in.^{2} = 6.9 kN/m^2. [Lee, K. L., and Seed, H. B. (1967). "Cyclic Stress Conditions Causing Liquefaction of Sand," *Journal of the Soil Mechanics and Foundations Division, ASCE,* 93 (SM1), Fig. 6, pp. 54–55.]

11.2.3 Influence of Various Parameters on Soil Liquefaction Potential

Influence of the Initial Relative Density

The effect of the initial relative density of a soil on liquefaction is shown in Figure 11.10. All tests shown in Figure 11.10 are for $\sigma_3 = 14.2$ lb/in.^{2} (98.1 kN/m^2). The *initial liquefaction* corresponds to the condition when the pore water pressure becomes equal to the confining pressure σ_3. In most cases, *20% double amplitude* strain is considered as failure. It may be seen that, for a given value of σ_d, the initial liquefaction and the failure occur simultaneously for loose sand (Figure 11.10a). However, as the relative density increases, the difference between the number of cycles to cause 20% double amplitude strain and to cause initial liquefaction increases.

Influence of Confining Pressure

The influence of the confining pressure σ_3 on initial liquefaction and 20% double amplitude strain condition is shown in Figure 11.11. For a given initial relative density and peak pulsating stress, the number of cycles to cause initial liquefaction or 20% strain increases with the increase of the confining pressure. This is true for all relative densities of compaction.

Influence of the Peak Pulsating Stress

Figure 11.12 shows the variation of the peak pulsating stress σ_d with the confining pressure for initial liquefaction in 100 cycles (Figure 11.2a) and

FIGURE 11.12 Influence of pulsating stress on the liquefaction of Sacramento River sand: **(a)** initial liquefaction in 100 cycles; **(b)** 20% strain in 100 cycles. [Lee, K. L., and Seed, H. B. (1967). "Cyclic Stress Conditions Causing Liquefaction of Sand," *Journal of the Soil Mechanics and Foundations Division, ASCE*, 93 (SM1), Fig. 7, p. 56.]

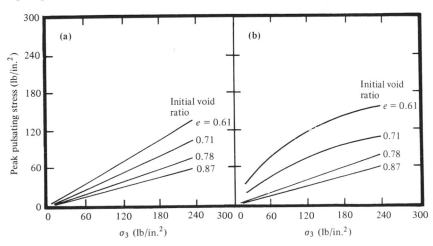

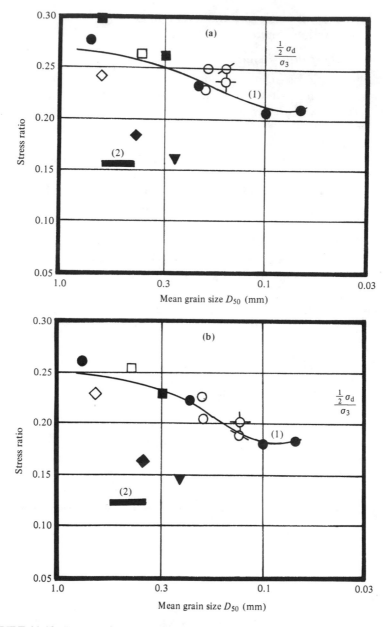

FIGURE 11.13 Stress ratio causing liquefaction of sands in **(a)** 10 and **(b)** 30 cycles: (1) triaxial compression data for $\frac{1}{2}\sigma_d/\sigma_3$ at liquefaction; (2) field value of τ_h/σ_v causing liquefaction estimated from the results of simple shear test. Relative density $\approx 50\%$. [Seed, H. B., and Idriss, I. M. (1971). "Simplified Procedure for Evaluating Soil Liquefaction Potential," *Journal of the Soil Mechanics and Foundations Division*, *ASCE*, 97 (SM9), Figs. 6 and 7, p. 1257.]

for 20% axial strain in 100 cycles (Figure 11.12b). Note that for a given initial void ratio (i.e., relative density R_D) and number of cycles of load application, the variation of σ_d for initial liquefaction with σ_3 is practically linear. A similar relation also exists for loose sand with a 20% axial strain condition.

11.2.4 Development of Standard Curves for Initial Liquefaction

By compilation of the results of liquefaction tests conducted by several investigators on various types of sand, average standard curves for initial liquefaction for a given number of load cycle application can be developed. These curves can then be used for evaluation of liquefaction potential in the field. Some of these plots developed by Seed and Idriss (1971) are given in Figure 11.13.

Figure 11.13a is a plot of $\frac{1}{2}\sigma_d/\sigma_3$ vs D_{50} to cause initial liquefaction in 10 cycles of stress application. The plot is for an initial relative density of compaction of 50%. Note that D_{50} in Figure 11.13a is the mean grain size, i.e., the size through which 50% of the soil will pass. It should be kept in mind that $\frac{1}{2}\sigma_d$ is the magnitude of the maximum cyclic shear stress imposed on a soil specimen (see plane $X-X$ and $Y-Y$ of Figure 11.2d, f). Figure 11.13b is a similar plot for initial liquefaction in 30 cycles of stress application. These curves are used in Section 11.4.4 for evaluation of liquefaction potential.

11.3 CYCLIC SIMPLE SHEAR TEST

11.3.1 General Concepts

Cyclic simple shear tests can be used to study liquefaction on saturated sands by using the simple shear apparatus developed by Roscoe (1953). In this type of test, the soil specimen is consolidated by a vertical stress σ_v. At this time, lateral stress is equal to $K_0\sigma_v$ (K_0 = coefficient of earth pressure at rest). The initial stress conditions of a specimen in a simple shear device are shown in Figure 11.14a; the corresponding Mohr's circle is shown in Figure 11.14b. After that, a cyclic horizontal shear stress of peak magnitude τ_h is applied (undrained condition) to the specimen as shown in Figure 11.14c. The pore water pressure and the strain are observed with the number of cycles of horizontal shear stress application.

Using the stress conditions on the soil specimen at a certain time during the cyclic shear test, a Mohr's circle is plotted in Figure 11.14d. Note that the maximum shear stress on the specimen in simple shear is not τ_h, but

$$\tau_{max} = \sqrt{\tau_h^2 + \left[\tfrac{1}{2}\sigma_v(1-K_0)\right]^2} \qquad (11.2)$$

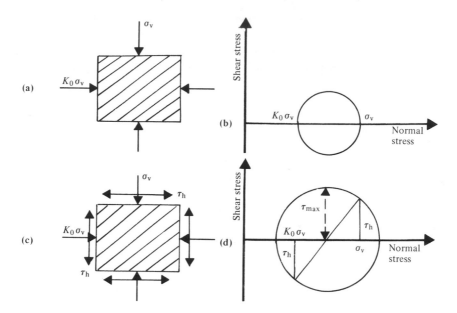

FIGURE 11.14 Maximum shear stress for cyclic simple shear test.

11.3.2 Typical Test Results

Typical results of some soil liquefaction tests on Monterey sand using simple shear apparatus are shown in Figure 11.15. Note that these are for the initial liquefaction condition. From the figure the following facts may be observed:

For a given value of σ_v and relative density R_D, a decrease of τ_h requires an increase of the number of cycles to cause liquefaction.

For a given value of R_D and number of cycles of stress application, a decrease of σ_v requires a decrease of the peak value of τ_h for causing liquefaction.

For a given value of σ_v and number of cycles of stress application, τ_h for causing liquefaction increases with the increase of the relative density.

Another important factor—the variation of the peak value of τ_h for causing initial liquefaction with the initial relative density of compaction (for a given value of σ_v and number of stress cycle applications)—is shown in Figure 11.16. For a relative density up to about 80%, the peak value of τ_h for initial liquefaction increases linearly with R_D.

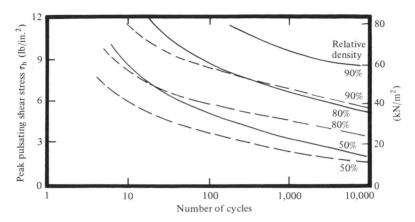

FIGURE 11.15 Initial liquefaction of cyclic simple shear test on Monterey sand: ——— $\sigma_v = 113.7$ lb/in.^{2} (784.8 kN/m^2); – – – $\sigma_v = 71.1$ lb/in.^{2} (490.5 kN/m^2). [Peacock, W. H., and Seed, H. B. (1968). "Sand Liquefaction Under Cyclic Loading, Simple Shear Conditions," *Journal of the Soil Mechanics and Foundations Division*, *ASCE*, 94 (SM3), Fig. 15, p. 701.]

FIGURE 11.16 Effect of relative density on cyclic stress causing initial liquefaction (in 100 cycles of pulsating stress) for Monterey sand. [Peacock, W. H., and Seed, H. B. (1968). "Sand Liquefaction Under Cyclic Loading Simple Shear Conditions," *Journal of the Soil Mechanics and Foundations Division*, *ASCE*, 94 (SM3), Fig. 16, p. 701.]

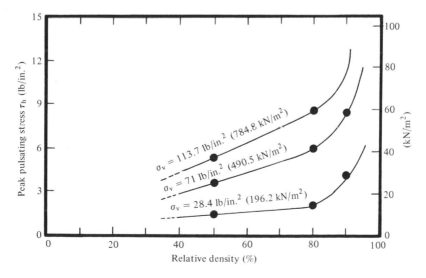

Influence of Test Condition

In simple shear test equipment, there is always some nonuniformity of stress conditions. This causes specimens to develop liquefaction under lower applied horizontal cyclic stresses as compared to that in the field. This happens even though care is taken to improve the preparation of the specimens and rough platens are used at the top and bottom of the specimens to be tested. For that reason, for a given value of σ_v, R_D, and number of cyclic shear stress applications, the peak value of τ_h in the field is about 15%–50% higher than that obtained from the cyclic simple shear test. This fact has been demonstrated by Seed and Peacock (1971) for a uniform medium sand ($R_D \approx 50\%$) in which the field values are about 20% higher than the laboratory values.

Influence of Overconsolidation Ratio on the Peak Value of τ_h Causing Liquefaction

For the cyclic simple shear test, the value of τ_h is highly dependent on the value of the initial lateral earth pressure coefficient at rest (K_0). The value of K_0 is in turn dependent on the overconsolidation ratio (OCR). The variation of τ_h/σ_v for initial liquefaction with the overconsolidation ratio as

FIGURE 11.17 Influence of overconsolidation ration (OCR) on stresses causing liquefaction in simple shear tests. [Seed, H. B., and Peacock, W. H. (1971). "The Procedure for Measuring Soil Liquefaction Characteristics," *Journal of the Soil Mechanics and Foundations Division*, *ASCE*, 97 (SM8), Fig. 6, p. 1104.]

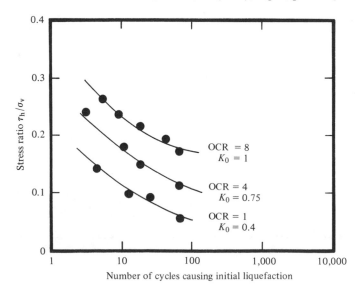

determined by the cyclic simple shear test is shown in Figure 11.17. For a given relative density and number of cycles causing initial liquefaction, the value of τ_h/σ_v decreases with the decrease of K_0. It needs to be mentioned at this point that *all* the *cyclic triaxial* studies for liquefaction are conducted for the initial value of $K_0 = 1$.

11.4 DEVELOPMENT OF PROCEDURE FOR DETERMINATION OF FIELD LIQUEFACTION

11.4.1 Correlation of the Liquefaction Results from Simple Shear and Triaxial Tests

The conditions for determination of field liquefaction problems are related to the ratio of τ_h/σ_v; this is also true for the case of cyclic simple shear tests. However, in the case of triaxial tests, the results are related to the ratio of $\frac{1}{2}\sigma_d/\sigma_3$. It appears that a correlation between τ_h/σ_v and $\frac{1}{2}\sigma_d/\sigma_3$ needs to be developed (for a given number of cyclic stress application to cause liquefaction). Seed and Peacock (1971) considered the following alternative criteria for correlation for the onset of soil liquefaction.

1. The *maximum ratio of the shear stress* developed during cyclic loading *to the normal stress* during consolidation on any plane of the specimen can be a controlling factor. For triaxial specimens, this is equal to $\frac{1}{2}\sigma_d/\sigma_3$, and for simple shear specimens it is about $\tau_h/(K_0\sigma_v)$. Thus

$$\tau_h/(K_0\sigma_v) = \frac{1}{2}\sigma_d/\sigma_3$$

or

$$[\tau_h/\sigma_v]_{\text{simple shear}} = K_0[\tfrac{1}{2}\sigma_d/\sigma_3] \tag{11.3}$$

2. Another possible condition for the onset of liquefaction can be the *maximum ratio of change in shear stress* during cyclic loading *to the normal stress* during consolidation on any plane. For simple shear specimens this is about $\tau_h/(K_0\sigma_v)$, and for triaxial specimens it is $\frac{1}{2}\sigma_d/\sigma_3$. This leads to the same equation as Eq. (11.3).

3. The third possible alternative can be given by the *ratio of the maximum shear stress* induced in a specimen during cyclic loading *to the mean principal stress* on the specimen during consolidation. For simple shear specimens:

$$\frac{\text{Maximum shear stress}}{\text{during cyclic loading}} = \sqrt{\tau_h^2 + [\tfrac{1}{2}\sigma_v(1 - K_0)]^2} \tag{11.2}$$

$$\frac{\text{Mean principal stress during}}{\text{consolidation (Figure 11.14a)}} = \tfrac{1}{3}(\sigma_v + K_0\sigma_v + K_0\sigma_v) = \tfrac{1}{3}\sigma_v(1 + 2K_0)$$

$$\tag{11.4}$$

For triaxial specimens, maximum shear stress during cyclic loading $= \frac{1}{2}\sigma_d$ and mean principal stress during consolidation $= \sigma_3$; so

$$\sqrt{\tau_h^2 + \left[\frac{1}{2}\sigma_v(1-K_0)\right]^2} / \left[\frac{1}{3}\sigma_v(1+2K_0)\right] = \frac{1}{2}\sigma_d/\sigma_3$$

$$\sqrt{(\tau_h/\sigma_v)^2 + \left[\frac{1}{2}(1-K_0)\right]^2} / \frac{1}{3}(1+2K_0) = \frac{1}{2}\sigma_d/\sigma_3$$

$$\left[\tau_h/\sigma_v\right]_{\text{simple shear}} = \sqrt{\left(\frac{1}{2}\sigma_d/\sigma_3\right)^2 \left[\frac{1}{3}(1+2K_0)\right]^2 - \left[\frac{1}{2}(1-K_0)\right]^2}$$

or

$$\left[\tau_h/\sigma_v\right] = \left(\frac{1}{2}\sigma_d/\sigma_3\right)\sqrt{\frac{1}{9}(1+2K_0)^2 - \frac{1}{4}(1-K_0)^2 / \left(\frac{1}{2}\sigma_d/\sigma_3\right)^2} \quad (11.5)$$

4. The fourth possible alternative may be the *ratio of the maximum change in shear stress* on any plane during cyclic loading *to the mean principal stress* during consolidation. Thus, for simple shear specimens, it is equal to $3\tau_h/[\sigma_v(1+2K_0)]$, and for triaxial specimens it is $\frac{1}{2}\sigma_d/\sigma_3$; so

$$\left[\tau_h/\sigma_v\right]_{\text{simple shear}} = \frac{1}{3}(1+2K_0)\left(\frac{1}{2}\sigma_d/\sigma_3\right) \quad (11.6)$$

Thus, in general, it can be written as

$$\left(\tau_h/\sigma_v\right)_{\text{simple shear}} = \alpha\left(\frac{1}{2}\sigma_d/\sigma_3\right)_{\text{triax}} \quad (11.7)$$

where $\alpha = K_0$ for Cases 1 and 2,

$$\alpha = \sqrt{\frac{1}{9}(1+2K_0)^2 - \frac{1}{4}(1-K_0)^2 / \left(\frac{1}{2}\sigma_d/\sigma_3\right)^2} \qquad \text{for Case 3}$$

and

$$\alpha = \frac{1}{3}(1+2K_0) \qquad \text{for Case 4}$$

The values of α for the four cases considered above are given in Table 11.1.

From Table 11.1 it may be seen that, for normally consolidated sands, the value of α is generally in the range of 45%–50%, with an average of about 47%.

Finn et al. (1971) have shown that, for initial liquefaction of normally consolidated sands, α is equal to $\frac{1}{2}(1+K_0)$. The value of K_0 can be given by the relation (Jaky, 1944)

$$K_0 = 1 - \sin\phi \quad (11.8)$$

Castro (1975) has proposed that the initial liquefaction may be controlled by the criteria of the *ratio of the octahedral shear stress* during cyclic loading *to the effective octahedral normal stress during consolidation*. The effective octahedral normal stress during consolidation σ'_{oct} is given by the relation

$$\sigma'_{\text{oct}} = \frac{1}{3}(\sigma'_1 + \sigma'_2 + \sigma'_3) \quad (11.9)$$

TABLE 11.1 Values of α [Eq. (11.7)][a]

K_0	Case 1	Case 2	Case 3	Case 4
0.4	0.4	0.4	—	0.6
0.5	0.5	0.5	0.25[b]	0.67
0.6	0.6	0.6	0.54[b]	0.73
0.7	0.7	0.7		0.80
0.8	0.8	0.8	0.83[b]	0.87
0.9	0.9	0.9		0.93
1.0	1.0	1.0	1.0	1.0

[a]Seed, H. B., and Peacock, W. H. (1971). "The Procedure for Measuring Soil Liquefaction Characteristics," *Journal of the Soil Mechanics and Foundations Division, ASCE*, 97 (SM8), Table 1, pp. 1110.
[b]For $\frac{1}{2}\sigma_d/\sigma_3 \approx 0.4$.

where σ_1', σ_2', σ_3' are, respectively, the major, intermediate, and minor *effective* principal stresses.

The octahedral shear stress τ_{oct} during cyclic loading is

$$\tau_{oct} = \frac{1}{3}\left[(\sigma_1 - \sigma_3)^2 + (\sigma_1 - \sigma_2)^2 + (\sigma_2 - \sigma_3)^2\right]^{1/2} \qquad (11.10)$$

where $\sigma_1, \sigma_2, \sigma_3$ are, respectively, the major, intermediate, and minor principal stresses during cyclic loading. For cyclic triaxial tests,

$$(\tau_{oct}/\sigma_{oct}')_{triax} = \frac{2}{3}\sqrt{2}\left(\frac{1}{2}\sigma_d/\sigma_3\right)_{triax} \qquad (11.11)$$

For cyclic simple shear tests,

$$(\tau_{oct}/\sigma_{oct}')_{simple\ shear} = \left[\sqrt{6}/(1+2K_0)\right](\tau_h/\sigma_v)_{simple\ shear} \qquad (11.12)$$

Thus

$$(\tau_{oct}/\sigma_{oct}')_{simple\ shear} = (\tau_{oct}/\sigma_{oct}')_{triax}$$

or

$$\left[\sqrt{6}/(1+2K_0)\right](\tau_h/\sigma_v)_{simple\ shear} = \frac{2}{3}\sqrt{2}\left(\frac{1}{2}\sigma_d/\sigma_3\right)_{triax}$$

or

$$\left(\frac{\tau_h}{\sigma_v}\right)_{simple\ shear} = \frac{2}{3}\sqrt{2}\left(\frac{1+2K_0}{\sqrt{6}}\right)\left(\frac{\frac{1}{2}\sigma_d}{\sigma_3}\right)$$

$$= \frac{\frac{2}{3}(1+2K_0)}{\sqrt{3}}\left(\frac{\frac{1}{2}\sigma_d}{\sigma_3}\right)_{triax} \qquad (11.13)$$

Comparing Eqs. (11.7) and (11.13),

$$\alpha = \frac{2}{3}(1+2K_0)/\sqrt{3} \qquad (11.14)$$

11.4.2 Correlation of the Liquefaction Results from Triaxial Tests to Field Conditions

Section 11.3.2 explained that the field value of τ_h/σ_v for initial liquefaction is about 15%–50% higher than that obtained from simple shear tests. Thus,

$$(\tau_h/\sigma_v)_{\text{field}} = \beta(\tau_h/\sigma_v)_{\text{simple shear}} \qquad (11.15)$$

The approximate variation of β with relative density of sand is given in Figure 11.18. Combining Eqs. (11.7) and (11.15), one obtains

$$\left(\frac{\tau_h}{\sigma_v}\right)_{\text{field}} = \beta\left(\frac{\tau_h}{\sigma_v}\right)_{\text{simple shear}} = \alpha\beta\left(\frac{\frac{1}{2}\sigma_d}{\sigma_3}\right)_{\text{triax}} = C_r\left(\frac{\frac{1}{2}\sigma_d}{\sigma_3}\right)_{\text{triax}} \qquad (11.16)$$

where $C_r = \alpha\beta$.

Using an average value of $\alpha = 0.47$ and the values of β given in Figure 11.18, the variation of C_r with relative density can be obtained. This is also shown in Figure 11.18.

Eq. (11.16) presents the correlations for initial liquefaction between the stress ratios in the field, cyclic simple shear tests, and cyclic triaxial tests for a given sand at the *same relative density*. However, when laboratory tests are conducted at relative density, say, $R_{D(1)}$, whereas the field conditions show the sand deposit to be at a relative density of $R_{D(2)}$, one has to convert the laboratory test results to correspond to a relative density of $R_{D(2)}$. It has been shown in Figure 11.16 that τ_h for initial liquefaction in the laboratory in a given number of cycles is approximately proportional to the relative

FIGURE 11.18 Variation of correction factor β and C_r with relative density [Eqs. (11.15) and (11.16)].

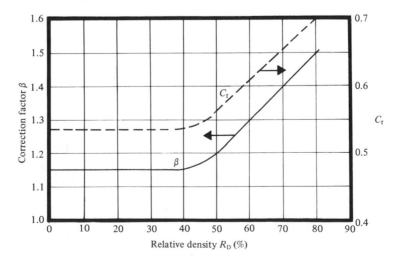

density (for $R_D \leqslant 80\%$). Thus

$$\tau_{h[R_{D(2)}]} = \tau_{h[R_{D(1)}]}\left[R_{D(2)}/R_{D(1)}\right] \tag{11.17}$$

where $\tau_{h[R_{D(1)}]}$ is the cyclic peak shear stress required to cause initial liquefaction in the laboratory for a given value of σ_v and number of cycles, by simple shear test; and $\tau_{h[R_{D(2)}]}$ is the cyclic peak shear stress required to cause initial liquefaction in the field for the same value of σ_v *and number of cycles*, by simple shear test.

Combining Eqs. (11.16) and (11.17),

$$\left(\frac{\tau_h}{\sigma_v}\right)_{field[R_{D(2)}]} = C_r\left(\frac{\frac{1}{2}\sigma_d}{\sigma_3}\right)_{triax[R_{D(1)}]} \cdot \frac{R_{D(2)}}{R_{D(1)}} \tag{11.18}$$

11.4.3 Zone of Initial Liquefaction in the Field

There are five general steps for determining the zone in the field where soil liquefaction due to an earthquake can be initiated:

1. Establish a design earthquake.
2. Determine the time history of shear stresses induced by the earthquake at various depths of sand layer.
3. Convert the shear stress–time histories into N number of equivalent stress cycles (see Section 8.8). These can be plotted against depth as shown in Figure 11.19.

FIGURE 11.19 Zone of initial liquefaction in field.

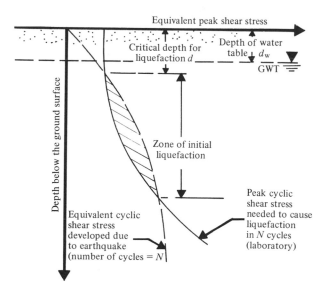

4. Using the laboratory test results, determine the magnitude of the cyclic stresses required to cause initial liquefaction in the field in N cycles (determined from Step 3) at various depths. Note that the cyclic shear stress levels change with depth due to change of σ_v. These can be plotted with depth as shown in Figure 11.19.

5. The zone in which the cyclic shear stress levels required to cause initial liquefaction (Step 4) are equal to or less than the equivalent cyclic shear stresses induced by an earthquake is the zone of possible liquefaction. This is shown in Figure 11.19.

11.4.4 Relation Between Maximum Ground Acceleration and the Relative Density of Sand for Soil Liquefaction

This section discusses a simplified procedure developed by Seed and Idriss (1971) to determine the relation between the maximum ground acceleration due to an earthquake and the relative density of a sand deposit in the field for the initial liquefaction condition. Figure 11.20a shows a layer of sand deposit in which we consider a column of soil of height h and unit area of cross section. Assuming the soil column to behave as a *rigid body*, the maximum shear stress at a depth h due to a maximum ground surface acceleration of a_{max} can be given by

$$\tau_{max} = (\gamma h/g) a_{max} \tag{11.19}$$

where τ_{max} is the maximum shear stress, γ is the unit weight of soil, and g is acceleration due to gravity.

However, the soil column is not a rigid body. For the deformable nature of the soil, the maximum shear stress at a depth h, determined by Eq. (11.19), needs to be modified as

$$\tau_{max(modif)} = C_D\left[(\gamma h/g) a_{max}\right] \tag{11.20}$$

where C_D is a stress reduction factor. The range of C_D for different soil profiles is shown in Figure 11.20b, along with the average value up to a depth of 40 ft (12.2 m).

It has been shown that the maximum shear stress determined from the shear stress–time history during an earthquake can be converted into an equivalent number of significant stress cycles. According to Seed and Idriss,

TABLE 11.2 Significant Number[a] of Stress Cycles N Corresponding to τ_{av}

Earthquake magnitude	N
7	10
7.5	20
8	30

[a]Seed, H. B., and Idriss, I. M. (1971). "Simplified Procedure for Evaluating Soil Liquefaction Potential," *Journal of the Soil Mechanics and Foundations Division, ASCE*, 97 (SM9), p. 1256.

377

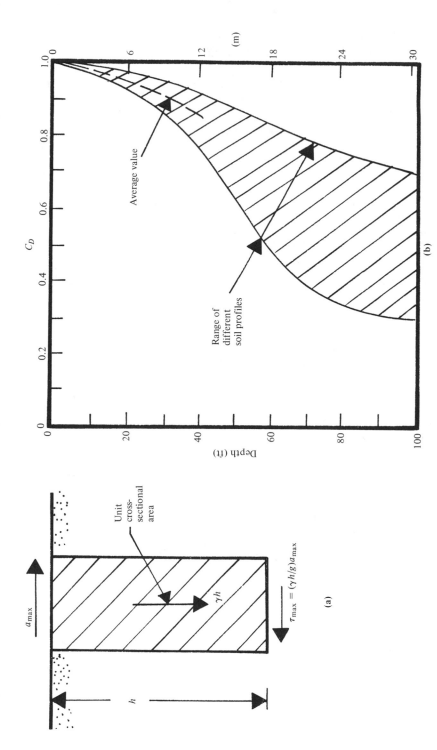

FIGURE 11.20 (a) Maximum shear stress at a depth for a rigid soil column; (b) range of the shear stress reduction factor C_D for the deformable nature of soil. [(b) is from Seed, H. B., and Idriss, I. M. (1971). "Simplified Procedure for Evaluating Soil Liquefaction Potential," *Journal of the Soil Mechanics and Foundations Division, ASCE*, 97 (SM9), Fig. 4, p. 1255.]

TABLE 11.3 Site Conditions and Earthquake Data for Known Cases of Liquefaction and Nonlinquefaction[a]

Earthquake	Date	Magnitude	Site	Approximate distance from source of energy release (miles)	Soil type	Depth of water table (feet)
Niigata	1802	6.6	Niigata	24	Sand	3
Niigata	1802	6.6	Niigata	24	Sand	3
Niigata	1887	6.1	Niigata	29	Sand	3
Niigata	1887	6.1	Niigata	29	Sand	3
Mino Owari	1891	8.4	Ogaki	20	Sand	3
Mino Owari	1891	8.4	Ginan West	20	Sand	6
Mino Owari	1891	8.4	Unama	20	Sand and gravel	6
Mino Owari	1891	8.4	Ogase Pond	20	Sand	8
Santa Barbara	1925	6.3	Sheffield Dam	7	Sand	≈ 15
El Centro	1940	7.0	Brawley	5	Sand	≈ 15
El Centro	1940	7.0	All-Am. Canal	5	Sand	≈ 20
El Centro	1940	7.0	Solfatara Canal	5	Sand	5
Tohnankai	1944	8.3	Komei	100	Sand	5
Tohnankai	1944	8.3	Meiko St.	100	Silt and sand	2
Fukui	1948	7.2	Takaya	4	Sand	11
Fukui	1948	7.2	Takaya	4	Sand	3
Fukui	1948	7.2	Shonenji Temple	4	Sand	4
Fukui	1948	7.2	Agr. Union	4	Sand and silt	3
San Francisco	1957	5.5	Lake Merced	4	Sand	8
Chile	1960	8.4	Puerto Montt	≈ 70	Sand	12
Chile	1960	8.4	Puerto Montt	≈ 70	Sand	12
Chile	1960	8.4	Puerto Montt	≈ 70	Sand	12
Niigata	1964	7.5	Niigata	32	Sand	3
Niigata	1964	7.5	Niigata	32	Sand	3
Niigata	1964	7.5	Niigata	32	Sand	3
Niigata	1964	7.5	Niigata	32	Sand	12
Alaska	1964	8.3	Snow River	60	Sand	0
Alaska	1964	8.3	Snow River	60	Sand	8
Alaska	1964	8.3	Quartz Creek	70	Sandy gravel	0
Alaska	1964	8.3	Scott Glacier	55	Sand	0
Alaska	1964	8.3	Valdez	35	Sand and gravel	5
Tokachioki	1968	7.8	Hachinohe	45–110	Sand	3
Tokachioki	1968	7.8	Hachinohe	45–110	Sand	3
Tokachioki	1968	7.8	Hachinohe	45–110	Sand	5
Tokachioki	1968	7.8	Hakodate	100	Sand	3

[a]Seed, H. B., and Idriss, I. M. (1971). "Simplified Procedure for Evaluating Soil Liquefaction Potential," *Journal of the Soil Mechanics and Foundations Division, ASCE*, 97 (SM9), Table 1, p. 1262.

TABLE 11.3 (*continued*)

Critical depth (feet)	Average penetration resistance at critical depth	Relative density (%)	Maximum ground surface acceleration (g)	Duration of shaking (sec)	Field behavior
20	6	53	0.12	≈ 20	No liquefaction
20	12	64	0.12	≈ 20	No liquefaction
20	6	53	0.08	≈ 12	No liquefaction
20	12	64	0.08	≈ 12	No liquefaction
45	17	65	≈ 0.35	≈ 75	Liquefaction
30	10	55	≈ 0.35	≈ 75	Liquefaction
25	19	75	≈ 0.35	≈ 75	No liquefaction
20	16	72	≈ 0.35	≈ 75	Liquefaction
25	—	40	≈ 0.2	15	Liquefaction
≈ 15	—	58	≈ 0.25	30	Liquefaction
≈ 25	—	43	≈ 0.25	30	Liquefaction
≈ 20	—	32	≈ 0.25	30	Liquefaction
13	4	40	≈ 0.08	≈ 70	Liquefaction
8	1	30	≈ 0.08	≈ 70	Liquefaction
23	18	72	≈ 0.30	≈ 30	Liquefaction
23	28	90	≈ 0.30	≈ 30	No liquefaction
10	3	40	≈ 0.30	≈ 30	Liquefaction
20	5	50	≈ 0.30	≈ 30	Liquefaction
10	7	55	≈ 0.18	18	Liquefaction
15	6	50	≈ 0.15	≈ 75	Liquefaction
15	8	55	≈ 0.15	≈ 75	Liquefaction
20	18	75	≈ 0.15	≈ 75	No liquefaction
20	6	53	0.16	40	Liquefaction
25	15	70	0.16	40	Liquefaction
20	12	64	0.16	40	No liquefaction
25	6	53	0.16	40	No liquefaction
20	5	50	≈ 0.15	180	Liquefaction
20	5	44	≈ 0.15	180	Liquefaction
≈ 25	40–80	100	≈ 0.12	180	No liquefaction
≈ 20	10	65	≈ 0.16	180	Liquefaction
≈ 20	13	68	≈ 0.25	180	Liquefaction
12	14	78	0.21	45	No liquefaction
12	6	58	0.21	45	Liquefaction
10	15	80	0.21	45	No liquefaction
15	6	55	0.18	45	Liquefaction

one can take

$$\tau_{av} = 0.65\tau_{max(modif)} = 0.65C_D\left[(\gamma h/g)a_{max}\right] \qquad (11.21)$$

The corresponding number of significant cycles N for τ_{av} is given in Table 11.2. Note that although the values of N given in the table are somewhat different from those given in Figure 8.24, it does not make a considerable difference in the calculations.

One can now combine Eq. (11.18), which gives the correlation of laboratory results of cyclic triaxial tests to the field conditions, and Eq. (11.21) to determine the relationships between a_{max} and R_D. This can be better shown with the aid of a numerical example.

In general, the critical depth of liquefaction (see Figure 11.19) occurs at a depth of about 20 ft (6.1 m) when the depth of water table d_w is 0–10 ft (0–3.05 m); similarly, the critical depth is about 30 ft (9.15 m) when the depth of the water table is about 15 ft. This can be seen from Table 11.3.

Liquefaction occurs in sands having a mean grain size D_{50} of 0.075–0.2 mm. Consider a case where

$D_{50} = 0.075$ mm

$d_w = 15$ ft (≈ 4.58 m)

$\quad \gamma =$ unit weight above the ground water table (GWT) $= 115$ lb/ft^3
$\qquad (18.08$ kN/m$^3)$
$\gamma_{sat} =$ unit weight of soil below GWT $= 122.4$ lb/ft^3
$\qquad (19.24$ kN/m$^3)$
$\quad \gamma' =$ effective unit weight of soil below GWT

$\qquad = (122.4 - 62.4) = 60.$ lb/ft^3 $(9.43$ kN/m$^3)$

significant number of stress cycles $= 10$

(earthquake magnitude $= 7$)

The critical depth of liquefaction d is about 30 ft (9.15 m), which at that depth the *total normal stress* is equal to

$$15(\gamma) + 15\gamma_{sat} = 15(115) + 15(122.4) = 3561 \text{ lb/ft}^2 \quad (170.63 \text{ kN/m}^2)$$

From Eq. (11.21)

$$\tau_{av} = 0.65C_D\left[(\gamma h/g)a_{max}\right]$$

The value of C_D for $d = 30$ ft is 0.925 (Fig. 11.20). Thus

$$\tau_{av} = (0.65)(0.925)(3561)a_{max}/g = 2141a_{max}/g \qquad (11.22)$$

Again, from Eq. (11.18),

$$\tau_{h(field)[R_{D(2)}]} = \sigma_v C_r \left(\frac{\frac{1}{2}\sigma_d}{\sigma_3}\right)_{triaxial[R_{D(1)}]} \frac{R_{D(2)}}{R_{D(2)}}$$

TABLE 11.4 Relation Between a_{max}/g vs $R_{D(2)}$ [Eq. (11.24)]

$R_{D(2)}$ (%)	Ratio[a] C_r	$\dfrac{a_{max}}{g}$
20	0.54	0.0572
30	0.54	0.0856
40	0.54	0.1144
50	0.565	0.1497
60	0.61	0.1938
70	0.66	0.2449
80	0.705	0.2989

[a]From Figure 11.18.

At a depth 30 feet below the ground surface, the *initial effective stress* σ_v is equal to $15(\gamma)+15(\gamma')=15(115)+15(60)=2625$ lb/ft² (125.78 kN/m²). From Figure 11.13, for $D_{50}=0.075$ mm, $(\tfrac{1}{2}\sigma_d/\sigma_3)_{triax,\,R_D=50\%} \approx 0.215$. Hence

$$\tau_{h(field)[R_{D(2)}]} = 2625\left[C_r(0.215)\right]R_{D(2)}/50 = 11.29C_r R_{D(2)} \quad (11.23)$$

For liquefaction, τ_{av} of Eq. (11.22) should be equal to $\gamma_{h(field)[R_{D(2)}]}$. Hence,

$$2141a_{max}/g = 11.29C_r R_{D(2)}$$

or

$$a_{max}/g = 0.0053C_r R_{D(2)} \quad (11.24)$$

It is now possible to prepare Table 11.4 to determine the variation of a_{max}/g with $R_{D(2)}$. Note that $R_{D(2)}$ is the relative density in the field.

FIGURE 11.21 Plot of a_{max}/g vs relative density from Table 11.4: $N=10$, $d_w=15$ ft (4.58 m), $D_{50}=0.075$ mm.

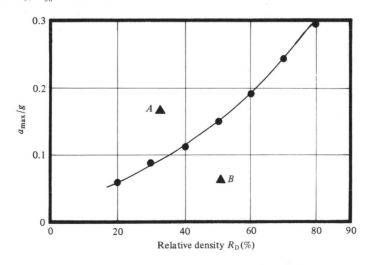

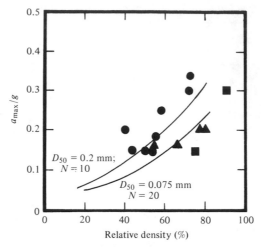

FIGURE 11.22 Evaluation of liquefaction potential for sand, water table 10 ft (3.05 m) below ground surface: ● liquefaction, a_{max} estimated; ■ no liquefaction, a_{max} estimated; ▲ no liquefaction, a_{max} recorded. [Seed, H. B., and Idriss, I. M. (1971). "Simplified Procedure for Evaluating Soil Liquefaction Potential," *Journal of the Soil Mechanics and Foundations Division*, *ASCE*, 97 (SM9), Fig. 12, p. 1261.]

FIGURE 11.23 Liquefaction potential evaluation chart, water table at depth of 5 ft (1.53 m), $a_{max} = 0.25g$. [Seed, H. B., and Idriss, I. M. (1971). "Simplified Procedure for Evaluating Soil Liquefaction Potential," *Journal of the Soil mechanics and Foundations Division*, *ASCE*, 97 (SM9), Fig. 16, p. 1269.]

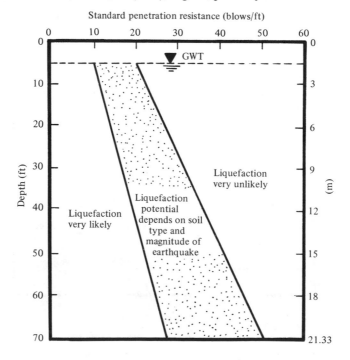

Figure 11.21 shows a plot of a_{max}/g vs the relative density as determined from Table 11.4. For this given soil (i.e., given D_{50}, d_w, and number of significant stress cycles N), if the relative density in the field and a_{max}/g are such that they plot as point A in Figure 11.21 (i.e., above the curve showing the relationship of Eq. (11.24)], then liquefaction would occur. On the other hand, if the relative density and a_{max}/g plot as point B [i.e., below the curve showing the relationship of Eq. (11.24)], then liquefaction would not occur.

Diagrams of the type shown in Figure 11.21 could be prepared for various combinations of D_{50}, d_w, and N. Since, in the field, for liquefaction the range of D_{50} is 0.075–0.2 mm and the range of N is about 10–20, one can take the critical combinations (i.e., $D_{50} = 0.075$ mm, $N = 20$; $D_{50} = 0.2$ mm, $N = 10$) and plot graphs as shown in Figure 11.22. These graphs provide a useful guide in the evaluation of liquefaction potential in the field.

11.4.5 Liquefaction Potential Evaluation Charts: Correlation with Standard Penetration Resistance

Another way of evaluating the soil liquefaction potential is to prepare correlation charts with the standard penetration resistance of soils. Some of these types of chart have been prepared by Seed and Idriss (1971). Relations

FIGURE 11.24 Penetration resistance values for which liquefaction is unlikely to occur under any earthquake condition. [Seed, H. B., and Idriss, I. M. (1971). "Simplified Procedure for Evaluating Soil Liquefaction Potential," *Journal of the Soil Mechanics and Foundations Division, ASCE,* 97 (SM9), Fig. 18, p. 1270.]

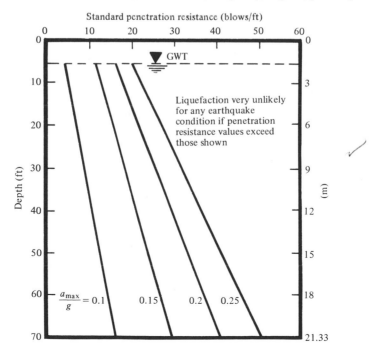

between standard penetration resistance, relative density, and effective overburden pressure σ_v have been developed by Gibbs and Holtz (1957). Relationships such as these can be combined with studies like that shown in Figure 11.22 to develop the liquefaction potential correlation with standard penetration resistance. Figure 11.23 is the liquefaction potential chart for $a_{max}/g = 0.25$ with water table located at a depth of 5 feet (1.53 m) below the ground surface. Figure 11.24 shows plots of the limiting soil liquefaction conditions with depth for $a_{max}/g = 0.1, 0.15, 0.2$, and 0.25 [depth of water table = 5 feet (1.53 m)]. For a given a_{max}/g, if the standard penetration resistance at any depth falls above the limiting line shown in Figure 11.24, liquefaction is very unlikely to occur.

11.5 STABILIZATION OF POTENTIALLY LIQUEFIABLE SAND DEPOSITS

A possible means of stabilizing a sand deposit which is potentially liquefiable is to use *gravel or rock drains* as shown in Figure 11.25. The purpose of the installation of gravel or rock drains is to dissipate the excess pore water pressure almost as fast as it is generated in the sand deposit due to cyclic loading. The design principles of gravel and rock drains have been developed by Seed and Booker (1977) and are described in the following sections.

11.5.1 Equation for Generation and Dissipation of Pore Water Pressure

Assuming that Darcy's law is valid, the continuity of flow equation in the sand layer may be written as

$$\frac{\partial}{\partial x}\left(\frac{k_h}{\gamma_w}\frac{\partial u}{\partial x}\right) + \frac{\partial}{\partial y}\left(\frac{k_h}{\gamma_w}\frac{\partial u}{\partial y}\right) + \frac{\partial}{\partial z}\left(\frac{k_v}{\gamma_w}\frac{\partial u}{\partial z}\right) = \frac{\partial \epsilon}{\partial t} \qquad (11.25)$$

where k_h is the coefficient of permeability of the sand in the horizontal direction; k_v is the coefficient of permeability of the sand in the vertical direction; u is the excess pore water pressure; γ_w is the unit weight of water; and ϵ is the volumatic strain (compression positive).

During a time interval dt, the pore water pressure in a soil element changes by du. However, if a cyclic shear stress is applied on a soil element, there is an increase of pore water. In a time dt, there are dN number of cyclic shear stresses; the corresponding increase of pore water pressure is $(\partial u_g/\partial N)\,dN$ (where u_g is the excess pore water pressure generated by cyclic shear stress). Thus, the net change in pore water pressure in time dt is equal to $[du - (\partial u_g/\partial N)\,dN)$, and

$$\partial \epsilon = m_{v_3}\left[\partial u - \left(\partial u_g/\partial N\right)dN\right]$$

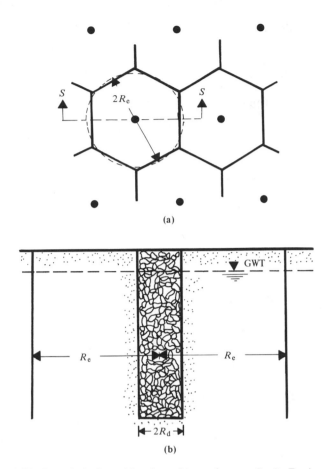

FIGURE 11.25 Gravel drains: (a) plan, (b) section at $S-S$. Drain radius, R_d; effective radius of drain, R_e.

or

$$\frac{\partial \epsilon}{\partial t} = m_{v_3}\left(\frac{\partial u}{\partial t} - \frac{\partial u_g}{\partial N}\frac{\partial N}{\partial t}\right) \qquad (11.26)$$

where m_{v_3} is the coefficient of volume compressibility.

Combining Eqs. (11.25) and (11.26),

$$\frac{\partial}{\partial x}\left(\frac{k_h}{\gamma_w}\frac{\partial u}{\partial x}\right) + \frac{\partial}{\partial y}\left(\frac{k_h}{\gamma_w}\frac{\partial u}{\partial y}\right) + \frac{\partial}{\partial z}\left(\frac{k_h}{\gamma_w}\frac{\partial u}{\partial z}\right) = m_{v_3}\left(\frac{\partial u}{\partial t} - \frac{\partial u_g}{\partial N}\frac{\partial N}{\partial t}\right)$$

$$(11.27)$$

If m_{v_3} is a constant and *radial symmetry* exists, then Eq. (11.27) can be written in cylindrical coordinates as

$$\frac{k_h}{\gamma_w m_{v_3}}\left(\frac{\partial^2 u}{\partial^2 r}+\frac{1}{r}\frac{\partial u}{\partial r}\right)+\frac{k_v}{\gamma_w m_{v_3}}\frac{\partial^2 u}{\partial z^2}=\frac{\partial u}{\partial t}-\frac{\partial u_g}{\partial N}\frac{\partial N}{\partial t} \qquad (11.28)$$

For the condition of purely radial flow, Eq. (11.28) takes the form

$$\frac{k_h}{\gamma_w m_{v_3}}\left(\frac{\partial^2 u}{\partial^2 r}+\frac{1}{r}\frac{\partial u}{\partial r}\right)=\frac{\partial u}{\partial t}-\frac{\partial u_g}{\partial N}\frac{\partial N}{\partial t} \qquad (11.29)$$

In order to solve Eq. (11.29), it is necessary to evaluate the terms k_h, m_{v_3}, $\partial N/\partial t$, and $\partial u_g/\partial N$. The value of k_h can be easily determined from field pumping tests. The coefficient of volume compressibility can be determined from cyclic triaxial tests (Lee and Albaisia, 1974). The term $\partial N/\partial t$ can be expressed as

$$\frac{\partial N}{\partial t}=N_S/t_d \qquad (11.30)$$

where N_S is the significant number of uniform stress cycles due to an earthquake and t_d is the duration of an earthquake.

The rate of the excess pore water pressure buildup in a saturated undrained cyclic simple shear tests is shown in Figure 11.26. The average value of u_g/σ_v (Seed et al., 1975) can be given by the relation

$$u_g/\sigma_v=(2/\pi)\arcsin(N/N_1)^{1/2\alpha} \qquad (11.31)$$

where u_g is the excess pore water pressure developed due to N number of cyclic shear stress applications, σ_v is the initial *consolidation* pressure, N_1 is the number of stress cycles needed for initial liquefaction, and α is a

FIGURE 11.26 Rate of pore water pressure buildup in cyclic simple shear test. [Seed, H. B., and Brooker, J. R. (1977). "Stabilization of Potentially Liquefiable Sand Deposits Using Gravel Drains," *Journal of the Geotechnical Engineering Division, ASCE*, 103 (GT7), Fig. 2, p. 758.]

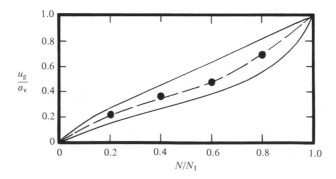

constant ≈ 0.7. Therefore,

$$\frac{\partial u_g}{\partial N} = \frac{2\sigma_v}{\alpha\pi N_1}\left[\sin^{2\alpha-1}\left(\tfrac{1}{2}\pi\frac{u_g}{\sigma_v}\right)\cos\left(\tfrac{1}{2}\pi\frac{u_g}{\sigma_v}\right)\right]^{-1} \tag{11.32}$$

11.5.2 Solutions for Gravel or Rock Drains

For radial flow conditions, the relation given by Eq. (11.29) has been solved by Seed and Brooker (1977). It has been shown that the ratio u/σ_v is a function of the following parameters:

$$\frac{R_d}{R_e} = \frac{\text{radius of rock or gravel drains}}{\text{effective radius of the rock or gravel drains}} \tag{11.33}$$

N_S/N_1, and

$$T_{ad} = \frac{k_h}{\gamma_w}\left(\frac{t_d}{m_{v_3}R_d^2}\right) \tag{11.34}$$

Using the above parameters, the solution to Eq. (11.29) is given in a nondimensional form in Figure 11.27 for design of rock or gravel drains. In Figure 11.27, the term r_g is defined as

$$r_g = \frac{\text{greatest limiting value of } u_g \text{ chosen for design}}{\sigma_v} \tag{11.35}$$

In obtaining the solutions given in Figure 11.27, it was assumed that the coefficient of permeability of the material used in the gravel or rock drains is infinity. However, in practical case, it would be sufficient to have a value of

$$k_{h(\text{rock or gravel})}/k_{(\text{sand})} \approx 200.$$

Example 11.1 For a sand deposit, it is given that

$$m_{v_3} = 2.8 \times 10^{-5} \text{ m}^2/\text{kN}$$

$$k_h = 0.02 \text{ mm}/\text{sec} = 2 \times 10^{-5} \text{ m}/\text{sec}$$

For a design earthquake, the equivalent number of uniform stress cycles (for uniform stress $= \tau_{av}$) was determined to be 30. The duration of the earthquake is about 65 sec.

From laboratory tests, it was determined that 12 cycles of cyclic stress application (the peak magnitude of the cyclic stress is equal to τ_{av}) would be enough to cause initial liquefaction in the sand.

Assuming that the radius of the gravel drains to be used is 0.25 m, and $r_g = 0.6$, determine the spacing of gravel drains.

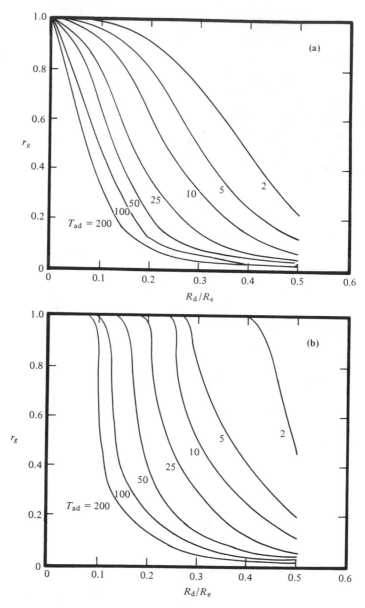

FIGURE 11.27 Relation between greatest pore water pressure ratio and drain system parameters: $N_S/N_1 =$ **(a)** 1, **(b)** 2, **(c)** 3, **(d)** 4. [Seed, H. B., and Brooker, J. R. (1977). "Stabilization of Potentially Liquefiable Sand Deposits Using Gravel Drains," *Journal of the Geotechnical Engineering Division*, *ASCE*, 103 (GT7), Fig. 5, p. 763.]

389

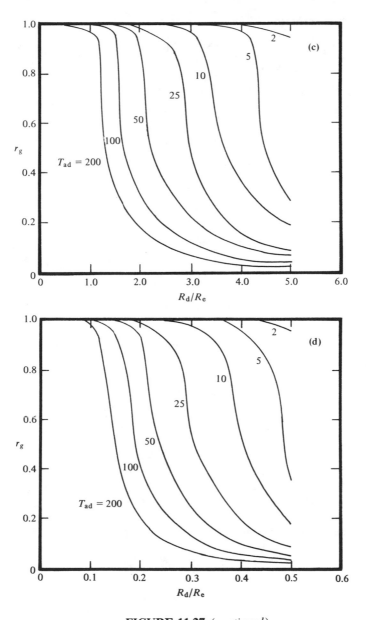

FIGURE 11.27 (*continued*)

Solution: From Eq. (11.34),

$$T_{ad} = \frac{k_h}{\gamma_w}\left(\frac{t_d}{m_{v_3}R_d^2}\right) = \frac{2\times 10^{-5}}{9.81}\left[\frac{65}{2.8\times 10^{-5}(0.25)^2}\right] = 75.72$$

$$N_S/N_1 = 30/12 = 2.5, \qquad r_g = 0.6$$

Referring to Figure 11.27b, for $T_{ad} = 75.72$, $N_S/N_1 = 2$, $r_g = 0.6$,

$$R_d/R_e \approx 0.17$$

From Figure 11.27c, for $T_{ad} = 75.72$, $N_s/N_1 = 3$, $r_g = 0.6$,

$$\frac{R_d}{R_e} \approx 0.2$$

Thus, for $N_S/N_1 = 2.5$, $R_d/R_e \approx \frac{1}{2}(0.17 + 0.2) = 0.185$. Hence,

$$R_e = R_d/0.185 = 0.25/0.185 = 1.35 \text{ m}$$

PROBLEMS

11.1. Explain the terms "initial liquefaction" and "cyclic mobility."

11.2. For a sand deposit with the following as given:

$$\text{mean grain size } (D_{50}) = 0.2 \text{ mm}$$
$$\text{depth of water table} = 10 \text{ ft}$$
$$\text{unit weight of soil above GWT} = 105 \text{ lb/ft}^3$$
$$\text{unit weight of soil below GWT} = 120 \text{ lb/ft}^3$$
$$\text{expected earthquake magnitude} = 7.5$$

make all calculations and prepare a graph showing the variation of a_{max}/g and the relative density in the field for liquefaction to occur.

11.3. Repeat Problem 11.2 for a mean grain size of 0.075 mm.

11.4. Repeat Problem 11.2 for the following conditions:

$$\text{mean grain size} = 0.075 \text{ mm}$$
$$\text{depth of water table} = 4.6 \text{ m}$$
$$\text{unit weight of soil above GWT} = 15 \text{ kN/m}^3$$
$$\text{unit weight of soil below GWT} = 18 \text{ kN/m}^3$$
$$\text{expected magnitude of earthquake} = 8$$

11.5. Repeat Problem 11.4 for a mean grain size of 0.2 mm.

11.6. Consider the soil and the ground water table condition given in Problem 11.2. Assume that the relative density in the field is 60%. The maximum expected intensity of ground shaking (a_{max}/g) is 0.2 and the magnitude of earthquake is 7.5.

a. Calculate and plot the variation of the shear stress τ_{av} induced in the sand deposit with depth (0–70 ft). Use Eq. (11.21).

b. Calculate the variation of the shear stress required to cause liquefaction with depth. Plot the shear stress determined in the same graph as used in (a). Use Eq. (11.18).

c. From the graph plotted above, determine the depths at which liquefaction is initiated.

11.7. Repeat Problem 11.6(a)–(c) for the data given in Problem 11.4. Assume the relative density of sand to be 60% and the maximum expected intensity of ground shaking to be 0.15g.

11.8. At a given site of sand deposit, the standard penetration resistance is 15 blows/ft at a depth of 30 ft. The ground water table is located at a depth of 5 ft. Is liquefaction very likely at that depth for

a. $a_{max} = 0.1g$?

b. $a_{max} = 0.15g$?

11.9. Solve the gravel drain problem given in Example 11.1 for $r_g = 0.7$.

11.10. Repeat Example 11.1 for the gravel drain with the following data:

$$m_{v_3} = 3.5 \text{ m}^2/\text{kN}$$

$$k_h = 1.4 \times 10^{-5} \text{ m/sec}$$

$$\frac{\text{equivalent number of uniform}}{\text{stress cycles due to earthquake}} = 20$$

$$\text{duration of earthquake} = 50 \text{ sec}$$

$$\frac{\text{number of uniform stress}}{\text{cycles for liquefaction}} = 12$$

$$\text{radius of gravel drains} = 0.3 \text{ m}$$

$$r_g = 0.7$$

REFERENCES

Casagrande, A. (1936), "Characteristics of Cohesionless Soils Affecting the Stability of Slopes and Earthfills," *Journal of the Boston Society of Civil Engineers*, January vol. 23, pp. 257–276.

Castro, G. (1975). "Liquefaction and Cyclic Mobility of Saturated Sands," *Journal of the Geotechnical Engineering Division, ASCE* 101 (GT6), 551–569.

Finn, W. D. L., Bransby, P. L., and Pickering, D. J. (1970), "Effect of Strain History on Liquefaction of Sands," *Journal of the Soil Mechanics and Foundations Division, ASCE* 96 (SM6), 1917–1934.

Finn, W. D. L., Pickering, D. J., and Bransby, P. L. (1971), "Sand Liquefaction in Triaxial and Simple Shear Tests," *Journal of the Soil Mechanics and Foundations Division, ASCE*, 97 (SM4), 639–659.

Gibbs, H. J., and Holtz, W. G. (1957). "Research on Determining the Density of Sand by Spoon Penetration Test," *Proceedings, Fourth International Conference on Soil Mechanics and Foundation Engineering*, 1, 35–37.

Ishibashi, I., and Sherif, M. A. (1974), "Soil Liquefaction by Torsional Simple Shear Device," *Journal of the Geotechnical Engineering Division, ASCE* 100 (GT8), 871–888.

Jaky, J. (1944), "The Coefficient of Earth Pressure at Rest," *Journal of the Society of the Hungarian Architectural Engineers* 21, 355–358.

Lee, K. L., and Albaisa, A. (1974), "Earthquake Induced Settlements in Saturated Sands," *Journal of the Geotechnical Engineering Division, ASCE* 100 (GT4), 387–406.

Lee, K. L., and Seed, H. B. (1967), "Cyclic Stress Conditions Causing Liquefaction of Sand," *Journal of the Soil Mechanics and Foundations Division, ASCE* 93 (SM1), 47–70.

Peacock, W. H., and Seed, H. B. (1968). "Sand Liquefaction Under Cyclic Loading Simple Shear Conditions," *Journal of the Soil Mechanics and Foundations Division, ASCE* 94 (SM3), 689–708.

Prakash. S., and Mathur, J. N. (1965). "Liquefaction of Fine Sand Under Dynamic Loads," *Proceedings, 5th Symposium of the Civil and Hydraulic Engineering Departments*, Indian Institute of Science, Bangalore, India.

Roscoe, K. H. (1953). "An Apparatus for the Application of Simple Shear to Soil Samples," *Proceedings, 3rd International Conference on Soil Mechanics and Foundation Engineering*, Vol. 1, pp. 186–191.

Seed, H. B. (1979). "Soil Liquefaction and Cyclic Mobility Evaluation for Level Ground During Earthquakes," *Journal of the Geotechnical Engineering Division, ASCE* 105 (GT2), 201–255.

Seed, H. B., and Booker, J. R. (1977). "Stabilization of Potentially Liquefiable Sand Deposits Using Gravel Drains," *Journal of the Geotechnical Engineering Division, ASCE* 103 (GT7), 757–768.

Seed, H. B., and Idriss, I. M. (1971). "Simplified Procedure for Evaluating Soil Liquefaction Potential," *Journal of the Soil Mechanics and Foundations Division, ASCE* 97 (SM9), 1249–1273.

Seed, H. B., and Lee, K. L. (1966). "Liquefaction of Saturated Sands During Cyclic Loading," *Journal of the Soil Mechanics and Foundations Division, ASCE* 92 (SM6), 105–134.

Seed, H. B., and Peacock, W. H. (1971). "The Procedure for Measuring Soil Liquefaction Characteristics," *Journal of the Soil Mechanics and Foundations Division, ASCE* 97 (SM8), 1099–1119.

Seed, H. B., Martin, P. P., and Lysmer, J. (1975). "The Generation and Dissipation of Pore Water Pressures during Soil Liquefaction," Report No. EERC 75-26, Earthquake Engineering Research Center, University of California, Berkeley, August.

Yoshimi, Y. (1970). "Liquefaction of Saturated Sand During Vibration Under Quasi-Plane-Strain Conditions," *Proceedings, 3rd Japan Earthquake Engineering Symposium*, Tokyo.

Yoshimi, Y., and Oh-Oka, H. (1973). "A Ring Torsion Apparatus for Simple Shear Tests," *Proceedings, 8th International Conference on Soil Mechanics and Foundation Engineering*, Vol. 1.2, Moscow, USSR.

PRIMARY AND SECONDARY FORCES OF SINGLE-CYLINDER ENGINES

Machineries involving a crank mechanism produce a reciprocating force. This mechanism is shown in Figure A.1a, in which

$$0A = \text{crank length} = r_1$$

$$AB = \text{length of the connecting rod} = r_2$$

Let the crank rotate at a constant angular velocity ω. At time $t = 0$, the vertical distance between 0 and B (Figure A.1b) is equal to $r_1 + r_2$. At time t, the vertical distance between 0 and B is equal to $r_1 + r_2 - z$, or

$$z = (r_1 + r_2) - (r_2 \cos \alpha + r_1 \cos \omega t) \tag{A.1}$$

But,

$$r_2 \sin \alpha = r_1 \sin \omega t \tag{A.2}$$

Now,

$$\cos \alpha = \sqrt{1 - \sin^2 \alpha} = \sqrt{1 - (r_1/r_2)^2 \sin^2 \omega t}$$

$$\approx 1 - \tfrac{1}{2}(r_1/r_2)^2 \sin^2 \omega t \tag{A.3}$$

Substituting Eq. (A.3) into Eq. (A.1),

$$z = (r_1 + r_2) - (r_2 \cos \alpha + r_1 \cos \omega t)$$

$$= r_2(1 - \cos \alpha) + r_1(1 - \cos \omega t)$$

$$= r_2 \left[1 - 1 + \tfrac{1}{2}(r_1/r_2)^2 \sin^2 \omega t \right] + r_1(1 - \cos \omega t)$$

$$= \left(\tfrac{1}{2} r_1^2 / r_2 \right) \sin^2 \omega t + r_1(1 - \cos \omega t) \tag{A.4}$$

However,

$$\sin^2 \omega t = \tfrac{1}{2}(1 - \cos 2\,\omega t) \tag{A.5}$$

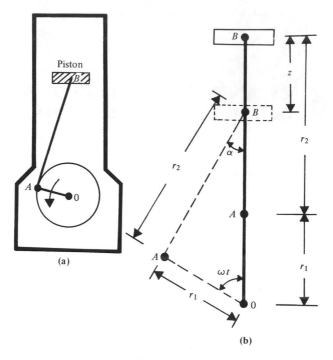

FIGURE A.1

Substituting Eq. (A.5) into Eq. (A.4), one obtains

$$z = \left(\tfrac{1}{4}r_1^2/r_2\right)\left(1 - \cos 2\,\omega t\right) + r_1\left(1 - \cos \omega t\right)$$
$$= \left(r_1 + \tfrac{1}{4}r_1^2/r_2\right) - \left[r_1\cos \omega t + \left(\tfrac{1}{4}r_1^2/r_2\right)\cos 2\,\omega t\right] \qquad (A.6)$$

The acceleration of the piston can be given by

$$\ddot{z} = r_1\omega^2\left[\cos \omega t + (r_1/r_2)\cos 2\,\omega t\right] \qquad (A.7)$$

If the mass of the piston is m, the force can be obtained as

$$F = m\ddot{z} = mr_1\omega^2\cos \omega t + m\left(r_1^2/r_2\right)\omega^2\cos 2\,\omega t \qquad (A.8)$$

The first term of Eq. (A.8) is the primary force, and the maximum primary force

$$F_{\max(\text{prim})} = mr_1\omega^2 \qquad (A.9)$$

Similarly, the second term of Eq. (A.8) is generally referred to as the secondary force, and

$$F_{\max(\text{sec})} = m\left(r_1^2/r_2\right)\omega^2 \qquad (A.10)$$

INDEX